PLANETS AROUND PULSARS

A SERIES OF BOOKS ON RECENT DEVELOPMENTS IN
ASTRONOMY AND ASTROPHYSICS

390 Ashton Avenue, San Francisco, California 94112

Printed by BookCrafters, Inc.

First published 1993

Library of Congress Catalog Card Number: 92-75905
ISBN 0-937707-55-4

D. Harold McNamara, Managing Editor of Conference Series
408 ESC Brigham Young University
Provo, UT 84602
801-378-2298

A SERIES OF BOOKS ON RECENT DEVELOPMENTS IN ASTRONOMY AND ASTROPHYSICS

Vol. 1-Progress and Opportunities in Southern Hemisphere Optical Astronomy: The CTIO 25th Anniversary Symposium
ed. V. M. Blanco and M. M. Phillips ISBN 0-937707-18-X

Vol. 2-Proceedings of a Workshop on Optical Surveys for Quasars
ed. P. S. Osmer, A. C. Porter, R. F. Green, and C. B. Foltz ISBN 0-937707-19-8

Vol. 3-Fiber Optics in Astronomy
ed. S. C. Barden ISBN 0-937707-20-1

Vol. 4-The Extragalactic Distance Scale: Proceedings of the ASP 100th Anniversary Symposium
ed. S. van den Bergh and C. J. Pritchet ISBN 0-937707-21-X

Vol. 5-The Minnesota Lectures on Clusters of Galaxies and Large-Scale Structure
ed. J. M. Dickey ISBN 0-937707-22-8

Vol. 6-Synthesis Imaging in Radio Astronomy: A Collection of Lectures from the Third NRAO Synthesis Imaging Summer School
ed. R. A. Perley, F. R. Schwab, and A. H. Bridle ISBN 0-937707-23-6

Vol. 7-Properties of Hot Luminous Stars: Boulder-Munich Workshop
ed. C. D. Garmany ISBN 0-937707-24-4

Vol. 8-CCDs in Astronomy
ed. G. H. Jacoby ISBN 0-937707-25-2

Vol. 9-Cool Stars, Stellar Systems, and the Sun. Sixth Cambridge Workshop
ed. G. Wallerstein ISBN 0-937707-27-9

Vol. 10-The Evolution of the Universe of Galaxies. The Edwin Hubble Centennial Symposium
ed. R. G. Kron ISBN 0-937707-28-7

Vol. 11-Confrontation Between Stellar Pulsation and Evolution
ed. C. Cacciari and G. Clementini ISBN 0-937707-30-9

Vol. 12-The Evolution of the Interstellar Medium
ed. L. Blitz ISBN 0-937707-31-7

Vol. 13-The Formation and Evolution of Star Clusters
ed. K. Janes ISBN 0-937707-32-5

Vol. 14-Astrophysics with Infrared Arrays
ed. R. Elston ISBN 0-937707-33-3

Vol. 15-Large-Scale Structures and Peculiar Motions in the Universe
ed. D. W. Latham and L. A. N. da Costa ISBN 0-937707-34-1

Vol. 16-Atoms, Ions and Molecules: New Results in Spectral Line Astrophysics
ed. A. D. Haschick and P. T. P. Ho ISBN 0-937707-35-X

Vol. 17-Light Pollution, Radio Interference, and Space Debris
ed. D. L. Crawford ISBN 0-937707-36-8

Vol. 18-The Interpretation of Modern Synthesis Observations of Spiral Galaxies
ed. M. Duric and P. C. Crane ISBN 0-937707-37-6

Vol. 19-Radio Interferometry: Theory, Techniques, and Applications, IAU Colloquium 131
ed. T. J. Cornwell and R. A. Perley ISBN 0-937707-38-4

Vol. 20-Frontiers of Stellar Evolution, celebrating the 50th Anniversary of McDonald Observatory
ed. D. L. Lambert ISBN 0-937707-39-2

Vol. 21-The Space Distribution of Quasars
ed. D. Crampton ISBN 0-937707-40-6

Vol. 22-Nonisotropic and Variable Outflows from Stars
ed. L. Drissen, C. Leitherer, and A. Nota ISBN 0-937707-41-4

Vol. 23-Astronomical CCD Observing and Reduction Techniques
ed. S. B. Howell ISBN 0-937707-42-4

Vol. 24-Cosmology and Large-Scale Structure in the Universe
ed. R. R. de Carvalho ISBN 0-937707-43-0

Vol. 25-Astronomical Data Analysis Software and Systems I
ed. D. M. Worrall, C. Biemesderfer, and J. Barnes ISBN 0-937707-44-9

Vol. 26-Cool Stars, Stellar Systems, and the Sun, Seventh Cambridge Workshop
ed. M. S. Giampapa and J. A. Bookbinder ISBN 0-937707-45-7

Vol. 27-The Solar Cycle
ed. K. L. Harvey ISBN 0-937707-46-5

Vol. 28-Automated Telescopes for Photometry and Imaging
ed. S. J. Adelman, R. J. Dukes, Jr., and C. J. Adelman ISBN 0-937707-47-3

Vol. 29-Workshop on Cataclysmic Variable Stars
ed. N. Vogt ISBN 0-937707-48-1

Vol. 30-Variable Stars and Galaxies, in honor of M. S. Feast on his retirement
ed. B. Warner ISBN 0-937707-49-X

Vol. 31-Relationships Between Active Galactic Nuclei and Starburst Galaxies
ed. A. V. Filippenko ISBN 0-937707-50-3

Vol. 32-Complementary Approaches to Double and Multiple Star Research, IAU Colloquim 135
ed. H. A. McAlister and W. I. Hartkopf ISBN 0-937707-51-1

Vol. 33-Research Amateur Astronomy
ed. S. J. Edberg ISBN 0-937707-52-X

Vol. 34-Robotic Telescopes in the 1990s
ed. A. V. Filippenko ISBN 0-937707-53-8

Vol. 35-Massive Stars: Their Lives in the Interstellar Medium
ed. J. P. Cassinelli and E. B. Churchwell ISBN 0-937707-54-6

Vol. 36-Planets and Pulsars
ed. J. A. Phillips, S. E. Thorsett, and S. R. Kulkarni ISBN 0-937707-55-4

Inquiries concerning these volumes should be directed to the:
Astronomical Society of the Pacific
CONFERENCE SERIES
390 Ashton Avenue
San Francisco, CA 94112-1722
415-337-1100

ASTRONOMICAL SOCIETY OF THE PACIFIC
CONFERENCE SERIES

Volume 36

PLANETS AROUND PULSARS

Edited by
J. A. Phillips, S. E. Thorsett, and S. R. Kulkarni

Table of Contents

Preface

The conference *Planets Around Pulsars* was held at Caltech on April 30 and May 1, 1992, only four months after the official announcement of the discovery of planets orbiting PSR B1257+12. The timing may seem precipitate to some, but they might be surprised to learn that the subject of the meeting is actually two decades old. In this Preface, we will briefly outline the history of a field that spent twenty years as one of the tiniest and least noticed niches in astronomy.

The first planet around around a pulsar was reported in 1970, by Richards *et al.* They had used the Arecibo radio telescope to measure pulse arrival times from the Crab Nebula pulsar. After subtracting a model of the pulsar's linear spin-down from their data, the timing "residuals" showed a remarkable thing: a quasi-sinusoidal oscillation with a period of about 77 days that they interpreted as evidence for an Earth-mass planet in a ~ 0.4 AU circular orbit. Their observations generated tremendous interest, and theorists wrote a flurry of papers explaining how planets could survive the supernova explosion or form in its aftermath. Some of those early works have been forgotten but they are very relevant today (see the bibliography at the end of this section).

Unfortunately, subsequent timing measurements showed that the oscillations were not really periodic. Richards *et al.* had discovered not a planet but timing noise—a subject less interesting to planetary scientists, perhaps, but of great importance to pulsar observers and condensed matter physicists studying neutron star interiors.

The "discovery" of planets in the Crab Nebula was wrong, but it prompted pulsar observers to inspect their data more carefully for evidence of planetary companions. In 1979, Demiański & Prószyński reported evidence for a $\sim 0.1 M_\oplus$ planet orbiting PSR B0329+54 every 3 years. By 1979, however, more was known about timing noise and many were suspicious of this result. Cordes and Downs (1985) examined independent timing measurements and found no evidence for coherent oscillations. More recently Bailes, Lyne, and Shemar (these proceedings) have analysed their own timing data and assert that the periodicity still exists. So the case of PSR B0329 is controversial.

Except for the modest attention given to PSR B0329, interest in planets around pulsars flagged from the mid-seventies until 1991. Then everything changed. On July 25th, 1991, Bailes, Lyne, and Shemar surprised astronomers with their announcement of a planet orbiting PSR B1829−10. In most respects PSR B1829 was a garden variety, slow pulsar, but its timing residuals exhibited a high-Q sinusoidal oscillation that Bailes *et al.* had tracked for five years. The putative planet was about $10 M_\oplus$ in a circular orbit ~ 0.7 AU from the neutron star. As had happened twenty years before, the discovery of a planet created a stir of activity. Approximately twenty scientific papers and popular articles were published on PSR B1829−10 in the first six months after the announcement.

The planet orbiting PSR B1829 had one striking peculiarity: its orbital period was within one percent of six months. In their discovery paper, Bailes *et al.* noted the coincidence between the period of the pulsar planet and that of Earth, but concluded that the planet was probably real. They pointed out that they had timed several hundred other pulsars with the same hardware and software and that similar oscillations were absent in all those cases. Furthermore, there was no similar periodicity in the residuals of PSR B1829−08, an object two degrees from PSR B1829−10. The discovery seemed secure.

Meanwhile, in 1990, Alex Wolszczan discovered a millisecond pulsar, PSR B1257+12, in an undirected search with the Arecibo radiotelescope (described in this volume). As he collected timing data over the next year, Wolszczan realized that there was something unusual about this pulsar: its period varied, with a timescale of months, in a highly ordered but indecipherable pattern. The class of millisecond pulsars, to which PSR B1257 belongs, is widely known for its extreme frequency stability—indeed, no intrinisic timing noise has been seen in any millisecond pulsar. In mid-1991, with the news of the Jodrell discovery fresh in the air and astrometric measurements made by Dale Frail at the VLA in hand, Wolszczan realized that the odd timing residuals he saw could be well explained by a sum of two nearly perfect sinusoids—the gravitational signature of a two-planet system. Wolsczcan and Frail had evidently discovered not just a single companion, but a whole planetary system. After months of waiting, while they checked that the orbital model correctly predicted future observations, Wolszczan and Frail scheduled nearly simultaneous announcements in the pages of *Nature* and at the 1992 winter meeting of the American Astronomical Society, held in Atlanta, Georgia.

The weeks before the AAS meeting in January found trouble for the Jodrell team, however. They found a bug their analysis software, by which the eccentricity of the Earth's orbit was neglected in part of the reduction of the pulsar arrival times from the reference frame of the observatory to the inertial reference frame of the Solar System center-of-mass. Fixing the code had the unfortunate side effect of eliminating the biannual oscillation that they had interpreted as evidence for a planetary system. PSR B1829 was the only pulsar affected because the error depended on the difference between the true position of the pulsar on the sky and the position assumed for the data reduction, and large intensity variations in the pulsar signal had led to misidentification of the pulsar's position with their single dish telescope. At the AAS meeting, Lyne and his group retracted their discovery, in a talk right before Wolszczan's announcement.

Given the checkered history of planets around pulsars we felt it valuable to organize a meeting where experts in the field could debate the evidence in favor of planets, propose critical tests, and put forward alternative models. It is not our purpose to summarize the meeting here, but we note that by the end of the conference most—if not all—of the participants were convinced of the reality of Wolszczan and Frail's signal. Don Backer delivered an especially important talk in which he presented Green Bank timing observations of PSR B1257+12 that agreed very well with the Arecibo results. The confirmation of sinusoidal timing residuals obtained with different telescopes and independent timing software (see also Backer *et al.* 1992) shows that the basic data are correct. But what about the interpretation? Some have suggested precession models to explain the data, but at present, planets appear to be the simplest explanation. Malhotra *et al.* and Rasio *et al.* (these proceedings) have proposed observational tests based on dynamical interactions between the planets that should prove or falsify the planetary hypothesis within about a decade.

We were surprised by the overwhelming number of models put forth at the meeting explaining the origin of the planets. While no such model yet seems to answer all its critics, the variety of plausible models suggests that planets can probably be formed in nearly any and all astrophysical sites. The fact that such low mass objects are now known to exist outside the solar system, and that their formation does not require extraordinary conditions found only in certain proto-stellar nebulae, is perhaps the biggest impact that the discovery of planets orbiting PSR B1257+12 has had. The discovery emphasizes the importance of new approaches and ideas in the search for planetary systems, while also giving new impetus to searches towards conventional stars using conventional techniques.

A secondary purpose of the meeting was tutorial. The issues raised by planets around pulsars are interdisciplinary, combining elements of pulsar astronomy, astrophysics, and planetary science. To bridge the gap between these diverse fields we scheduled review talks on the fundamentals: pulsar timing, neutron star formation, protoplanetary disks, and theories of planet formation. We wish to thank those speakers who worked to make their talks educational for newcomers to their discipline.

The 1992 LA riots broke out on the first day of the meeting, and we want to thank all the participants for their understanding of the resulting logistical difficulties. We especially thank the speakers, all of whom showed up for their talks. We owe a giant debt of gratitude to several people who helped out during the crisis. Rachel Akeson and Leila Belkora shuttled many participants safely to and from their hotels, Dhiraj Thakkar provided emergency lodging for one participant, and Helen Knudsen offered her hospitality to 12 others. The conference set-up and registration were accomplished with the help of Rachel Akeson, Leila Belkora, José Navarro, Gita Patel, Jagmit Sandhu, and Todd Small. Claire Chandler was a valuable source of editorial advice in preparing these proceedings and she helped affix headers to many of the papers in the volume. Alex Wolszczan provided the cover illustration. We extend our sincere thanks to everyone who helped.

We would also like to thank the Scientific Organizing Committee for their advice on the scientific agenda: David Black, Phil Nicholson, Steve Ostro, Sterl Phinney, Ed Salpeter, Anneila Sargent, Dave Stevenson, and Joe Taylor. Special thanks are due to Sterl Phinney and Anneila Sargent who, with the three of us, set the program, schedule, and invited speaker list.

As interest in the meeting grew, we discovered just how expensive an international conference can be. This one would not have been possible without very generous support from the NASA "Origins of the Solar System" program, through the Lunar and Planetary Institute. We also gratefully acknowledge Tony Readhead and Dave Stevenson, who supported the meeting with funds from the Astronomy Department and the Department of Geology and Planetary Sciences at Caltech.

Tony Phillips
Steve Thorsett
Shri Kulkarni

REFERENCES

Richards, D.W., Pettengill, G.H., Counselman, C.C. & Rankin, J.M. 1970, "Quasi-sinusoidal oscillation in arrival times of pulses from NP 0532," ApJ, 160, L1.

Colgate, S.A. 1970, "Ejection of companion objects by supernovae," Nature, 225, 247.

Hills, J.G. 1970, "Planetary companions of pulsars," Nature, 226, 730.

Michel, F.C., 1970, "Pulsar planetary systems," ApJ, 159, L25.

Rees, M.J., Trimble, V.L. & Cohen, J.M. 1971, "Planet, pulsar, glitch, and wisp," Nature, 229, 395.

Treves, A. 1971, "Importance of pulsar observations for planetary physics," A& A, 15, 471.

Demiańsky, M. & Proszyński, M., 1979, "Does PSR0329+54 have companions?" Nature, 282, 383.

Cordes, J.M. & Downs G.S. 1985, "JPL pulsar timing observations III. Pulsar rotation fluctuations", ApJ Supp, 59, 343.

Bailes, M., Lyne, A.G. & Shemar, S.L. 1991, "A planet orbiting the neutron star PSR1829-10," Nature, 352, 311.

Wolszczan, A. & Frail, D.A. 1992, "A planetary system around the millisecond pulsar PSR 1257+12," Nature, 355, 145.

Lyne, A.G. & Bailes, M. 1992, "No planet orbiting PSR 1829-10," Nature, 355, 213.

Backer, D.C., Sallmen, S. & Foster, R. 1992 "Pulsar's double period confirmed," Nature, 358, 24.

Registered Participants

Rachel Akeson
Caltech 105-24
Pasadena, CA 91125
rla@deimos.caltech.edu

Thomas Ahrens
Caltech 252-21
Pasadena, CA 91125
tja@seismo.gps.caltech.edu

Stuart Anderson
Caltech 206-49
Pasadena, CA 91125
sba@ccsf.caltech Edu

Arkady Dolginov
3600 Bay Area Blvd
Lunar and Planetary Institute
Houston, TX
dolginov@lpi.nasa.gov

Jonathan Arons
Astronomy Department
601 Campbell Hall
University of California
Berkeley, CA 94720
arons@astroplasma.berkeley.edu

Don Backer
601 Campbell Hall
University of California
Berkeley, CA 94720
dbacker@bkypsr.berkeley.edu

Matthew Bailes
NRAL Jodrell Bank
Macclesfield
Cheshire SK119DL UK
mxb@star.jb.man.ac.uk

Menashe Banit
Columbia University
Columbia Astrophysics Lab
Physics Department
538 West 170th Street
New York, NY 10027
mbi@cuthry.bitnet

Mary Barsony
MS-78
60 Garden Street
Harvard-Smithsonian CFA
Cambridge, MA 02138
fun@cfa.harvard.edu

Leila Belkora
Caltech 264-33
Pasadena, CA 91125
leila@sundog.caltech.edu

Garrett Biehle
Caltech 130-33
Pasadena, CA 91125
biehle@tapir.caltech.edu

Lars Bildsten
Caltech 130-33
Pasadena, CA 91125
bildsten@tapir.caltech.edu

Geoffrey A. Blake
Caltech 170-25
Pasadena, CA 91125
gab@csardas.gps.caltech.edu

Peter Bodenheimer
Lick Observatory
University of California
Santa Cruz, CA 95060
peter@helios.ucsc.edu

Leigh Brookshaw
LLNL L-413
P.O. Box 808
Livermore, CA 94550
leigh@sunlight.llnl.gov

A. G. W. Cameron
Harvard College Observatory
60 Garden Street
Cambridge, MA 02138
cameron@cfa.harvard.edu

Pat Cassen
NASA — Ames
245-3
Moffett Field, CA 94035
gal::cassen (Span)

Sandip Chakrabarti
Theoretical Astrophysics
TIFR
Homi Bhabha Rd.
Bombay 400005 India
chakraba@tifrvax.bitnet

Claire Chandler
Caltech 105-24
Pasadena, CA 91125
cjc@milli.caltech.edu

James Chiang
Dept. of Physics
Stanford Univ.
Stanford, CA 94305-4060
jwc@egret0.stanford.edu

Chris Chyba
Laboratory for Extraterrestrial Physics
Code 693
NASA Goddard Space Flight Center
Greenbelt, MD 20771
chyba@pan.asc.nasa.gov

Bill Cochran
Dept of Astronomy
Univ. of Texas
Austin, TX 78712
wdc@astro.as.ateras.edu

Marshall Cohen
Caltech 105-24
Pasadena, CA 91125

Jim Cordes
520 Space Science Bldg
Cornell University
Ithaca, NY 14853
cordes@astrosun.tn.cornell.edu

Ken Croswell
2430 Prospect St, #310
Berkeley, CA 94704

Melvyn B. Davies
MS 10
60 Garden St
Cambridge, MA 02138
melvyn@cfa160.harvard.edu

Will Deich
Caltech 105-24
Pasadena, CA 91125
will@surya.caltech.edu

W. F. Detlefs
Caltech 105-24
Pasadena, CA 91125
wfd@surya.caltech.edu

Rachel Dewey
JPL
MS 301-125
Pasadena, CA 91109
rjd@sgra.jpl.nasa.gov

Rosanne Distefano
Physics 6-201
MIT
Cambridge, MA 02139
rd@space.mit.edu

Dale A. Frail
National Radio Astronomy Obs
Socorro, NM 87801
dfrail@nrao.edu

Andy Fruchter
601 Campbell Hall
Astronomy Dept
Berkeley, Ca 94709
asf@orestes.berkeley.edu

Andrea Ghez
Caltech 320-47
Pasadena, CA 91125
andrea@mop.caltech.edu

Alan Harris
JPL
MS 183-501
Pasadena, CA 91109
awharris@jplgp.jpl.nasa.gov

Lee Hartmann
Center for Astrophysics
60 Garden Street
Cambridge, MA 02138
hartmann@cfa.harvard.edu

Carl Heiles
Department of Astronomy
University of California
Berkeley, CA 94720
heiles@bkyast.berkeley.edu

Ed van den Heuvel
Astronomical Institute
Univ. of Amsterdam
Kruislaan 403
1098 SJ Amsterdam
The Netherlands
edvdh@astro.uva.nl

Jan Gil
MPI für Radioastronomie
Auf Dem Hugel 69
5300 Bonn 1
Germany
v584jag@mpirbn.mpifr-bonn.mpg.de

Helen Johnston
Sterrekundig Institute
University of Utrecht
Utrecht
The Netherlands
johnston@fys.ruu.nl

Simon Johnston
ATNF — CSIRO
PO Box 76
Epping, NSW 2121
Australia
sjohnston@atnf.csiro.au

Vicky Kaspi
Physics Dept
Princeton University
Princeton, NJ 08544
vicky@puppsr.princeton.edu

Jocilyn Keene
Caltech 320-47
Pasadena, CA 91125

John F. Kerridge
Inst. Geophysics
UCLA
Los Angeles, CA 90024-1567

David Koerner
Caltech 170-25
Pasadena, CA 91125
david@satur1.gps.caltech.edu

Julian Krolik
Dept. of Physics & Astronomy
Johns Hopkins Univ.
Baltimore, MD 21218
jhk@gauss.pha.jhu.edu

Shri Kulkarni
Caltech 105-24
Pasadena, CA 91125
srk@surya.caltech.edu

Jack Lissauer
Astronomy Program
SUNY
Stony Brook, NY 11794
jlissauer@sbccmail.bitnet

Scott Lundgren
518 Space Sciences
Cornell University
Ithaca, NY 14850
lundgren@astrosun.tn.cornell.edu

Tim Lynch
Caltech 105-24
Pasadena, CA 91125
tlynch@juliet.caltech.edu

Renu Malhotra
Lunar & Planetary Institute
3600 Bay Area Blvd
Houston, TX 77058
renu@lpi.jsc.nasa.edu

Erik A. Martinez
Institute for Theoretical Physics
University of Berne
Sidlerstrasse 5
CH-3012 Bern
Switzerland
martinez@butp.unibe.ch

Stuart Mcmuldroch
Caltech 170-25
Pasadena, CA 91125
sxm@sol1.caltech.edu

Larry Molnar
Dept of Physics & Astronomy
University of Iowa
Iowa City, IA 52242
lam@lamsun.physics.uiowa.edu

Norman Murray
Caltech 130-33
Pasadena, CA 91125
murray@tapir.caltech.edu

Tadashi Nakajima
Caltech 105-24
Pasadena, CA 91125
tn@miranda.caltech.edu

Jose Navarro
Caltech 105-24
Pasadena, CA 91125
jn@surya.caltech.edu

David Nice
Physics Dept, Princeton Univ
P.O. Box 708
Princeton, NJ 08544
david@pulsar.princeton.edu

Phil Nicholson
Space Sciences
Cornell University
Ithaca NY 14853
nicholson@astrosun.tn.cornell.edu

Michael Nowak
Dept. of Physics
Stanford, CA 94305-4060
man@mensh.stanford.edu

Joe Nuth
Code 691
NASA/GSFC
Greenbelt, MD 20771
lepvax::uijan

Steve Ostro
JPL
300-233
Pasadena, CA 91109
berlin::ostro
ostro@echo.jpl.nasa.gov

Stan Peale
Department of Physics
University of California
Santa Barbara, CA
peale@io.ucsb.edu

Chris A. Perez
Physics Dept
Stanford University
Stanford, CA 94305
perez@mensch.stanford.edu

J. A. Phillips
Caltech 105-24
Pasadena, CA 91125
jap@milli.caltech.edu
phillips@surya.caltech.edu

Sterl Phinney
Caltech 130-33
Pasadena, CA 91125
esp@tapir.caltech.edu

Philipp Podsiadlowski
Institute of Astronomy
Cambridge CB3 OHA
U.K.
ppodsi@starlink.astronomy.cambridge.ac.uk

Tom Prince
Caltech 220-47
Pasadena, CA 91125
prince@citsrl.srl.caltech.edu

Fred Rasio
Institute for Advanced Study
Olden Lane
Princeton, NJ 08540
rasio@guinness.ias.edu

Paul S. Ray
Caltech 206-49
Pasadena, CA 91125
paulr@caltech.edu

Andreas Reisenegger
Caltech 130-33
Pasadena, CA 91125
andreas@iago.caltech.edu

Martin Rees
Institute of Astronomy
Madingley Road
Cambridge CB3 OHA
England

Paul Robinson
JPL 180-500
Pasadena, CA 91109
probinson@vlsi.jpl.nasa.gov

Steve Ruden
Physics Department
University of California
Irvine, CA 92717
sruden@nebula.ps.uci.edu

Shauna Sallmen
Department of Astronomy
University of California
Berkeley, CA 94720
sallmen@bkyast.berkeley.edu

Jagmit Sandhu
Caltech 105-24
Pasadena, CA 91125
jss@surya.caltech.edu

Basilio X. Santiago
Caltech 105-24
Pasadena, CA 91125
basilio@deimos.caltech.edu

Anneila Sargent
Caltech 105-24
Pasadena, CA 91125
afs@mmstar.caltech.edu

Steinn Sigurdsson
Lick Observatory
University of California
Santa Cruz, CA 95064
steinly@helios.ucsc.edu

Nick Scoville
Caltech 105-24
Pasadena, CA 91125

Todd Small
Caltech 105-24
Pasadena, CA 91125
tas@elgar.caltech.edu

Tomasz Stepinski
Lunar and Planetary Institute
3600 Bay Area Bldg
Houston, TX 77058
stepinski@lpi.jsc.nasa.gov

Karen M. Strom
Five College Astronomy
University of Mass
Amherst, MA 01003
kstrom@hanksville.phast.umass.edu

Stephen Strom
Five College Astronomy
University of Mass
Amherst, MA 01003
sstrom@donald.phast.umass.edu

Curt Struck-Marcell
Department of Physics & Astronomy
Iowa State University
Ames, IA 50011
s1.csm@isumvs.iastate.edu

Dhiraj Thakkor
Caltech 105-24
Pasadena, CA 91125
ddt@surya.caltech.edu

Stephen Thorsett
Caltech 105-24
Pasadena, CA 91125
steve@surya.caltech.edu

Susan Terebey
IPAC 100-22
Pasadena, CA 91125
st@ipac.caltech.edu

Scott Tremaine
CITA
Mclennan Labs
60 St George St
Univ. of Toronto
Toronto, M5S 1A7
Canada
tremaine@cita.utoronto.ca

David Van Buren
Caltech 130-33
Pasadena, CA 91125
dave@groucho.caltech.edu

Gautam Vasisht
Caltech 105-24
Pasadena, CA 91125
gv@surya.caltech.edu

Mark Voit
Caltech 130-33
Pasadena, CA 91125

Dan Werthimer
Space Science Labs
University of California
Berkeley, CA 94720
danw@ssp.berkeley.edu

Alex Wolszczan
Penn. State University
Dept. of Astronomy & Astrophysics
515A Davey Lab
University Park, PA 16802
alex@astro.psu.edu

Dotty Woolman
Physics Department
California State Univ.
Fullerton, CA 92634
woolman@fullerton.edu

Stan Woosley
Lick Observatory
Astronomy Department
UCSC
Santa Cruz, CA 95064
woosley@helios.ucsc.edu

Ben Zuckerman
Astronomy Dept
UCLA
Los Angeles, CA 90024
zuckerman@uclastro.bitnet

Ellen Zweibel
APAS Dept Box 391
University of Colorado
Boulder, CO 80309
solarz::zweibel

Section 1:

PSR 1257+12 and its Planetary Companions

PSR1257+12 AND ITS PLANETARY COMPANIONS

A. WOLSZCZAN[1]
Joseph Henry Laboratories and Physics Department, Princeton University, Princeton, New Jersey 08544

ABSTRACT The most recent results of timing observations of the 6.2–ms radio pulsar PSR1257+12 are discussed. As demonstrated earlier by Wolszczan and Frail (1992), postulating the presence of two Earth–like mass companions around the pulsar provides the most plausible explanation of its unusual timing behavior. A timing model including the standard pulsar parameters and two non–interacting, Keplerian orbits gives an excellent fit to the observed pulse arrival times. A firm upper limit to the total amplitude contributed by any presently unmodeled effects is set at ~3 μs by this fit. Consequently, any alternative models to explain the timing results must be able to mimic Keplerian orbits with the above accuracy over many cycles of the observed periodicities. The two planet–like objects have minimum masses of $3.4M_{\oplus}$ and $2.8M_{\oplus}$, and orbit the pulsar at distances similar to that of Mercury from the Sun with the respective periods of 66.5 and 98.2 days.

INTRODUCTION

Very high rotational stability of radio pulsars makes them excellent "cosmic clocks" with a variety of important applications in fundamental physics and astrophysics. A potential of the pulse timing method to detect planets or planetary systems around neutron stars has been indicated in the past by reports on possible planet–sized companions to the Crab pulsar (Richards et al. 1970), PSR0329+54 (Prószyński and Demiański 1979) and, most recently, to PSR1829-10 (Bailes, Lyne and Shemar 1991). In these and other similar cases, further studies have demonstrated that rather than originating from orbital motion, the observed periodic behavior of the pulse arrival times was caused by timing noise or by inaccuracies in the data analysis.

PSR1257+12 was discovered with the 305–m Arecibo radiotelescope during the first Arecibo survey specifically designed to detect millisecond pulsars at high galactic latitudes (Wolszczan 1991). The discovery has been followed by systematic timing measurements of the new pulsar at Arecibo and by astrometric observations with the Very Large Array (VLA). Two strict periodicities that could not be accounted for by the standard timing model have been extracted from timing residuals and interpreted in terms of the orbital motion of planet–like objects around the pulsar (Wolszczan and Frail 1992). So far, this finding

[1]On leave from Arecibo Observatory, Arecibo, PR 00613

constitutes the most suggestive available evidence for the existence of a planetary system other than our own.

The purpose of this paper is to present a summary of observations and analysis that have led to the above conclusion and to describe the updated timing model of PSR1257+12 based on almost two years of continuing measurements of the pulsar with the Arecibo telescope.

OBSERVATIONS AND ANALYSIS

A recent survey of the sky at high galactic latitudes with the 305–m Arecibo radiotelescope has revealed two new millisecond radio pulsars: PSR1257+12 and PSR1534+12. The characteristics of this survey and a description of the relativistic binary pulsar PSR1534+12 are given in Wolszczan (1991). In contrast with PSR1534+12, the initial study of the 6.2–ms pulsar, PSR1257+12, was not conclusive and long–term observations were clearly necessary, to understand the pulsar's timing properties. A description of these observations can be found in Wolszczan and Frail (1992). Here, the analysis of TOA measurements made over a period of 18 months is summarized and supplemented with the results based on the larger data set, whose time span has now grown up to almost two years.

Systematic measurements of the times of arrival (TOA) of pulses from PSR1257+12 have been made at Arecibo since July 1990. The results published by Wolszczan and Frail (1992) were based on observations made at 430 MHz with the 40–MHz, 3–level autocorrelation spectrometer utilized as a pulsar backend. In this case, a typical uncertainty of TOA measurements was $\sim 15\ \mu$s. Later in 1990, adding the Princeton Mark III pulsar processor to the backend configuration, in parallel with the 40–MHz correlator, has improved the accuracy of TOA determination by almost a factor of two. A timing model including the pulsar's rotational parameters (period, P and its derivative, $\dot{P}$) and the celestial coordinates, when fitted to the data, has resulted in large post–fit residuals shown in Figure Ia. These residuals measure deviations of the arrival times predicted by the model from the actual TOAs and clearly indicate a quasi–periodic "wandering" of TOAs with an amplitude of about $\pm$ 2.5 ms. A timing "noise" of this magnitude is entirely unknown in very old, millisecond pulsars, which are known to be extraordinarily stable rotators (e.g. Davis et. al 1985). Moreover, further analysis has shown that the quasi–periodic TOA variations of PSR1257+12 consist of two strictly periodic processes with periods of 66.5 and 98.2 days, as illustrated in Figure Ib,c. Careful tests have eliminated instrumental problems as a possible source of these periodicities.

The observed residuals are difficult to explain in terms of the phenomena such as precession of the neutron star axis (Nelson, Finn and Wasserman 1990) and interstellar or circumpulsar propagation effects (Foster and Cordes 1991; Fruchter, Stinebring and Taylor 1988). Precession would have to account for two periodicities and the evident lack of systematic changes in the pulse profile morphology down to a 1% level (Kaspi and Wolszczan, these proceedings). Any propagation–related phenomena are ruled out by the observed frequency independence of TOA variations. Consequently, the most plausible remaining possibility is that the observed residuals are caused by a reflex motion of the pulsar in response to the presence of orbiting low–mass companions.

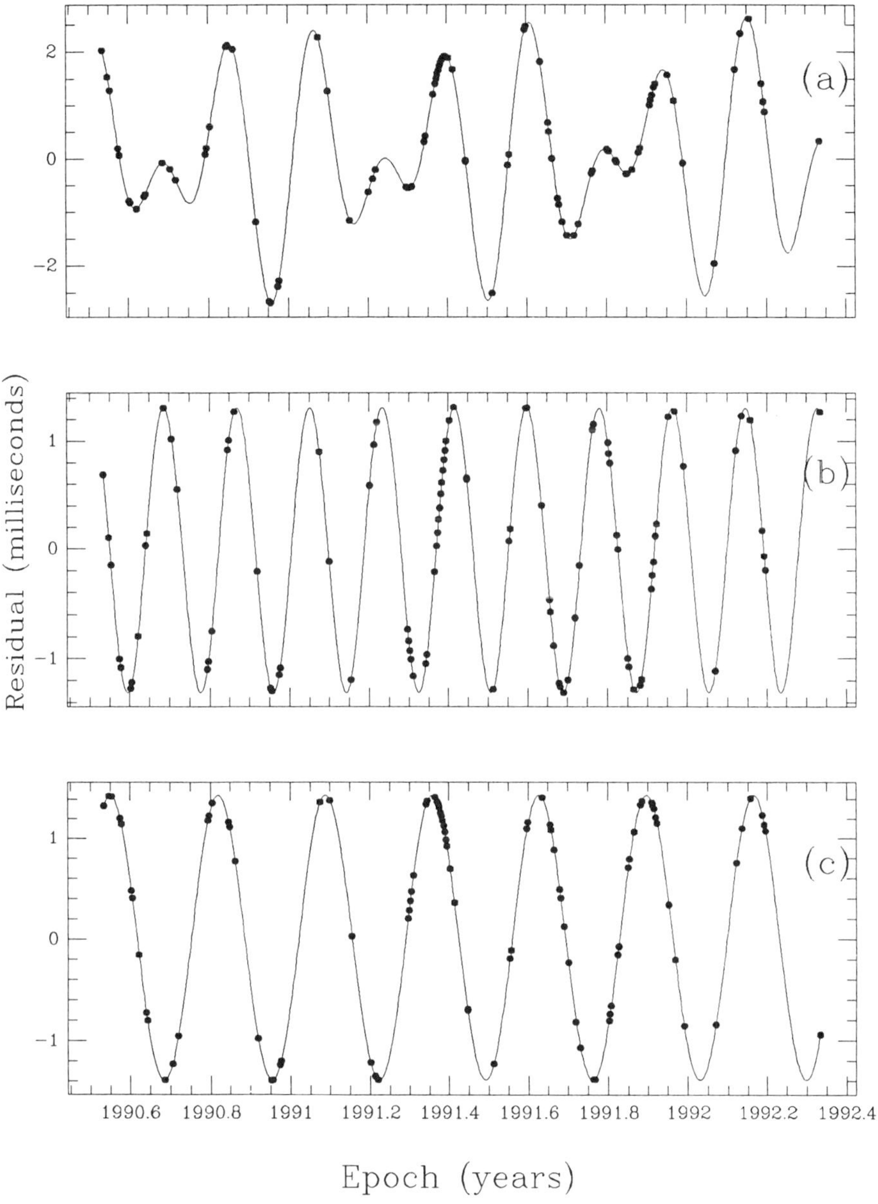

FIGURE I The post–fit residuals of pulse arrival times from PSR1257+12. (a) Fit for rotational and positional parameters only, (b) fit for a 98.2–day Keplerian orbit (leaves a 66.5–day periodicity as residual), (c) fit for a 66.5–day Keplerian orbit (leaves a 98.2–day periodicity as residual).

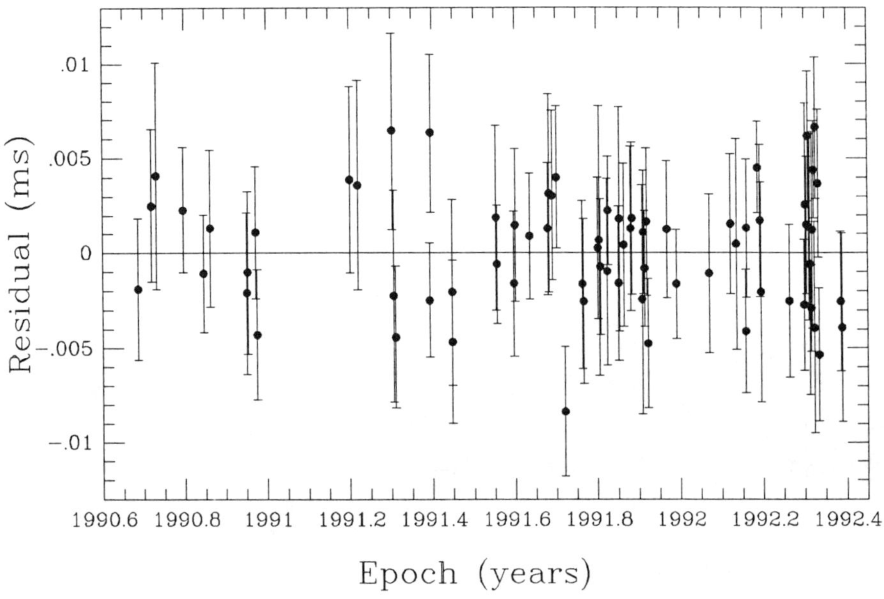

FIGURE II The post–fit residuals of pulse arrival times from PSR1257+12 for the timing model including the standard pulsar parameters and two non–interacting Keplerian orbits.

The planetary system hypothesis has been tested by constructing a timing model that contained two non–interacting Keplerian orbits, in addition to a set of standard parameters (phase, P, $\dot{P}$, celestial coordinates and proper motion). The result of a fit of this model to the TOAs measured between July 1990 and May 1992 is shown in Figure II. The rms magnitude of the post–fit, daily–averaged residuals for the two–planet model is ~ 3 μs with no discernible systematic variation. The model parameters of the pulsar and its assumed two companions are listed in Table I. In most cases, the accuracy of parameter estimates has improved, as compared to the original values given by Wolszczan and Frail (1992). The derived parameters have been obtained from standard considerations involving very low–mass objects in Keplerian orbits around a $1.4M_{\odot}$ neutron star.

One important difference between the old model and its updated version presented here is that the proper motion must be added to the new model, in order to fully account for the measured TOA variations over an increased period of time (Table I). Assuming the distance to PSR1257+12 to be about 400 pc, one obtains the pulsar's transverse motion of the order of $\sim$200 km s^{-1}, directed approximately toward the galactic plane. Although this result is entirely reasonable, the fit for proper motion parameters does not eliminate a $0.7''$ discrepancy between the VLA and the timing positions, which is considerably greater than the conservative error estimates (Table I). This discrepancy has been taken by Wolszczan and Frail (1992) as an indication that a presence of a third planet with the orbital period comparable to one year could have biased the timing position. Since the proper motion fit to the presently available data may be affected in a similar way, the above result must be treated with caution. Hopefully, a few more years of timing observations aided by the timing-independent astrometry of the pulsar with the VLA and VLBI will help to resolve these ambiguities. Since the currently measured value of the second period derivative, $\ddot{P}$,

TABLE I Parameters of the PSR1257+12 System

Pulsar parameters		
Rotational period, P	0.006218531931908(5) s	
Period derivative, $\dot{P}$	$1.143(2) \times 10^{-19}$ s s^{-1}	
α (B1950.0)	$12^h\ 57^m\ 33^s.121(1)$	
δ (B1950.0)	$12^\circ\ 57'\ 06''.57(1)$	
μ_α (B1950.0)	45(5) mas yr^{-1}	
μ_δ (B1950.0)	$-86(9)$ mas yr^{-1}	
Epoch	JD 2448088.9	
Dispersion measure	10.188(5) pc cm^{-3}	
Flux density (430 MHz)	20(5) mJy	
Flux density (1400 MHz)	1.0(2) mJy	
Surface magnetic field, B	8.8×10^8 G	
Characteristic age, τ_c	0.8×10^9 yr	
Keplerian orbital parameters		
Projected semi-major axis, $a_1 sin\ i$	1.312(2) light ms	1.413(2) light ms
Eccentricity, e	0.020(2)	0.024(2)
Epoch of periastron, $T_\circ$	JD 2448104(1)	JD 2447998(1)
Orbital period, P_b	5749068(500) s	8484936(1000) s
Longitude of periastron, ω	247°(10)	105°(10)
Parameters of the planetary system		
Planet mass, $m_{2,3}$ ($M_\oplus$)	$3.4/sin\ i$	$2.8/sin\ i$
Distance from the pulsar, d (AU)	0.36	0.47
Orbital period, P_b (days)	66.54	98.20

of the pulsar is not significant, the effect of any more distant objects that could be present in the 1257+12 system would be entirely absorbed in $\dot{P}$.

The pulsar period variations predicted by this model are displayed in Figure III together with the observed period values derived from timing data. This analysis is analogous to the single-line optical spectroscopy. Note that the presence of the planet-sized bodies orbiting the pulsar results in the apparent period variations of only ±15 picoseconds. This is caused by the pulsar's reflex motion which is characterized by the projected maximum velocity and displacement amplitudes of ±0.7 m/s and ±900 km, respectively. Given the timing accuracy

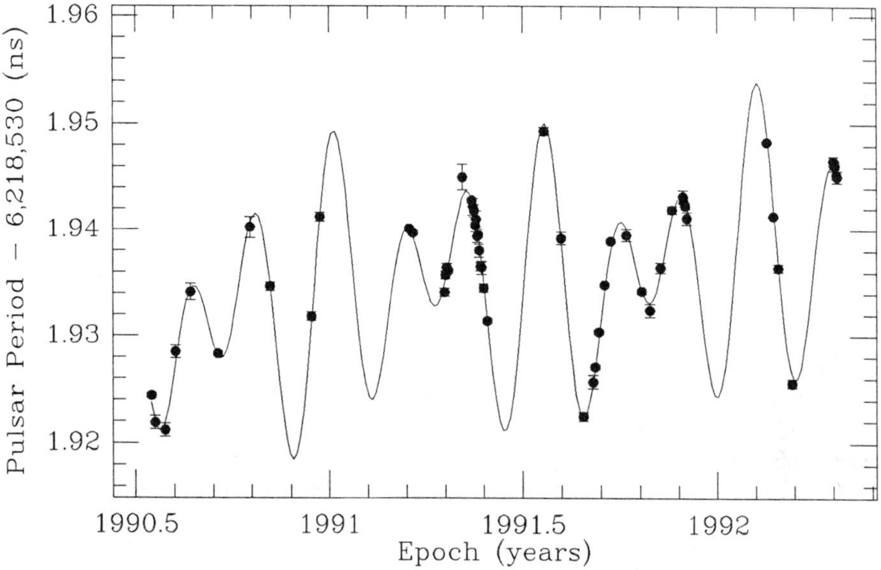

FIGURE III Period variations of PSR1257+12. Each period measurement is based on observations made on at least two consecutive days. The solid line denotes changes in period predicted by a two–planet model of the 1257+12 system.

of PSR1257+12, the pulsar's reflex motion caused by planetary masses smaller than that of the Moon could easily be detected with this technique.

DISCUSSION

The high quality of the two–companion model fit to the pulsar timing data covering many orbital cycles provides compelling evidence that PSR1257+12 possesses a planetary system consisting of at least two planet–sized bodies. Indeed, a comparison of post–fit residual amplitudes of Figure Ia and Figure II shows that the planetary model reduces the residuals by a factor of nearly 500 with respect to the model restricted to the standard parameters of a single pulsar. The resulting 3 μs rms residual places an upper limit to the magnitude of any effects that have not been accounted for by the present model. In particular, a detection of the consequences of neglecting a gravitational interaction between the two planets would be of great interest. As pointed out by Rasio et al. (1992) and Malhotra et al. (1992), a near–3:2 resonance condition of the 1257+12 system should lead to precisely calculable perturbations of the planetary orbits. In time, the effect of these perturbations should begin to manifest itself in the form of predictable departures from the timing model discussed above.

PSR1257+12, with its old age and a low surface magnetic field (Table I), belongs to the class of neutron stars which are believed to evolve in low mass binary systems and are spun up to the observed millisecond periods by accretion of matter from their stellar companions (e.g., Bhattacharya and van den Heuvel 1991). Since it is not very likely that any primordial planets would survive this kind of evolution (Nakano 1987), the observed 1257+12 system probably consists of "second generation" planets created at or after the end of the pulsar's binary history. Because a second supernova explosion is not expected to occur in a low–mass binary, such planets could form a stable system of bodies orbiting an old neutron star. Another important evolutionary constraint is provided by the observed very low eccentricities of the planetary orbits. This suggests the creation of planets from some form of a circumpulsar disk that would naturally provide means to circularize the orbits. Consequently, it appears that any plausible mechanism for the creation of a planetary system around a millisecond pulsar must provide a way to remove its stellar companion in a manner different from the disruption of a binary system caused by a supernova explosion and to retain enough circum–pulsar matter to form planet–sized objects. For example, one can envision detailed scenarios of planet formation, based on the evaporation and disruption of the pulsar's stellar companion (Stevens, Rees and Podsiadlowski 1992; Tavani and Brookshaw 1992).

The most recent results of timing analysis of PSR1257+12 described here strengthen the evidence presented by Wolszczan and Frail (1992) that this object is accompanied by a system of two or more planet–sized bodies. As outlined by Rasio et al. (1992) and Malhotra et al. (1992), an "ultimate test" of this hypothesis can be performed by detecting gravitational perturbations of the planetary orbits. Since the interactions between the two planets can be accurately predicted and incorporated in the timing model, their effect on the observed pulse arrival times is, in principle, measurable. Preliminary simulation analysis of this problem indicates that the perturbations should be detectable within a few years of continuing timing observations with a $\sim$1 μs precision.

The existence of the 1257+12 planetary system seems to indicate that planet–sized bodies can form under surprisingly diverse conditions. Moreover, our present understanding of the millisecond pulsar evolution does not contradict the idea that planetary companions to very old, "recycled" neutron stars may be not entirely uncommon. Since the new surveys at Arecibo and elsewhere are beginning to uncover more such pulsars (e.g., D. Nice, private communication; Wolszczan, Foster and Camilo, in preparation), one can realistically predict that this truly exciting possibility can be verified. Also, it appears that much more analysis of the existing timing data and new observations of the so–called "normal" pulsars have to be performed, to properly assess detectability of planets around these objects (Thorsett and Philips 1992). A positive outcome of these efforts would undoubtedly have very interesting consequences for our understanding of the formation and evolution of planetary systems and for future strategies of searches for planets outside the Solar System.

ACKNOWLEDGMENTS

I would like to thank J. Taylor for his contribution to data analysis and for making available the Princeton Mark III processor. Arecibo Observatory is part of the National Astronomy and Ionosphere Center, which is operated by Cornell University under contract with the National Science Foundation.

REFERENCES

Bailes, M., Lyne, A. G. and Shemar, S. L. 1991, *Nature* **352**, 311.
Bhattacharya, D. and van den Heuvel, E. P. J. 1991, *Phys. Rep.* **203**, 1.
Davis, M. M., Taylor, J. H., Weisberg, J. M. and Backer, D. C. 1985, *Nature* **315**, 547.
Demiański, M. and Prószyński, M. 1979, *Nature* **282**, 383.
Foster, R. S. and Cordes, J. M. 1990, *Astrophys. J.* **364**, 123.
Fruchter, A. S., Stinebring, D. R. and Taylor, J. H. 1988, *Nature* **333**, 237.
Malhotra, R., Black, D., Eck, A. and Jackson, A. 1992, *Nature* **356**, 583.
Nakano, T. 1987, *Mon. Not. R. astr. Soc.* **224**, 107.
Nelson, R. W., Finn, L. S. and Wasserman, I. 1990, *Astrophys. J.* **348**, 226.
Rasio, F. A., Nicholson, P. D., Shapiro, S. L. and Teukolsky, S. A. 1992, *Nature* **355**, 325.
Richards, D. W., Pettengill, G. H., Counselman, C. C., III and Rankin, J. M. 1970, *Astrophys. J. (Letters)* **160**, L1.
Stevens, I. R., Rees, M. J. and Podsiadlowski, P. 1992, *Mon. Not. R. astr. Soc.* **254**, 19P.
Tavani, M. and Brookshaw, L. 1992, *Nature* **356**, 320.
Thorsett, S. E. and Phillips, J. A. 1992, *Astrophys. J. (Letters)* **387**, L69.
Wolszczan, A. 1991, *Nature* **350**, 688.
Wolszczan, A. and Frail, D. A. 1992, *Nature* **355**, 145.

Planets Around Pulsars
ASP Conference Series, Vol. 36, 1993
J.A. Phillips, J.E. Thorsett, and S.R. Kulkarni

A PULSAR TIMING TUTORIAL AND NRAO GREEN BANK OBSERVATIONS OF PSR 1257+12

D. C. BACKER
Astronomy Department
601 Campbell Hall
University of California
Berkeley, CA 94720

ABSTRACT The steps required in precise timing of pulsars, including pulsars with orbital perturbations, are outlined. Observations of PSR 1257+12 confirm the double periodicity reported by Wolszczan and Frail (1992) with independent data and independent analysis. Observations of PSR 1855+09 compare favorably with a timing model based on independent data (Ryba & Taylor 1991), and thus serve as a clock and system check. The timing of PSR 1620-26, an 11-ms pulsar in the globular cluster M4 (Lyne *et al.* 1988), show a large period second derivative which, if gravitational, indicates a low mass companion or an encounter at distances of 10-100 AU.

PULSAR TIMING MODEL

I will briefly outline the model and steps involved in pulsar timing. Particular attention is given to the size of various effects which indicates their relative importance. The accuracies of some effects are mentioned, but these can not reflect any systematic errors in applying our knowledge in an analysis system. Some recent references about pulsar timing are: Backer & Hellings (1986); Backer (1988); Taylor (1988).

The arrival of the Nth pulse at the observatory is a space-time event that is recorded by a reading of the 'time' on a local atomic clock T_N. In practice such arrival times are obtained by averaging together, in synchronism with the apparent pulse period, the signals from a band of the spectrum for many rotation periods. Observations are continuous on a given day, but are typically obtained only at intervals ranging from a day to weeks or months. Careful modeling is required to ensure that the rotation number N is unambiguous during observation gaps.

The modeling proceeds in two steps. First one performs a relativistic space-time transformation of intervals at the observatory $(T_N - T_0)$ to intervals at the pulsar $(t_N - t_0)$. Then the intervals at the pulsar are compared to a rotation model that includes terms related to the slow decay of its period and possibly other factors. The transformation can be approximated by a list of additive 'corrections':

$$(t_N - t_0) = (T_N - T_0) + \Delta T_C + \Delta T_R + (\mathbf{R_e} - \mathbf{R_p}) \cdot \hat{\mathbf{n}}/c$$
$$+ \Delta T_S + \Delta T_{n_e}$$

The terms in the first line effect the space-time transformation within the framework of special relativity and with no interstellar medium. ΔT_C corrects the observatory clock to terrestrial proper time (Guinot 1988) and has a typical amplitude of microseconds. The most accurate atomic time scale has an uncertainty of a fraction of a microsecond over a time interval of one year. The second clock term ΔT_R removes the time varying effects of gravitational redshift and transverse Doppler between the terrestrial standard and a clock in the reference frame of the pulsar. The amplitude of the part of this effect that arises from the motion of the Earth in the variable solar system potential has periodic terms with maximum amplitude of 1.66 milliseconds. The variation of the solar system clock with respect to a clock in the pulsar system barycenter (PSB) is slowly varying and indistinguishable from a shift in the star's rotation period. The transformation from the PSB to the pulsar for a nonisolated pulsar will depend on the mass(es) and period(s) of the companion object(s).

Table 1

Planetary Mass Errors

System i	Δm_i^1 (10^{22} gm)	P_i (y)	r_i (AU)	$\Delta R_e/c$ (ns)	Reference
Mercury	1.4	0.25	0.39	1	Mariner 10
Venus	1.5	0.62	0.72	3	Mariner 5/10
Mars	0.2	1.88	1.52	1	Mariner 4
Jupiter	150	11.9	5.20	1950	Voyager 1/2
Saturn	300	29.5	9.52	7140	Voyager 1/2
Uranus	15	84.0	19.2	720	Voyager 2
Neptune	30	165	30.0	2250	Voyager 2

[1] from Anderson (1990) JPL Interoffice Memorandum.

The solar system barycenter (SSB) is taken as the origin of space. In this frame the Earth distance, $|\mathbf{R_e}|/c$, is 500 seconds, the Astronomical Unit, and its scalar product with the direction of the pulsar $\hat{\mathbf{n}}$ can vary with the full amplitude of $\pm$ 500 seconds. Hence one needs to know pulsar positions with great precision in the reference frame defined for the planetary ephemeris to prevent pulse number ambiguity over an entire year. The reference frames of the two ephemerides in common use, CfA 740R and JPL DE118, have a $0.4''$ arc seconds rotation between them. Ephemeris errors, $\Delta R_e/c$, are dominated by uncertainties in the outer planet masses (Δm_i, Table 1) and other orbital elements which cause only small errors in the true pulsar rotation parameters.

An arcsecond error in the pulsar position leads to an annual arrival time error with an amplitude as large as 2.5 milliseconds. The proper motion, μ, of the pulsar produces a time growing error that allows its detection from pulse timing analysis. A transverse velocity of 100 km s^{-1} for a pulsar at 200 pc corresponds to a motion of $0.1''$ y^{-1}, and a timing error after one year as large as 250 microseconds. Even the annual parallactic motion, π, can be detected for precisely timed pulsars at distances less than 1 kpc.

$$\hat{\mathbf{n}} = \hat{\mathbf{n}}(T_0) + \mu(T_N - T_0) + \pi(T_N).$$

The second spatial correction, $\mathbf{R_p}\cdot\hat{\mathbf{n}}/c$, consists of a slowly varying term for the SSB to PSB step and, in the case of a nonisolated pulsar, a periodic term, or terms, from the pulsar motion around the PSB whose amplitude, $a_1 \sin i/c$, depends on the orbital elements of a companion object.

General Relativity enters in the correction ΔT_S, which is often referred to as the Shapiro delay, because the photons traverse the gravitational potential in the solar system (annual variation with amplitude of 10 to 100 microseconds), the interstellar field (a slowly varying yet large term) and the potential in the pulsar system (detectable in a few cases). While this is part of the fundamental space time transformation, it is effectively a propagation term in Euclidean space. A second propagation term, ΔT_{n_e}, comes from the interstellar plasma and is proportional to the thermal electron column density, DM, and the inverse square of the observing frequency, ν. The total delay at 1 GHz relative to the infinite frequency arrival time is 0.03 to 3.0 seconds. Variations in the column density by parts in 10^5 over a year lead to microsecond arrival time variations at 1 GHz. Figure 1 displays the DM variations for PSR 1821-24, a 3-ms pulsar in the globular cluster M28 (Backer *et al.* 1992). Studies of the column density variations of a number of pulsars has led to the following estimate for temporal gradients in the dispersive delay:

$$\frac{\Delta T_{n_e}}{\Delta T} \approx (1 \text{ to } 100)\ \mu\text{s y}^{-1} \left(\frac{DM}{100 \text{ pc cm}^{-3}}\right)^{0.5} \left(\frac{v}{100 \text{ km s}^{-1}}\right) \left(\frac{\nu}{\text{GHz}}\right)^{-2},$$

where v is the transverse velocity from the proper motion.

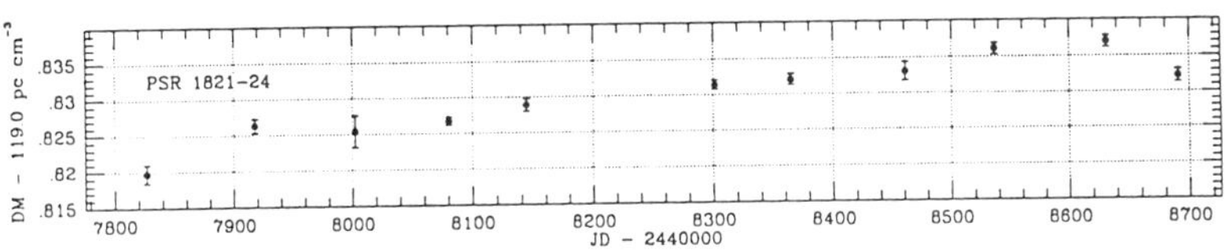

Fig. 1 Dispersion Measure (DM) record for PSR 1821-24.

The time interval in the pulsar frame can then be compared to a polynomial model for the pulsar's rotation evolution:

$$(t_N - t_0) = NP_\circ + N^2 \dot{P}_\circ P_\circ + \cdots$$

This modeling must proceed iteratively. Pulse periods are a rapidly varying effect while the unknown model parameters required to transform the intervals from the observer frame to the pulsar frame are, comparatively, slow effects. While these model parameter effects produce slowly varying perturbations of the timing residuals, they are not small in relation to the pulse period. Model extrapolations from one observing session to another must proceed very cautiously so that there is no pulse number (N) ambiguity. Pulsars in binary systems are first modeled strictly by Doppler variations of the apparent period. After sufficient accuracy in the model is obtained, the pulse arrival times, or phases, are used to obtain higher precision. If the orbital perturbations are small with respect to the period, as could be true for planet(s), then there is no ambiguity in tying together observations.

GREEN BANK OBSERVATIONS OF PSR 1257+12

In 1990 September the 6-ms pulsar PSR 1257+12 was added to our bimonthly Pulsar Timing Array experiment (see, *e.g.*, Foster & Backer 1990). In this experiment we are monitoring a spatial array of millisecond pulsars to create a grid of natural clocks to probe a number of fundamental scientific questions. PSR 1257+12 was added eagerly owing to its location which is far from the other strong millisecond pulsars in our Array. Wolszczan and Frail (1992; hereafter WF) have now demonstrated that PSR 1257+12 is questionable as a clock owing to a double periodicity in its arrival times. These perturbations have an obvious interpretation in a pair of 10 Earth mass planets orbiting about the pulsar.

Figure 2a displays the timing residuals between our observations and the rotation plus astrometry parameters published by WF (Backer, Sallmen & Foster 1992). An adjustment between the 1330 MHz and 800 MHz residuals is made to allow for a small inconsistency in the *DM* for this pulsar owing to small systematic effects in absolute *DM* determination with different observing systems at different observing frequencies. The residuals are obtained by using a version of the same program, TEMPO, that is used by WF, and the same ephemeris, CfA 740R. The perturbation of the residuals by a pair of planets with elements given by WF was independently calculated, and is shown in Figure 2a with an overall adjustment in phase to match the curve to the data in the mean.

Figure 2b displays the difference between the observations and the calculated perturbation. The agreement is excellent over the interval for which the WF model applies, up to 1991.8. In 1992 the model diverges from the data as can be expected for the extrapolation of a many parameter model beyond the domain of the fit. There is no ambiguity in pulse number in the 1992 data. A fit of our data for a revised period, period derivative, right ascension and declination reduces the residuals considerably for the full range of our observations. Similar parameter adjustment has been done by Wolszczan (this meeting).

Two further checks were done. First simultaneous observations were done as part of the Pulsar Timing Array of PSR 1855+09, a well-studied 5.3-ms pulsar in a 12-d binary system. These data show excellent agreement at the 10-μs level with a model derived from independent data (Ryba & Taylor 1991). Also, A. Lyne analyzed the PSR 1257+12 data with an independent software package (PSRTIME) using an independent ephemeris (JPL DE118). Results similar to those presented in Figure 2 were obtained. While these results confirm the effects reported by WF, complete confirmation of the planetary hypothesis will require detection of the orbital interaction effects described by Rasio *et al.* (1992).

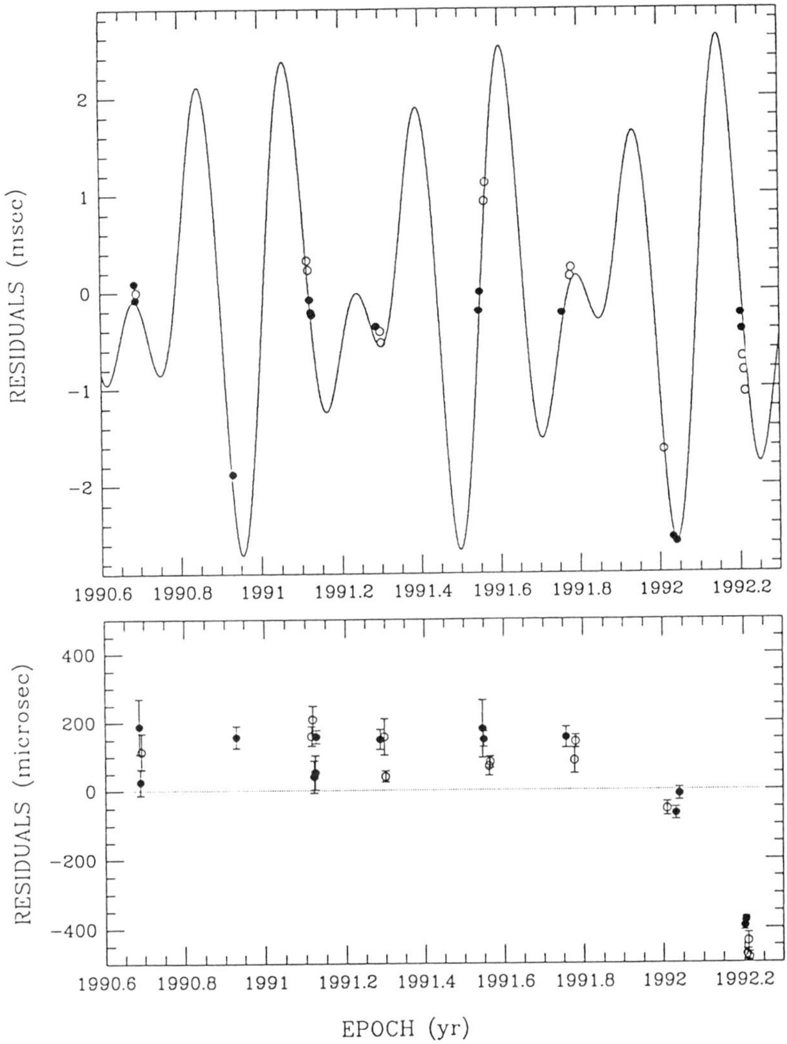

Fig. 2 Timing residuals for the 6ms pulsar PSR 1257+12 from NRAO Green Bank observations. The residuals in *(a)* are a result of a model that includes only the spin and astrometry parameters published by WF. The solid symbols are 1330 MHz data and the open symbols are 800 MHz data. The solid line is the calculated effect of the dual planet perturbations. The residuals in *(b)* are the difference between the data and the curve in *(a)*. The deviation of the 1992 points is the result of the expected imperfect extrapolation of the WF model beyond the range of the fit.

GLOBULAR CLUSTER PULSAR PSR 1620-26

In late 1987 Brinklow *et al.* announced the discovery of the second pulsar in a globular cluster – PSR 1620-26, an 11-ms pulsar in M4. We added this object to our Pulsar Timing Array experiment in 1988 January. Lyne *et al.* (1988) and McKenna and Lyne (1988) report that this pulsar is in a binary system with 0.3 $M_\odot$ companion and a 191-d orbital period. Their first paper presents a fit to the Doppler variations, and the second contains the results of a phase fit. The orbital elements include a highly significant eccentricity, 0.025.

Reduction of our data using the McKenna and Lyne model always fell short of reducing the residuals to white noise. Perhaps there was another effect, or possibly a slip in the pulse numbering scheme. However, the models obtained at various points in time never convincingly lost phase from one observation to the next. A switch of observing equipment in 1989 added to the confusion owing to a lack of data overlap. In late 1991 our conclusion was that the neutron star had perhaps experienced a "glitch" which is unexpected given the excellent timing record of the other millisecond pulsars. In 1992 March a solution was obtained that satisfied all data well with the addition of a single parameter, a large period second derivative, $\ddot{P}$. Figure 3a presents the results of a timing analysis without a $\ddot{P}$ term that shows a large cubic residual whose amplitude is about half of the rotation period. Figure 3b presents the residuals after a model fit that includes $\ddot{P}$ is done for data extending to early 1991. This displays the ability of the model to predict the arrival times of the late 1991 and early 1992 observations, more than one orbit, to within 1% of a period.

The period second derivative is $\ddot{P} = -2.3 \times 10^{-27}$ s s^{-2}. This gives a time scale for the change in sign of the period derivative of only 14 years. If the second derivative arises from the gravitational influence of a companion, then the acceleration derivative, or jerk, will be $\dot{a} \simeq 7 \times 10^{-15}$ cm s^{-3}. This $\dot{a}$ is larger by 5 orders of magnitude than the jerks expected from either nearby stars in the cluster or the overall cluster potential. The $\ddot{P}$ could be the result of a third body in a long period orbit about the binary. Periods in the range of 10 to 100 years would be allowed by the present size of $\dot{P}$ without appealing to special geometry. The corresponding companion mass would be 3.3×10^{29} g $(P_{o2}/10\ \mathrm{y})^{7/3}$, where P_{o2} is the second orbital period. However this explanation may run into difficulty owing to the weak binding of such an object to the inner binary in the moderately dense core of M4 (see poster by S. Sigurdsson). Future observations will allow even higher derivatives to be measured.

ACKNOWLEDGEMENTS

The Pulsar Timing Array experiment has been conducted with R. Foster, L. Fairhead, S-I. Hama, S. Van Hook and S. Sallmen. J. Taylor provided the initial hardware and software to start the experiment. I thank A. Wolszczan and A. Lyne for their contributions to this talk.

REFERENCES

Backer, D. C., & Hellings, R. W. 1986, *Ann. Rev. Astron. Astrophys.*, **24**, 537.

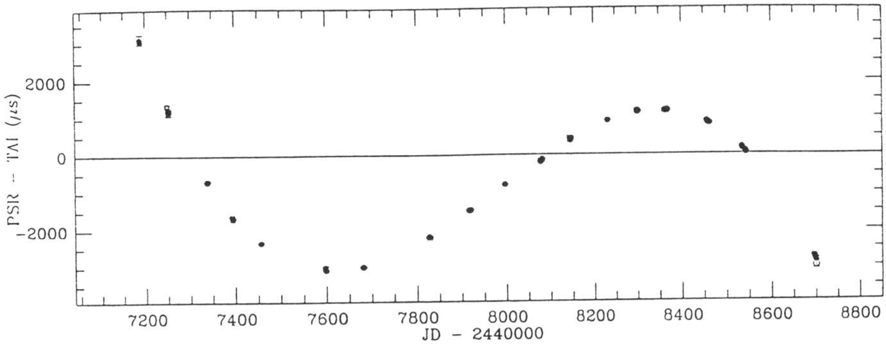

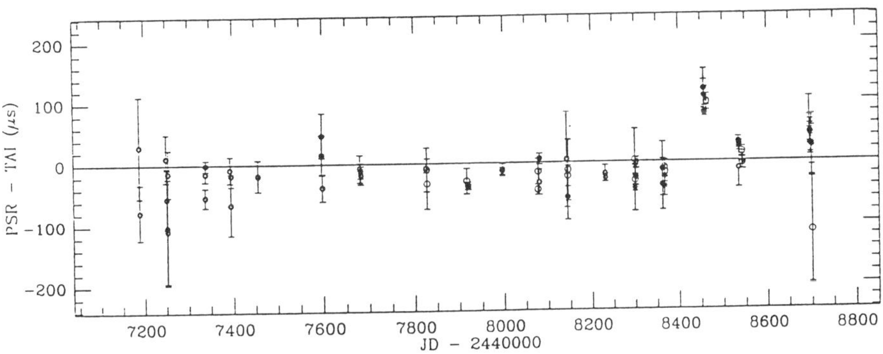

Fig. 3 Timing residuals for PSR 1620-26, an 11 ms pulsar in the globular cluster M4. In *(a)* the residuals are shown without a fit for a period second derivative. In *(b)* a period second derivative is fit along with spin, astrometry and orbital parameters through JD 2448400. The model extrapolates reasonably well to the end of the data set.

Backer, D. C. 1988, *NATO ASI*, **C:262**, 3.

Backer, D. C., Van Hook, S., Hama, S.-H., & Foster, R. S. 1992, *Ap. J.*, submitted.

Backer, D. C., Sallmen, S., & Foster. R. S. 1992, *Nature*, submitted.

Brinklow, A., Lyne, A. G., Biggs, J. D., & Ashworth, M. 1987, *IAU Circular* No. 4470.

Foster, R. S., & Backer, D. C. 1990, *Ap. J.*, **361**, 300.

Guinot, B. 1988, *Astron. Asrophys.*, **192**, 370.

Lyne, A. G., Biggs, J. D., Brinklow, A., Ashworth, M., & McKenna, J. 1988, *Nature*, **332**, 45.

McKenna, J., & Lyne, A. G. 1988, *Nature*, **336**, 226.

Rasio, F. A., Nicholson, P. D., Shapiro, S. L., & Teukolsky, S. A. 1992, *Nature*, **355**, 325.

Ryba M. F., & Taylor, J. H. 1991, *Ap. J.*, **371**, 739.
Taylor, J. H. 1988, *NATO ASI*, **C:262**, 17.
Wolszczan, A., & Frail, D. A. 1992, *Nature*, **355**, 145.

Planets Around Pulsars
ASP Conference Series, Vol. 36, 1993
J.A. Phillips, J.E. Thorsett, and S.R. Kulkarni

LIMITS ON PULSAR PLANETARY SYSTEMS FROM THE JODRELL BANK TIMING DATABASE

M. BAILES, A. G. LYNE AND S. L. SHEMAR
Nuffield Radio Astronomy Laboratories, University of Manchester, Cheshire, SK11 9DL, UK

ABSTRACT A large pulsar timing database exists at Jodrell Bank which we have used to search for the signature of planetary systems around pulsars. Approximately 80 % of the pulsars in our database cannot possess planets with mass greater than about ten times that of the Earth in orbits of period less than a few years. However, the limits on Earth-mass planets, and planets with long orbital periods are poor, partly due to the property of pulsars known as "timing noise", and partly because of the limited time-span of timing data on many pulsars.

With more recent data we verify the existence of a sinusoidal feature with a period of ~1100 days in the pulse timing residuals of the pulsar PSR 0329+54 claimed in 1979 by Demianski and Proszynski (1979). For almost all of the data we examined, there are no obvious signatures of planetary systems. A planetary system orbiting a pulsar would cause the residuals of the pulse arrival times to behave in an erratic manner which would resemble noise. One of our young pulsars (PSR 1828–11) contains an apparently large amount of timing noise, and has post-fit residuals that can be modelled quite successfully by two sinusoids, but only after the removal of a significant period second derivative.

INTRODUCTION

Over the past year, interest in pulsar planetary systems has made a resurgence despite the ill-fated report of the planet orbiting PSR 1829–10 (Bailes, Lyne and Shemar 1991), due to the timely discovery of two planets orbiting the millisecond pulsar PSR 1257+12 (Wolszczan and Frail 1992). Unlike PSR 1829–10, the planets orbiting PSR 1257+12 have survived further scrutiny with independent hardware and software (Backer, these proceedings). Although one cannot rule out exotic mechanisms in this pulsar which mimic planets, theorists are more comfortable with planets accompanying millisecond, rather than normal pulsars (see van den Heuvel in these proceedings for a review of millisecond and normal pulsar evolution). This is because it is now widely accepted that millisecond pulsars are recycled neutron stars which once possessed an accretion disk from which the planets could condense, (see Phinney, and Podsiadlowski, these proceedings) whereas normal pulsars experience great difficulty retaining planets during their birth and proceeding phases of stellar evolution (Bailes, Lyne and Shemar 1991). Nevertheless, there now exists a large amount of radio pulsar timing data on several hundred "normal" pulsars which can be searched for planetary signatures. Early limits on pulsar planetary systems were derived by Lamb and Lamb (1976), and more recently, Thorsett and Phillips (1992) have derived strong limits on some selected pulsars, particularly the original millisecond pulsar, PSR 1937+21.

THE PROGRAMME

The Jodrell Bank timing Programme has been developed over the last 17 years. Observations are carried out at 408, 610, 1400, and 1660 MHz. The history of the programme is as follows:

- 1975: Some sporadic polarization observations.
- 1984: 60 pulsars observed regularly to measure positions with respect to VLA observations.
- 1986: Another 80 mainly young pulsars added to determine positions, periods, period derivatives and for monitoring glitches.
- 1989: 160 pulsars observed every 2-3 months with another 160 being observed every 6 months for the purpose of GRO timing.

For about 30 pulsars, the measurements extend those of Downs and Reichley (1983). We also monitor several globular cluster pulsars discovered at Jodrell Bank and Parkes. Altogether, there are some 2000 pulsar years of monitoring. Over the next year or so it is hoped that all the data can be reduced and compiled into a flexible database. At present, just over 50% of the data has been reduced. Even in its incomplete form, the data provides an excellent opportunity to look for planetary systems around pulsars.

PLANETS OR TIMING NOISE?

In general, our ability to predict the arrival times of a pulsar is limited not only by signal-to-noise, but also by what pulsar astronomers have come to call "timing noise". After a simple slow-down model has been fitted to the pulse arrival times, there is often a large residual which can resemble a random walk. A planetary system orbiting a pulsar could mimic timing noise, as each individual planet would perturb the pulsar making it difficult to clearly identify a single planet's contribution. Can we discount the hypothesis that timing noise is solely due to the perturbations caused by planets?

Timing noise has been well-studied by Cordes and Downs (1985) amongst others. Dewey and Cordes (1989) demonstrated that timing noise is much more prevalent in young pulsars with large period derivatives than in older pulsars. Thus, the only way to retain our hypothesis is if pulsars lose their planets as they age. This seems most unlikely, as planets should be rather immune to the radiation impinging upon them from pulsars with periods > 100 ms or so unless they are particularly close to the pulsar and/or very gaseous. Furthermore, very young pulsars exhibit sharp discontinuities in their rotation period (glitches) which are clearly not related to orbiting bodies. It seems almost certain that normal pulsars are not perfect clocks, and, that for whatever reason have intrinsic timing noise. This is not to say that pulsars cannot possess planets as well, but it just makes it more difficult to place stringent limits on their existence.

"EASY" LIMITS

There are two simple methods for placing limits on the masses of planetary companions to radio pulsars. The first involves their period derivatives, which would be affected by acceleration due to any planetary bodies. The other involves their pulse arrival times which are extremely sensitive to any displacement of the pulsar from its mean position.

Period derivatives

All radio pulsars slow down due to rotational kinetic energy losses which power the pulsar. Negative period derivatives have only been observed for pulsars which are associated with globular clusters where acceleration towards the Earth in the gravitational field of the cluster core is taking place. Binary pulsars are extremely rare, with less than two percent of the disk population being members of binary systems. The pulse periods and period derivatives of over 400 radio pulsars have now been measured. The observed period of a radio pulsar ($P_{\rm obs}$) is related to its true period P, by the approximate relation

$$P_{\rm obs} = P(1 - v/c) \tag{1}$$

where v is the radial velocity of the pulsar and c is the speed of light. Differentiating (1) yields

$$\dot{P}_{\rm obs} = \dot{P}(1 - v/c) - Pa/c \tag{2}$$

where a is the radial acceleration of the pulsar. The maximum acceleration of a pulsar of mass $M_{\rm psr}$ in a binary system with a planet of mass $M_{\rm planet}$ with an orbital separation of R is given by

$$a_{\rm psr} = \frac{M_{\rm planet}}{M_{\rm psr} + M_{\rm planet}} \frac{GM_{\rm planet}}{R^2}. \tag{3}$$

Hence the maximum period derivative induced in the pulsar by the planet is given by:

$$\dot{P}_{\rm obs}/P_{obs} = \frac{a_{\rm psr}}{c} = \frac{M_{\rm planet}}{M_{\rm psr} + M_{\rm planet}} \frac{GM_{\rm planet}}{cR^2} \tag{4}$$

The fact that no galactic pulsars have negative period derivatives enables us to place an immediate limit on abundant planetary systems. Figure I shows a histogram of $\dot{P}/P$ for the pulsars detected in one of the major pulsar surveys. In table I, the *maximum* $\dot{P}/P$ that could be induced in a pulsar of mass 1.4 $M_\odot$ by planets with masses and orbital periods the same as each of the solar planets are given.

A "Jupiter" would induce at most a $\dot{P}/P$ of $10^{-15.3}$ s^{-1}. When one allows for random inclination angles and orbital phases then even if all pulsars possessed

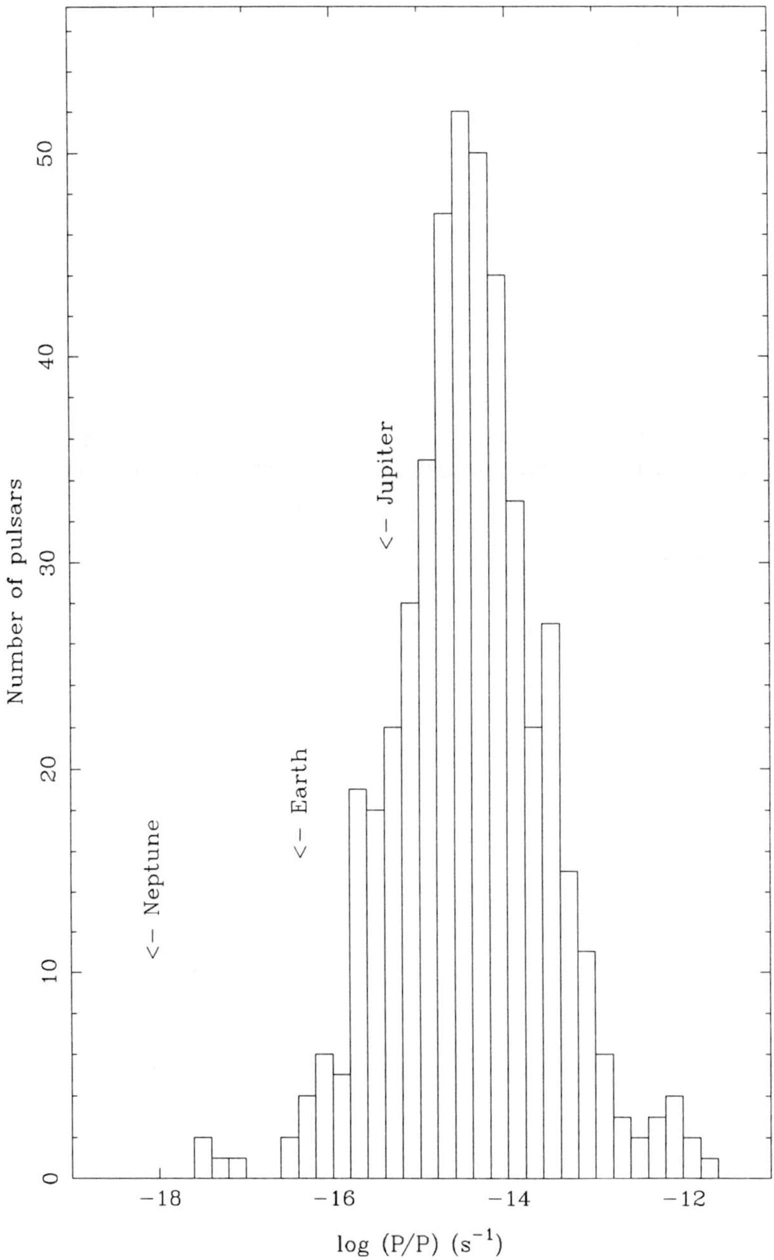

FIGURE I Histogram of $\dot{P}/P$ for all the pulsars which were discovered in one of the major pulsar surveys. Planets can contribute to a pulsar's value of $\dot{P}$ due to its gravitational pull on the pulsar. A Jupiter could induce at most a $\dot{P}/P$ of $10^{-15.3}$ s^{-1}. See Table I.

TABLE I Effects of planets orbiting pulsars.

Planet	Orbital Period (years)	$a \sin i$(max) (milliseconds)	$\log_{10}\dot{P}/P$ (s^{-1})
Mercury	0.24	0.02	−16.8
Venus	0.61	0.71	−16.1
Earth	1.00	1.20	−16.3
Mars	1.88	0.196	−17.7
Jupiter	11.86	1979.6	−15.3
Saturn	29.45	1086	−16.3
Uranus	84.0	335.5	−17.7
Neptune	164.8	614.9	−18.0
Pluto	248.2	0.111	−22.1

a "Jupiter" there are only about 10 (out of 450) where we might expect it to manifest itself as a negative $\dot{P}$. Nevertheless, the fact that all $\dot{P}$'s are positive suggests that unless there is some bias against pulsars with low $\dot{P}/P$, planets more massive than Jupiter cannot exist around more than about 10 % of the pulsar population. This method is not very useful for placing limits on the number of Earth-like planets orbiting pulsars, as there are only a few "normal" pulsars with $\dot{P}/P < 10^{-16.3}$, the *maximum* that could be induced by an Earth-mass planet with a period of a year.

Timing residuals

Much more restrictive limits can be placed on pulsar planetary systems from the size of their timing residuals relative to a simple slow-down model. If a pulsar possesses only a solitary companion, and it has been regularly observed for at least one planetary "year", then the planet will induce an rms residual in the pulse arrival times of 0.707 times the projected semi-amplitude of the pulsar's orbit due to the orbiting planet. For those pulsars currently in the Jodrell Bank timing database the rms residual obtained from the best-fitting timing solution is plotted against the length of time that the pulsar has been observed in Figure II. In Table I, the maximum values of $a \sin i$ that each of the solar planets would induce in a pulsar are presented. For some pulsars the limits are excellent, but for many, entire solar-like planetary systems cannot be excluded. This is because for most of the pulsars in our sample the timespan is insufficient to exclude Jovian-like planets, and the timing residual is greater than that which would be induced by the inner planets of our solar system. We can exclude planets about ten times that of the Earth's mass with periods less than a few years in about 80 % of the pulsars in our sample as they would induce an rms residual of about 8.5 $\sin i$ milliseconds in the post-fit residuals. Although one might expect to be able to exclude Jupiter-like planets well before the data span encompasses a complete orbit, this is not the case. When arrival times are fitted to obtain the period and period derivative, short segments of an orbit would simply be absorbed into the fitted parameters.

One obvious feature of Figure II is the tendency for pulsars with the largest dataspan to have systematically higher RMS residuals than those with shorter dataspans. This is precisely what one would expect if pulsars possessed planetary systems. Before the dominant planet had completed one orbit, its effect on the pulsar's timing would be absorbed into the fit. However, after it had completed one orbit, the period and period derivative would no longer be able to absorb the sinusoid and it would begin to manifest itself in the timing residuals. Unfortunately, any other long-term effects in the timing residuals due to "timing noise" would also make the residuals increase with time.

SEARCHING FOR PLANETS

We have demonstrated that for many pulsars there are only poor limits available on potential orbiting bodies, particularly for Earth-sized planets. With the constraints that timing noise place upon us, in order to convincingly detect a planet, one needs to observe at least 2, and preferably many, coherent planetary orbits, where the amplitude of the oscillation remains constant. Furthermore, after removing the "planet", there should be a statistically significant reduction in the RMS residuals of the fit given the number of additional free parameters in the model.

There are three methods we employed when looking for pulsar planets:

Fitting Explicitly for all Planetary Periods

Thorsett and Phillips (1992) developed a method for placing limits on pulsar planetary systems. It involves attempting to fit a sinusoid to pulsar timing residuals with all possible periods from some minimum, by advancing the period from some minimum value in steps which result in one radian of orbital phase over the data span. To do this properly requires adjusting the initial phase of the sinusoid and performing a dozen least squares fits for each trial period, recording the best-fit asini which yields the minimum rms at each period and plotting the results. We introduced this facility into PSRTIME, the Jodrell Bank pulsar timing package. Whilst we found that the method was very good at placing limits on pulsar planetary systems, it was quite poor at detecting sinusoids that we had introduced into the data even though the limits it produced were correct. However, if one plots the minimum rms residual as a function of orbital period then any planet with a significant amplitude is immediately visible as a sharp reduction in the fitted residuals. In Figure III we demonstrate how well this method works for the pulsar PSR 1829–10, in which we intentionally induce a 6-month sinusoid of amplitude 7 milliseconds into the residuals. The Figure demonstrates that although the signature of the 6-month orbit was clearly detected by the order of magnitude reduction in the rms, the feature associated with it in Figure IIIa is relatively obscure.

Fourier Transforms

Although the method of Thorsett and Phillips is very powerful for placing limits on simple planetary systems, (in particular it highlights the limitations that pulsar timing experiences in removing the effects of planets in the fitting process) it is also very time consuming. Our analysis of the pulsar PSR 0329+54 using this method took two sparc II CPU days, in which we performed several thousand

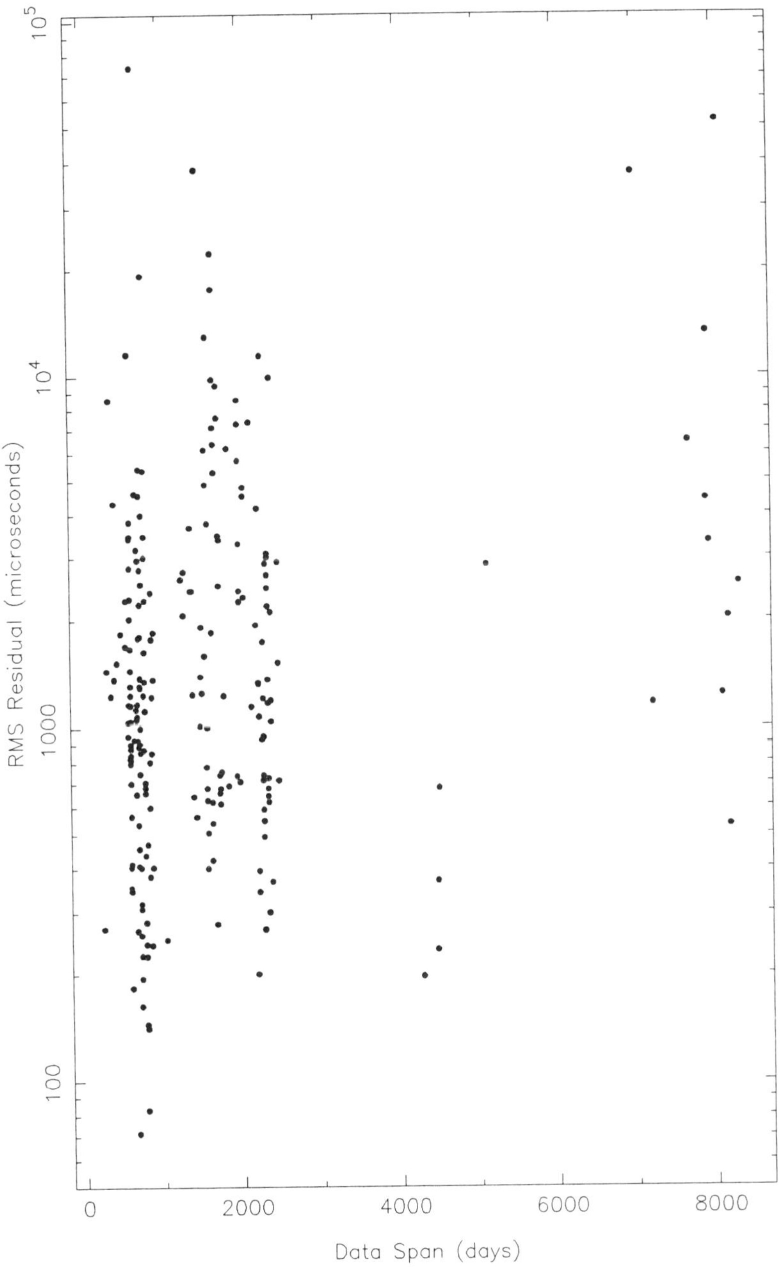

FIGURE II RMS timing residuals vs the length of time that the each of pulsars in the Jodrell Bank timing database has been observed. Any planet in a circular orbit will induce an rms in the timing residuals of the same order as the asini it induces in the planet provided that the data span is at least a planetary orbital period. See Table I for the asini induced by each of the solar planets. We can exclude planets with periods less than a few years and ten times that of the Earth in about 80 % of the sample.

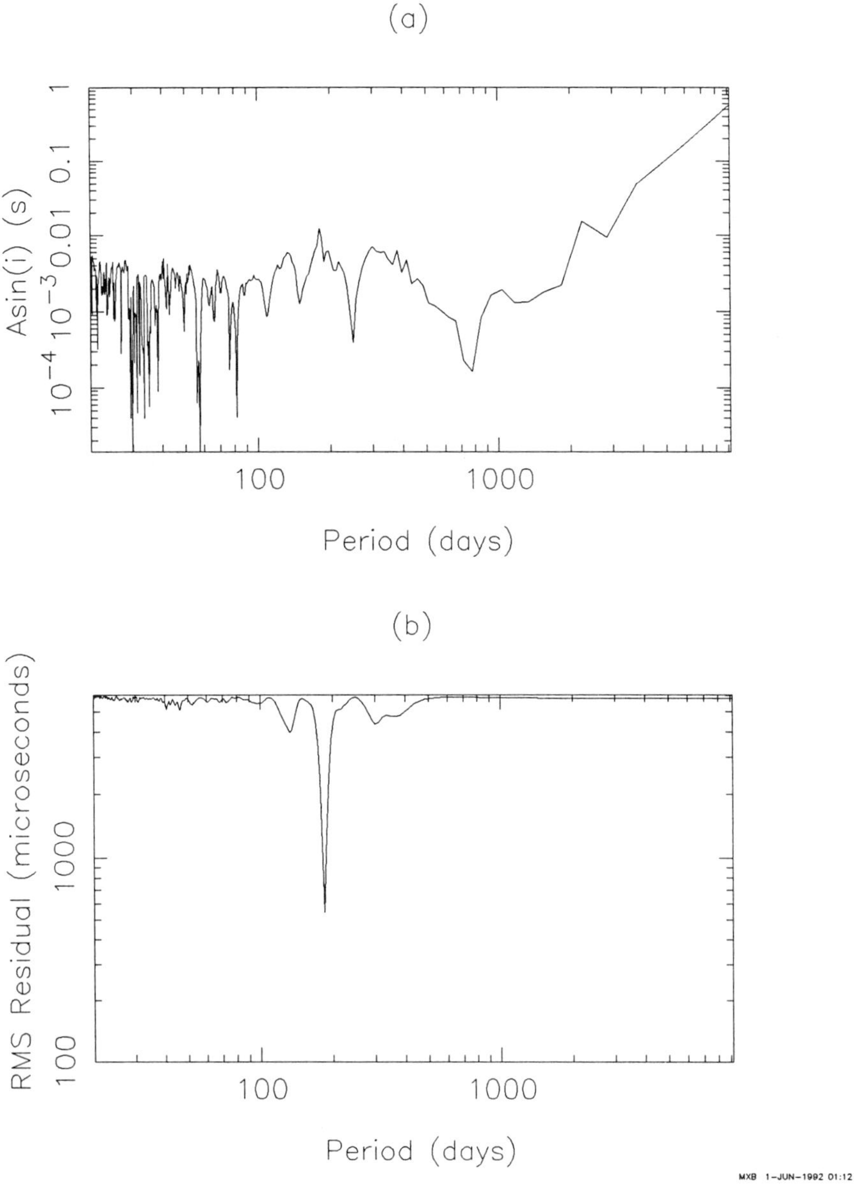

FIGURE III (a) Limits on asini for the pulsar PSR 1829–10 due to potential planetary companions with a six-month sinusoid introduced into the data. (b) RMS residuals for the best-fitting sinusoid at each trial period demonstrating the detection of the planet at a period of 6 months.

fits. A much quicker way to look for sinusoidal features is to Fourier transform the data. Unfortunately, the uneven sampling involved in pulsar timing makes use of the fast Fourier transform impossible without resampling the data. Good limits can still be obtained provided that the maximum spacing between data points is much less than the period down to which one is searching for planets. However this method does not take into account the possibility that the fitting process may have absorbed some of the periodicities which we are looking for.

Visual examing of Residuals

The quickest and easiest method was to simply examine the residuals by eye and look for any obvious periodic features.

RESULTS

Targets

The prohibitive amount of computer time involved in performing a Thorsett-Phillips type analysis on each of the 200+ pulsars at our disposal caused us to only perform fits to selected pulsars. We performed fits to all of the pulsars for which we had close to 8000 days of timing data, and also for all pulsars which were significantly below the spin-up line in the hope that some of them may have accreted matter in a similar way to PSR 1257+12, and therefore possess planets. Visual examinations of all the pulsar timing residuals were also performed for any obvious planets. We also studied the pulsar PSR 0329+54 because in the past Demianski and Proszynski (1979) had suggested that there existed an 1100 day periodicity in the timing residuals.

The vast majority of pulsars exhibited no obvious reduction in their residuals when we used the Thorsett-Phillips method of searching for planets. In fact, of all of our target sources, only PSR 0329+54 showed anything of interest by utilising the Thorsett-Phillips approach.

PSR 0329+54

Demianski and Proszynski (1979) noticed that PSR 0329+54 exhibited sinusoidal variations in its timing residuals after fitting for the normal pulsar parameters and a second period derivative. Although the sinusoid could not be verified by Cordes and Downs (1986), we now have data which extends the dataset to 1992. A Thorsett-Phillips analysis was performed on the complete dataset which now extends over more than 20 years. The results are displayed in Figure IV. At 1101 days, there is a significant decrease in the residuals with a further decrease in the residuals at 2370 days, suggestive of a second sinusoid with a longer period. The dataset was also split into two parts and the analysis redone on both halves of the data in case the decrease in the residuals at 1100 days was only due to the first few years of timing data. In both datasets the 1100 day periodicity was maintained, along with the 2370 day period. What remains apparent, is that there is a persistent 1100 day periodicity in the timing residuals of this pulsar. Whether this is due to an orbiting body, or is an artifact of the internal dynamics of the neutron star is not clear.

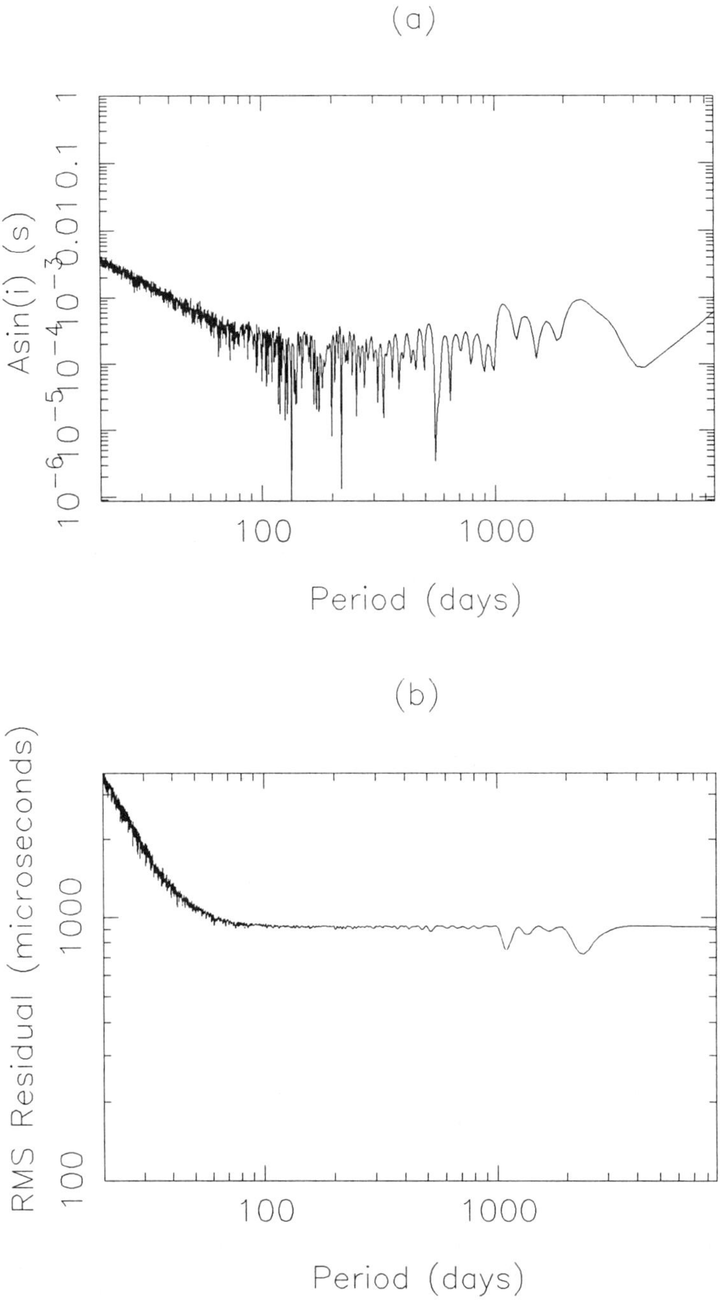

FIGURE IV Results of a Thorsett-Phillips type analysis on the pulsar PSR 0329+54 with over 20 years of data present. A clear depression in the RMS residuals exists at 1101 days, consistent with the periodicity claimed by Demianski and Proszynski in 1979.

PSR 1828–11

Young pulsars with large period derivatives often exhibit large amounts of timing noise. One of the young pulsars from the Clifton and Lyne survey was PSR 1828–11. This pulsar has a period of 405 ms and a $\dot{P}$ of ~60. Although the residuals for this pulsar can be modelled by two sinusoids with periods of 492 and 1043 days, and amplitudes of 13 and 22 milliseconds respectively, the model fitted to the data presented at the conference failed to correctly predict the arrival times of more recent data without the inclusion of a substantial period second derivative. Whilst we shall continue to monitor this pulsar in the future, until the model can accurately predict pulse arrival times, a planetary interpretation must be greeted with some caution, especially considering the large period derivative of the pulsar.

DISCUSSION

Whilst we have been able to place good limits on the number of pulsars with planets an order of magnitude larger than that of the Earth around normal pulsars in orbits of a few years, there are two factors which make limits on lighter and longer-period pulsars difficult to achieve. The first of these is the pulsar's intrinsic timing noise, which can mask or even immitate the signature of a pulsar planetary system. The second is the difficulty of maintaining a pulsar timing programme over more than a few years. Strong limits on pulsar planets with orbital periods greater than the length of an average timing programme are always going to be difficult to obtain.

The phenomenal rotational stability of millisecond pulsars, along with more plausible formation scenarios for orbiting planets makes them the most likely sites for the future detection of pulsar planets, particularly if the current surveys for millisecond pulsars detect a substantial number of them.

ACKNOWLEDGMENTS

MB would like to thank the LOC for travel support. SLS acknowledges receipt of a Fentham's trust postgraduate grant.

REFERENCES

Bailes, M., Lyne, A. G. and Shemar, S. L., (1991) *Nature*, **352**, 311-313.

Cordes, J. M. and Downs, G. S., (1985) *Astrophys. J. Supp. Series.*, **59**, 343-382

Demianski, M. and Prosznski, M., (1979) *Nature*, **282**, 383-385.

Dewey, R. J., and Cordes, J. M., (1989) *In Timing Neutron Stars, eds Ogelman, H and van den Heuvel, E. P. J., Kluwer, Dordrecht* 119-124

Downs, G. S., and Reichley, P. E., (1983) *Astrophys. J. Supp. Series.*, **53**, 169-240

Lamb, D. Q., and Lamb, F. K. (1976) *Astrophys. J.*, **204**, 168-186.

Thorsett, S. E., and Phillips, J. A., (1992) *Astrophys. J. (Letters)*, **387**, L69

Wolszczan, A. and Frail, D. A., (1992) *Nature*, **355**, 145-147.

Planets Around Pulsars
ASP Conference Series, Vol. 36, 1993
J.A. Phillips, J.E. Thorsett, and S.R. Kulkarni

A SEARCH FOR PLANETS IN ARCHIVAL TIMING DATA: THE JPL DATA SET

S. E. THORSETT AND J. A. PHILLIPS
Radio Astronomy 105-24, California Institute of Astronomy, Pasadena, CA 91125

J. M. CORDES
Department of Astronomy, Space Sciences Building, Cornell University, Ithaca, NY 14853

ABSTRACT Using the data analysis method of Thorsett & Phillips (1992), we have used the timing data of Downs and Reichley (1983) taken between 1970 and 1982 to set limits on the masses of possible planetary companions to 23 bright pulsars.

INTRODUCTION

The discovery by Wolszczan and Frail (1992) of two planetary companions to the pulsar PSR 1257+12 has renewed interest in the use of pulsar timing data to detect pulsar planetary systems. Thorsett & Phillips (1992) described a sensitive algorithm for detecting low-mass companions in pulsar timing data. Here we have applied the algorithm to search for planets in the timing data of 23 bright field pulsars. The data consist of pulse arrival times obtained by Downs and Reichley (1983) at 2390 MHz using the NASA Deep Space Network facility. The monitoring program for these pulsars was remarkable for its length, ≈ 13 yr. For this reason the data sets are very sensitive to *long-period companions*.

The search algorithm is based on standard timing techniques. The 'Tempo' software package (e.g., Rawley et al. 1987) was used to fit a model to the pulsar period, period derivative, position, and – in some cases – period second derivative and proper motion. In addition, a barycentric sinusoidal term with fixed period P_b was included in the fit, with amplitude x and phase ω_o allowed to vary. We repeated the fitting procedure for values of ω_o between 0 and 360° in units of 30°. Since our intent is to find the most conservative limits possible, we chose the maximum value of $x(\omega_o)$ for a given P_b.

RESULTS

Table I summarizes the objects we studied and the results obtained. All of the pulsars listed are relatively slow ($P \sim 1$s), and the rms timing residuals are about 3 orders of magnitude larger than those of the brightest millisecond pulsar, PSR 1937+21. An Earth-mass planet would be only marginally detectable in the data of most of the pulsars, however, Jupiter would be detectable at a level

Pulsar	Span (yrs)	Points	RMS (ms)	$\ddot{P}$	Proper Motion	M_1	M_2
0031−07	10.8	300	3.1			1.5	0.1
0329+54	12.0	327	1.7	√		1.1	0.9
0355+54	7.6	205	4.6			2.6	1.3
0525+21	12.0	294	5.9	√		1.7	9.4
0628−28	10.8	304	8.1			4.2	4.4
0736−40	8.9	225	1.7			1.2	7.1
0823+26	11.9	388	11.3	√		3.5	21.5
0950+08	12.3	285	3.6			2.1	0.4
1133+16	11.9	365	1.6	√	√	0.8	0.4
1237+25	11.3	348	2.2			1.1	0.2
1604−00	8.9	263	3.5			1.6	0.8
1642−03	11.5	336	9.7			4.4	3.9
1706−16	9.0	196	15.3			11.3	14.2
1749−28	11.6	279	5.9			3.2	10.2
1818−04	10.9	294	22.2			9.0	18.3
1911−04	10.8	261	4.0			1.6	3.6
1929+10	11.6	376	4.1			3.1	0.5
1933+16	12.0	352	4.6			2.4	0.6
2016+28	12.9	237	4.0			2.1	0.7
2021+51	10.9	305	4.8	√		2.5	2.3
2045−16	12.0	278	2.4	√		1.7	1.4
2111+46	11.0	280	6.2			2.6	0.4
2217+47	11.0	214	4.7			3.4	0.9

TABLE I Summary of JPL data set for 23 pulsars. Pulsar fits include pulse phase, period, period-derivative, and pulsar right ascension and declination in all cases, and period-second-derivative and proper motion where indicated. Mass limits 1 and 2 are the largest allowable planets with period within 10% of 98 days (the period of PSR 1237+12's second planet) and 11.9 years (the period of Jupiter), respectively, in Earth masses.

$\gtrsim 100\sigma$ in all cases. Figures 1 through 5 display the mass limits as a function of orbital period for all 23 pulsars. The basic features of the plots are discussed by Thorsett & Phillips (1992). Of special note is the spike at 365 days. The loss of sensitivity for periods close to one year is a result of fitting for the pulsar position, which removes a 1-yr sinusoid from the timing data.

SUMMARY

- A analysis of arrival times from the 13-yr JPL pulsar monitoring program is especially sensitive to long-period ($P_b \sim 1 - 10$ yr) planetary companions.
- We do not detect planets more massive than $\sim 1 M_\oplus$ in the timing data of 23 slow, bright pulsars.
- If Jupiter-size objects existed around any of the 23 pulsars, they would have been detected at a level $\gtrsim 100\sigma$.

ACKNOWLEDGMENTS

We thank David Nice for helpful discussions.

REFERENCES

Downs, G.S. & Reichley, P.E. 1983, ApJ Supp, 53, 169.

Rawley, L.A., Taylor, J.H., Davis, M.M. & Allan, D.W. 1987, Science, 238, 761.

Thorsett, S.E. & Phillips, J.A., 1992, ApJ, 387, L69.

Wolszczan, A. & Frail, D., 1992, Nature, 355, 145.

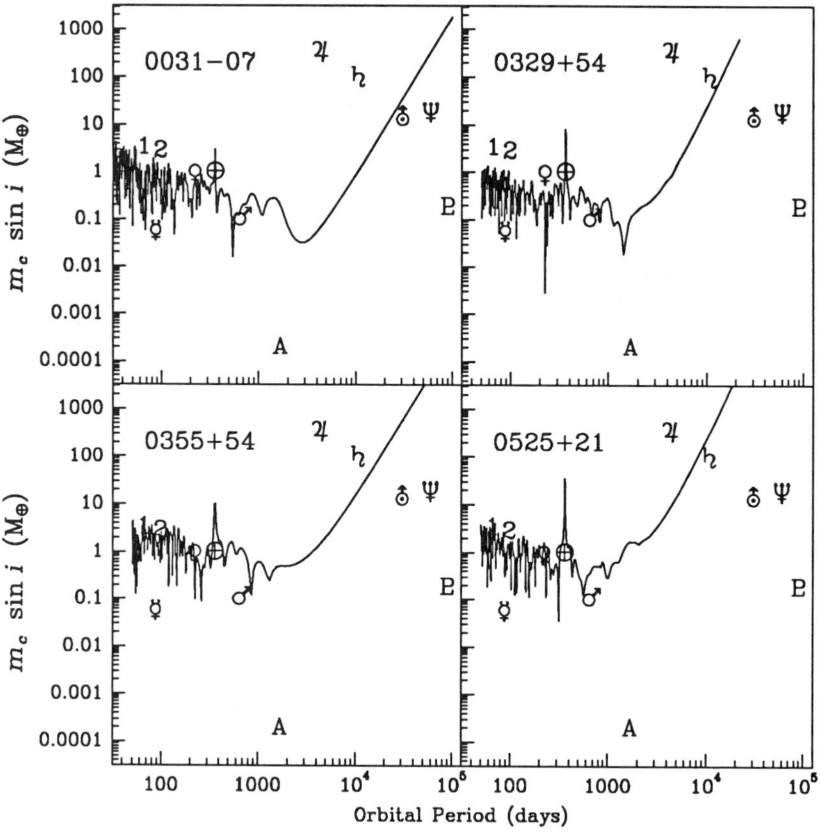

FIGURE I Limits on planets around four pulsars. The phase space above the curve is excluded by timing observations. For comparison, the planets of the Solar System are indicated with their usual symbols. The largest Solar System asteroids are marked with an 'A,' while the planets of Wolszczan and Frail are indicated by '1' and '2.' A full description of the plot format can be found in Thorsett and Phillips (1992).

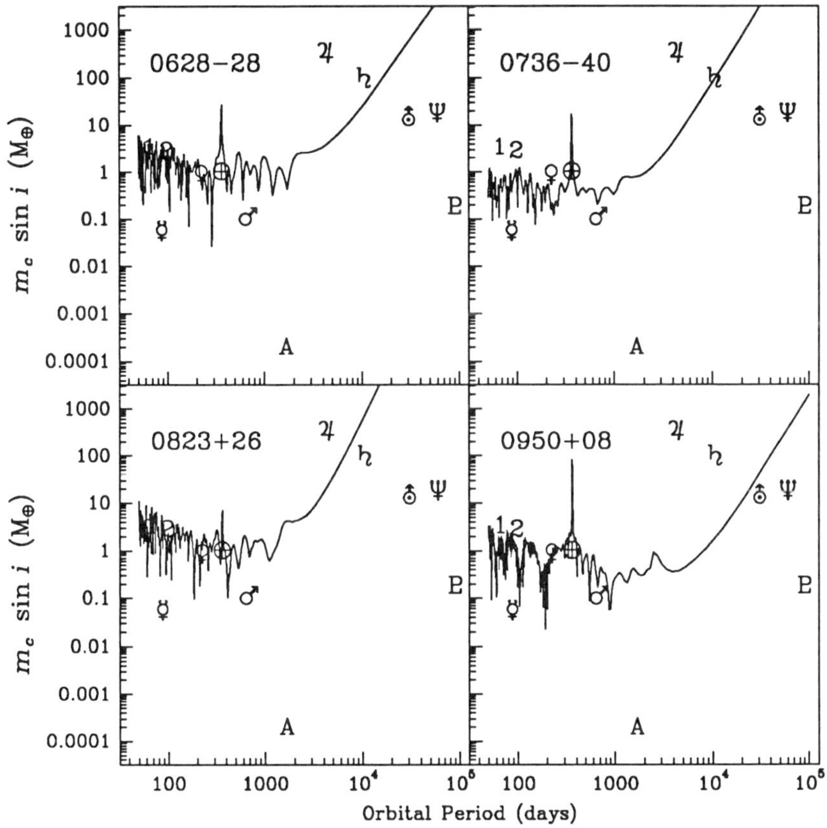

FIGURE II See caption for Figure I

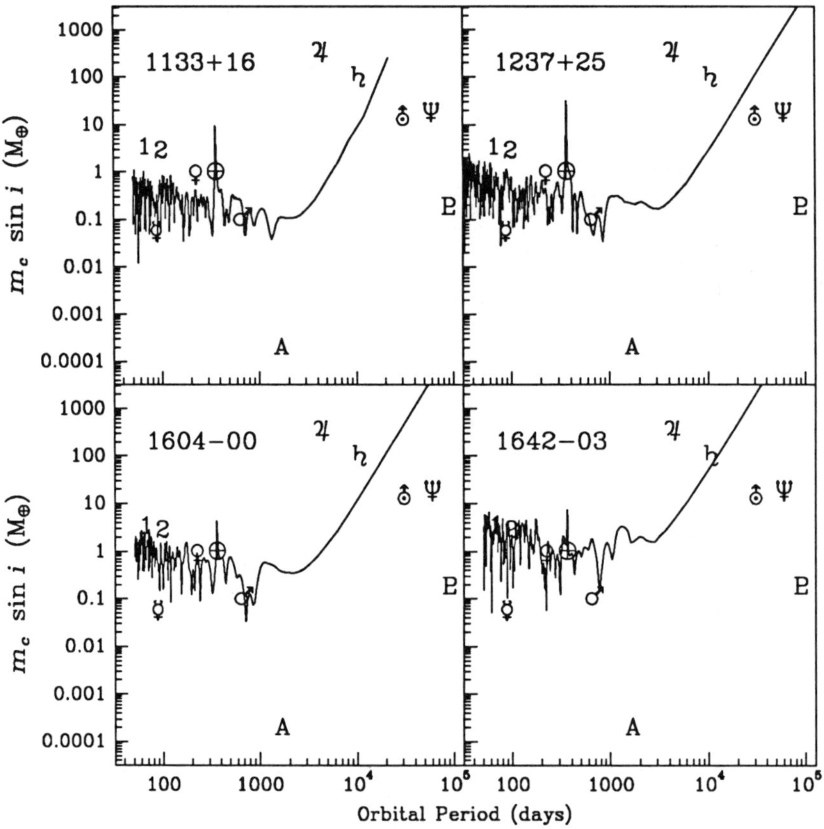

FIGURE III See caption for Figure I

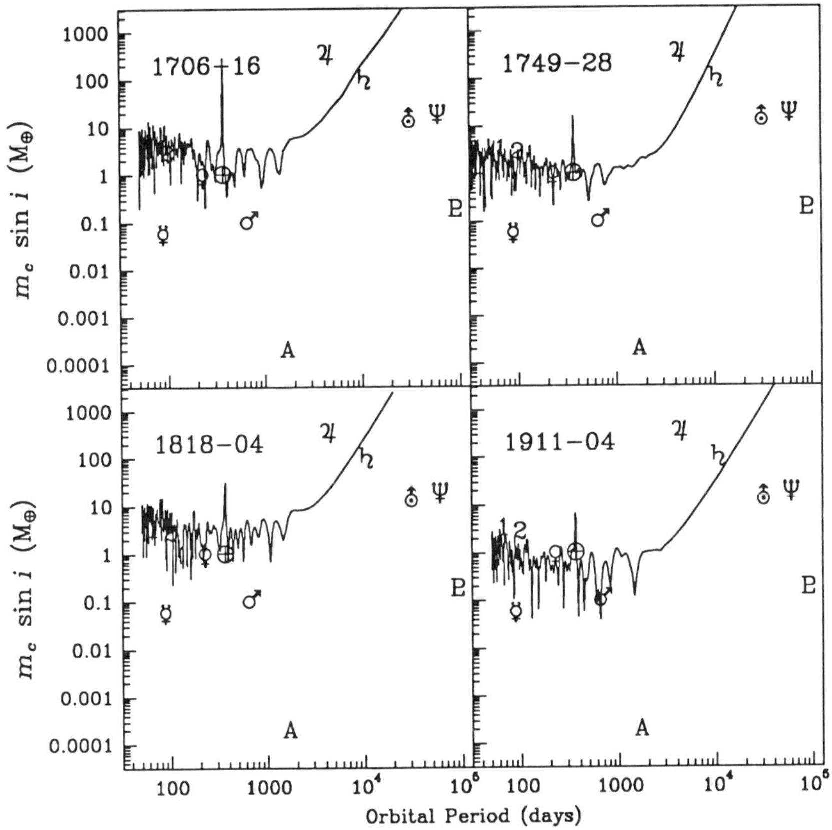

FIGURE IV See caption for Figure I

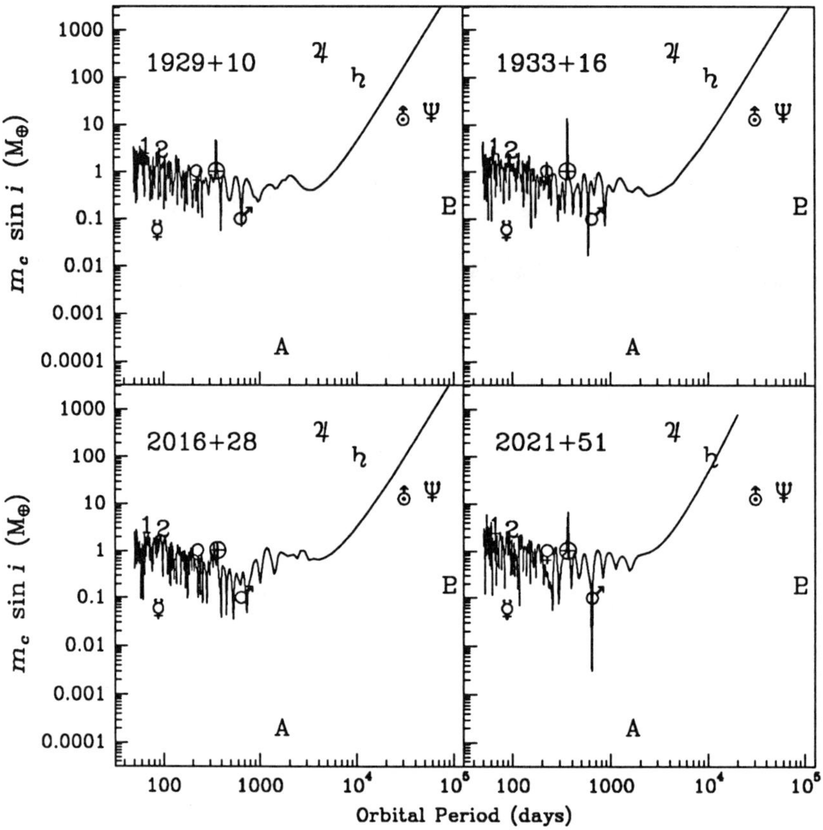

FIGURE V See caption for Figure I

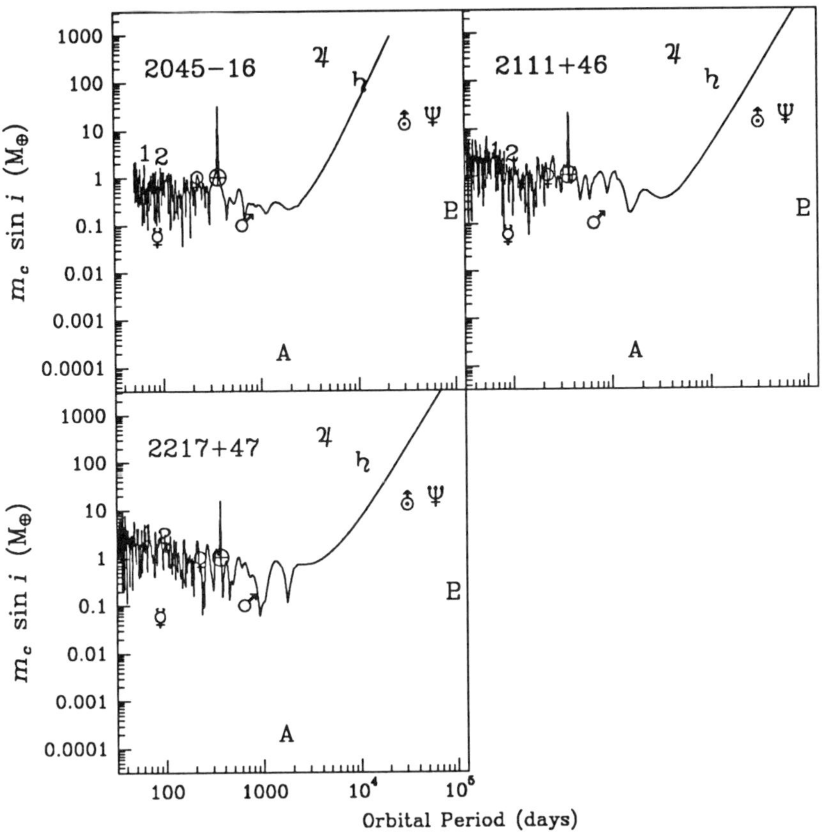

FIGURE VI See caption for Figure I

Section 2:

Timing Noise in Pulsars and Alternative Explanations

Planets Around Pulsars
ASP Conference Series, Vol. 36, 1993
J.A. Phillips, J.E. Thorsett, and S.R. Kulkarni

THE DETECTABILITY OF PLANETARY COMPANIONS TO RADIO PULSARS

JAMES M. CORDES
Astronomy Department, 520 Space Sciences Building, Cornell University, Ithaca, NY 14853

ABSTRACT I discuss pulsar phenomenology that is relevant to the detection of orbiting planetary companions. Pulse timing measurements track the pulsar spin to within a precision limited by intrinsic pulse phase jitter. We quantify the detectability of coherent oscillations in timing residuals amidst white noise (phase jitter, etc.) and red noise (spin fluctuations). Free precession may be driven by spin fluctuations that counter damping processes. Precession, in turn, can drive torque fluctuations that are manifest as cyclical (but most likely not coherent) timing variations. Precession-related timing variations need not be accompanied by detectable pulse shape changes and are most prominent in young, Crab-like pulsars with large spin-down rates. Considering all effects, planetary companions are easiest to detect in short period pulsars that have attenuated spindown rates, such as the millisecond pulsar population and other objects that have endured accretion-induced quenching of the magnetic field. Detectability of multiple planets is considered when the ranges of mass and orbital period are large, as with the solar system. Bayesian methods are advocated to compare planetary vs. spin noise interpretations of the timing data from a given pulsar.

INTRODUCTION

Reports of planets around neutron stars (NS) have been made five times since radio pulsars were discovered in 1967. All are based on perturbations in the apparent pulse phase measured over data spans of years. The first case, the Crab pulsar, was later reinterpreted as random spin variations ('timing noise') rather than coherent perturbations from orbital motion (Richards *et al.* 1970). The second case, PSR0329+54, also appears to be due to timing noise (Demianski & Prószyński 1979). The third case, a planet around an optical pulsar in SN1987a (Kristian *et al.* 1989), was retracted along with the pulsar detection. The fourth case, PSR1829−10, misidentified improperly reduced timing data that left residual traces of the Earth's orbital motion (Bailes, Lyne, & Shemar 1991). By contrast, the latest case, PSR1257+12 (Wolszczan & Frail 1992), appears to believably possess two perturbing planetary companions because (1) the timing signature shows a degree of coherence unlike that seen in the other cases (except 1829−10, for which, after all, a planet was certainly identified in the data); (2) timing noise, phenomenologically, is weak or nonexistent in the class of millisecond pulsars, of which 1257+12 is a member; and (3) the result has been confirmed with independent data obtained from another observatory and analyzed independently (Backer *et al.* 1992).

In this paper I first outline those phenomena that limit the detectability of doppler effects on the observed pulse train. Next, I discuss the occurrence of precession in pulsars. Free precession may be occurring in a few pulsars but at levels too small or too incoherent to be misidentified as an orbital effect. However, precession and other timing noise make detections of orbital signatures more difficult. Because precession is damped on times much less than the spindown times of most pulsars, it must be driven by crust cracking or differential rotation. Therefore it is more likely to appear in objects with large spindown rates $\dot{\nu}$ that show timing noise. I also show that, in such cases, precession is much more likely to be manifest as induced torque fluctuations rather than in changes in pulse shape.

Finally, I discuss the detectability of planetary companions around pulsars taking into account all limitations on the 'clock purity' of the pulsar signal. Planets are easiest to detect around millisecond pulsars. Planetary companions with masses about that of Earth's around more slowly spinning objects cannot be excluded without more sophisticated analysis of the extant timing data. The difficulty in identifying planetary signatures results not only from intrinsic noise but also from the likely possibility that multiple planetary systems have a wide range of masses and orbital periods. I advocate a Bayesian approach for comparing the alternative hypotheses that can apply to the data from a given object.

PULSAR BEAMS AND PHASE JITTER

Pulses from radio pulsars show widths that are consistent with random orientation angles between the pulsar beam and spin axis and a scaling of the beam width (in angular units) as $\theta_b \propto P^{-1/2}$ (Rankin 1990). The pulse width therefore scales, in time units, as $\theta_p P \propto P^{1/2}$. The scaling with spin period P is in accord with the Goldreich-Julian (1969) definition of the magnetic polar cap. Individual pulses show large ($\sim$ 100%) variations in amplitude and phase about the mean that, evidently, are due to motions of emitting plasma clumps within the open field line region of the pulsar magnetosphere. As such, single pulses represent the pulsar 'weather' that, when averaged, converges to a stable waveform 'climate' as the number of pulses N becomes large. Thus, the pulsar beam appears to be locked stably to the spin of the neutron star's crust.

Pulse phase jitter of individual pulses, combined with finite N, implies an error in assigning times of arrival (TOAs)

$$\sigma_J \approx 13.4\mu s\, f\,(1+m_I^2)^{1/2} N_3^{-1/2}\left(\frac{W_{FWHM}}{1\ ms}\right), \qquad (1)$$

where W_{FWHM} is the pulse width (full width at half maximum); $f \approx 1$ is the amount of jitter in units of the observed pulse width; $m_I \approx 1$ is the intensity modulation index (rms/mean); and $N_3 \equiv N/10^3$. Additive noise contributes a statistically independent error

$$\sigma_n \approx \left[\frac{21.9\mu\, s}{(S/N)_1}\right]\left(\frac{W_{FWHM}}{1\ ms}\right)\left(\frac{\Delta}{W_{FWHM}N_3}\right)^{1/2}, \qquad (2)$$

where Δ is the temporal resolution (~receiver time constant) and $(S/N)_1$ is the signal-to-noise ratio of a single pulse.

Timing precision is radiometer-noise limited for small signal-to-noise ratios and jitter-limited for large S/N, the cross-over occurring for $(S/N)_1 \approx 1$. TOAs from non-overlapping averages will show white noise errors associated with jitter and radiometer noise. Another contribution to white noise comes from diffractive interstellar scintillations, which scales strongly with pulsar distance and observation frequency but enters in, typically, at the sub-microsecond level.

There are no known long term perturbations to TOAs from magnetospheric effects, such as might arise if the magnetic field wandered or the emission altitude varied secularly. Long term interstellar propagation effects— refractive interstellar scintillations and DM variations— produce wavelength dependent perturbations that can be removed, largely, through multiple wavelength observations. In the remainder of this paper we focus on pulse phase jitter as the limiting factor in arrival time precision.

ROTATION VARIATIONS OF PULSARS

Identification of Spin Fluctuations

Spin rate variations may be identified as departures from smooth spindown. The rotational phase over a time span $T \ll \nu/|\dot{\nu}|$ is

$$\phi(t) = \phi_0 + \nu t + \dot{\nu} t^2/2 + \ddot{\nu} t^3/6 + \delta\phi_{TN}(t) + \phi_{WN}, \tag{3}$$

where $\nu, \dot{\nu}, \ddot{\nu}$ describe the spinrate and spindown from the magnetospheric torque and $\delta\phi_{TN}$ is 'timing noise' associated with spin-rate fluctuations and ϕ_{WN} is white noise. For most pulsars, the fourth, cubic term is negligible. Thus TN may be recognized as residuals $\mathcal{R}_2(t)$ from a second-order polynomial that exceed those due to white noise. Figure I shows phase residuals for 14 objects in the JPL timing program of Downs and Reichley (1983; see also Cordes and Downs 1985). The amount of timing noise is markedly different from object to object. Some show obvious changes in slope that indicate the occurrence of discrete events in the spin rate or its derivative. The bulk of timing variations are independent of observation frequency, so interstellar propagation effects may be ruled out. Current thinking is that timing noise is caused by a noisy interaction between the crust of the neutron star and the quantized array of superfluid vortices that permeate the inner crust about 1 km below the surface. The figure suggests that objects with the largest period derivatives $\dot{P}$ have the largest amount of timing noise. This is borne out by a quantitative analysis (Dewey and Cordes 1989) that shows that a timing noise 'activity parameter' correlates more strongly with $\dot{P}$ than with P. Some of the phase residuals in figure I are quite smooth; it is debatable whether these are due to intrinsic spin variations or due to orbital effects with $P_{orb} \gg T$. We discuss this issue later in the paper.

Timing Activity Parameter

The activity parameter is defined as

$$A_t \equiv \log\left[\frac{\sigma_2(T)}{\sigma_{2,Crab}(T)}\right] \tag{4}$$

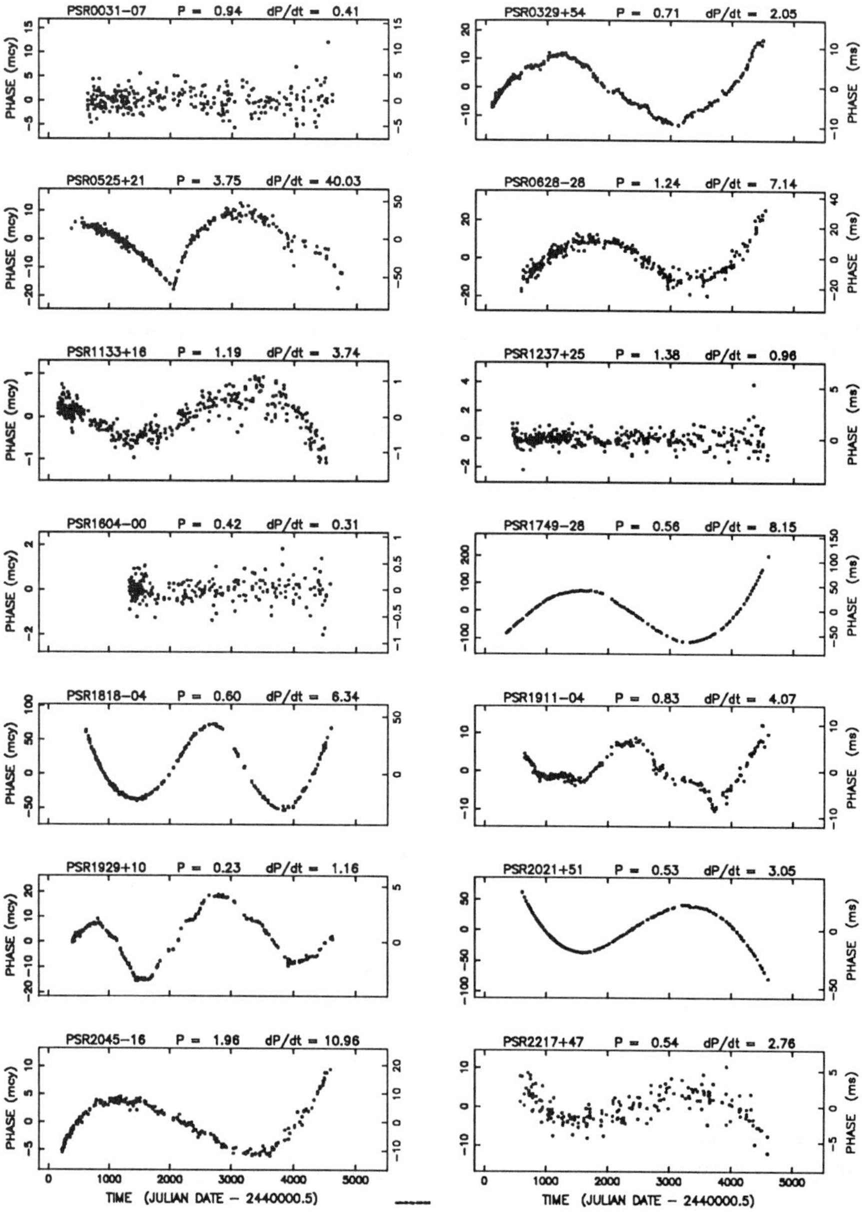

FIGURE I Phase residual curves $\mathcal{R}_2(t)$ for 14 pulsars from the JPL sample of Downs and Reichley (1983). Spin periods P (seconds) and derivatives $\dot{P}$ (in units of 10^{-15} s s^{-1}) are shown at the top of each panel.

where

$$\sigma_2(T) \equiv \langle \mathcal{R}^2(t) \rangle_T^{1/2} \tag{5}$$

is the rms phase residual after removal of a quadratic over a time span T and the timing noise of the Crab pulsar, consistent with a random walk in ν (see below), scales as (Boynton *et al.* 1972; Groth 1975; Cordes 1980)

$$\sigma_{2,Crab}(T) \approx 1.28 \; ms \, T_{yr}^{3/2}. \tag{6}$$

Dewey & Cordes (1989) found a best fit model for the activity parameter

$$\hat{A}_t = \alpha \log P + \beta \log \dot{P}_{-15} + \gamma \tag{7}$$

in terms of the spin period and its derivative (in units of 10^{-15} s s^{-1}). The parameters α, β, γ were calculated using a maximum liklihood method that includes upper limits on A_t as well as actual measurements. Another parameter is δ, the rms departure of the predicted from the actual activity parameter, also accounted for in the fitting process. The results are:

$$\alpha = -0.4 \pm 0.8, \qquad \beta = 0.9 \pm 0.4, \qquad \gamma = -2.0 \pm 0.4, \qquad \delta = 0.8 \pm 0.2. \tag{8}$$

Figure II shows activity parameter plotted against the best fit estimate. Although the trend with increasing $P^\alpha \dot{P}^\beta$ is evident, there is scatter about this relation by more than an order of magnitude, suggesting that other factors, such as temperature, composition, or evolutionary history, are also important. The important point here is that:

> *Young objects with large magnetic fields (eg. the Crab and Vela pulsars) tend to have the largest timing noise, whereas objects with small fields, particularly the millisecond pulsars, have small or negligible timing noise.*

Well timed, small-field objects show timing noise less than 20μs over time scales of years; the millisecond pulsar 1937+214 shows residuals of only 200 ns over a few years (Ryba 1991).

Random Walk Processes

Random walk processes have been proposed as a statistical description for timing noise (Boynton *et al.* 1972; Groth 1975; Cordes 1980). In particular, random walks in pulse phase, spin frequency, or frequency derivative are due, respectively, to step functions in the respective variable. The corresponding phase fluctuations have 'red' power spectra that are increasingly steeper in going from a 1st order random walk (in ϕ) to a 3rd order walk in ($\dot{\nu}$). Referring to these as PN (phase noise), FN (frequency noise), and SN (slowing-down noise), the corresponding power spectra are power laws scaling as f^{-2}, f^{-4}, and f^{-6}. The *total* (pre-fit) RMS phase variation over a time span T is

$$\sigma_{TN}^2(T) = P^2 \times \begin{cases} \frac{1}{2} S_{PN} T & \text{PN} \\ \frac{1}{12} S_{FN} T^3 & \text{FN} \\ \frac{1}{120} S_{SN} T^5 & \text{SN} \end{cases}, \tag{9}$$

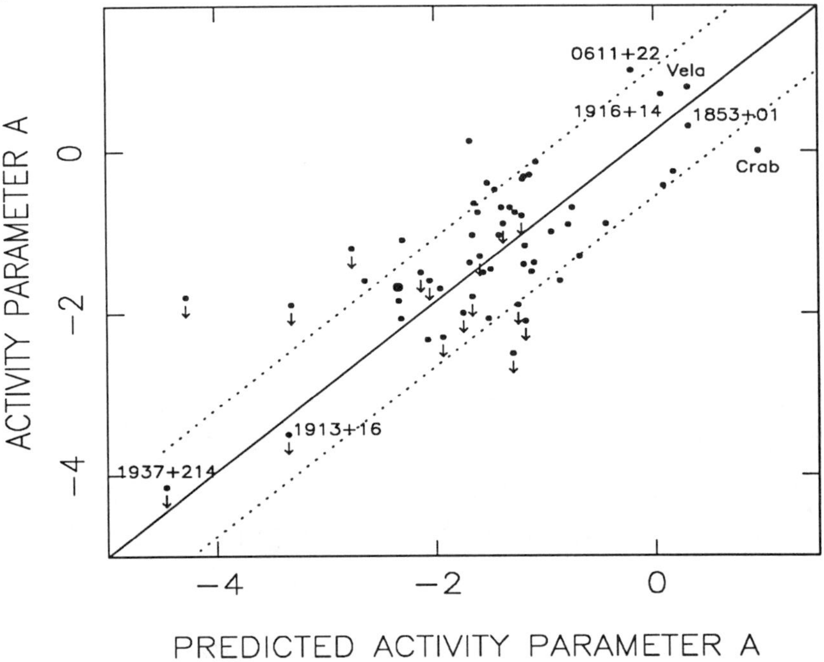

FIGURE II Timing noise activity parameter plotted against predicted activity parameter $\hat{A} = \log(\gamma P^{\alpha} \dot{P}^{\beta})$. The dotted lines are $\pm 1\delta$ departures from the model (solid line).

where S_{PN}, etc. are strength parameters for the random walks. Subtraction of a polynomial from the data (to remove spindown terms) removes a fraction of the random walk that is increasingly larger in going from PN→SN.

Detailed studies (Cordes & Downs 1985; Alpar, Nandkumar, & Pines 1986; Cordes, Downs, & Krause-Polstorff 1988) show that timing noise is often a *mixture* of idealized random walks and that there is a wide range of step rates in the walk. In some cases, individual events can be identified (eg. PSR 1929+10 in Figure I). Nonetheless, idealized random walks may be used to provide scaling laws in data span length T. Later in the paper, we will use the scaling for FN to address the detectability of planets around pulsars. Adopting FN, the strength parameters of various objects are

$$S_{FN}(\mathrm{Hz^2\ s^{-1}}) \approx \begin{cases} 5 \times 10^{-23} & \text{Crab pulsar} \\ 10^{-28} \text{ to } 10^{-25} & \text{`vanilla' pulsars} \\ < 10^{-28.5} & \text{millisecond pulsars.} \end{cases} \tag{10}$$

Alternatively, combining eqn (7)-(9) and applying a correction factor $C_{FN,2} \approx 15.5$ (cf. Cordes 1980; Blandford, Narayan, and Romani 1984; Deeter 1984) to account for a quadratic fit, we obtain

$$S_{FN} \approx 10^{-28.8} \mathrm{Hz^2\,s^{-1}} P^{-2.8} \dot{P}_{-15}^{1.8}. \tag{11}$$

PRECESSION

Precession of neutron stars is as yet poorly studied, though it is well recognized that study of free precession can yield important information about the internal structure of NS (eg. Pines and Shaham 1972) and that geodetic precession can reveal the cross sectional structure of pulsar beams and information about the kinematics of binary pulsars (Smarr & Blandford 1976; Cordes & Wasserman 1984). The influence of superfluid vortices on free-precession is controversial. With 'absolute' pinning of vortices on crustal nuclei, precession periods are probably limited to no more than $\sim 100P$ (Shaham 1977). However, the degree of pinning is uncertain (Jones 1988; Alpar & Ögelman 1987), so there may not be any theoretical obstacles to interpreting some long-term phenomena in terms of free precession.

Geodetic Precession

Geodetic precession, a General Relativistic effect where the pulsar's spin vector precesses around the orbital angular momentum, occurs at a rate

$$\Omega_p = \frac{3\pi G m_2(1 + m_1/3M)}{ac^2(1-e^2)P_{orb}} \tag{12}$$

where m_1, m_2 are the pulsar and companion masses, $M \equiv m_1 + m_2$, and we have used values for the masses and orbital elements from Taylor and Weisberg (1989). NS-NS binaries are devoid of Newtonian precessional effects owing to the compact nature of the stars. With orbital elements and masses determined from timing observations, the precession periods of two known NS-NS binaries are 300 yr (1913+16) and 600 yr (1534+12). Because the pulsar beam also precesses, pulse shape changes are expected whose amplitude depends on the spin-orbit angle λ. Such pulse-shape changes are seen from 1913+16 at the level of about 1% yr^{-1} (Weisberg, Romani, and Taylor 1989; unpublished data), consistent with the 300-yr precession period and $\lambda \approx$ few degrees.

Free Precession

Free precession (figure III) also yields pulse-shape changes from wobble of the beam, but differs from geodetic precession in that the angle α between the (instantaneous) spin axis and the magnetic moment varies over the precessional cycle. The torque almost certainly depends on α. Magnetic dipole radiation into a vaccuum yields $\dot{\nu} \propto \sin^2\alpha$, but homopolar generator models (Goldreich & Julian 1969) and empirical studies (eg. Bhattacharya 1992) suggest a much weaker dependence. If we assume $\dot{\nu} \propto g(\alpha)$, with g an unspecified function that changes by order unity, the induced torque variation is sinusoidal with frequency $\omega_p \sim (\delta I/I)\Omega$ (I is the moment of inertia) and amplitude of order

$$\frac{\delta\dot{\nu}}{\dot{\nu}} \sim \theta_p \frac{\partial \ell n g(\alpha)}{\partial\alpha} \sim \theta_p, \tag{13}$$

where θ_p is the wobble angle (the angle between the instantaneous spin vector and a symmetry axis of the neutron star crust). The corresponding phase

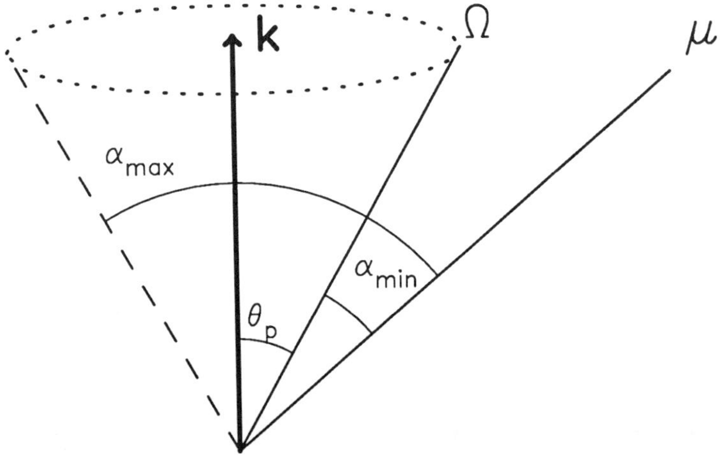

FIGURE III Configuration of axes for free precession in the neutron star frame. The instantaneous angular velocity Ω precesses around the symmetry axis k with frequency ω_p and by an angle θ_p. The magnetic moment μ is fixed with respect to k, while the angle between μ and Ω varies by $\alpha_{max} - \alpha_{min} = 2\theta_p$.

residuals have amplitude

$$\delta\phi_{torque} \approx \frac{\dot{\nu}\theta_p}{\omega_p^2} \approx 10^{-5.5} \text{ cycles} \left(\frac{\theta_p}{0.1^\circ}\right) \left(\frac{\dot{\nu}}{10^{-15}\,\text{Hz/s}}\right) \left(\frac{P_p}{100\,d}\right)^2. \tag{14}$$

Timing variations due to beam wobble are (eg. Nelson, Finn, & Wasserman 1990)

$$\delta\phi_{wobble} \sim \frac{\theta_p}{2\pi} \sim 10^{-3.6}\, cycles \left(\frac{\theta_p}{0.1^\circ}\right). \tag{15}$$

with accompanying (though not necessarily observable) changes in pulse shape

$$\frac{\delta S}{S} \sim \frac{\theta_p}{W} \sim 0.01 \left(\frac{\theta_p}{0.1^\circ}\right) \left(\frac{10^\circ}{W}\right), \tag{16}$$

where W is the pulse width.

Free precession in neutron stars, like the Chandler wobble of the Earth, must be excited continuously because it otherwise will damp on a time scale $\tau_p \approx (\Omega/\omega_p)\tau_{cc}$ (Shaham 1977), where the crust-core coupling time $\tau_{cc} \sim 100$ sec, consistent with glitches from the Vela pulsar (eg. Flanagan 1989; McCulloch *et al.* 1990). For precession periods short enough to be seen in existing data sets (eg. $P_p = 2\pi/\omega_p < 10$ yr), τ_p is much less than the spindown time $P/2\dot{P}$. Free precession may be excited continuously by crustal cracking that is itself driven by spindown of the neutron star. The wobble angle θ_p and precession frequency ω_p will both be random functions of time, because $\delta I/I$ is random, yielding stochastic precession. As shown in figure IV, free precession induces

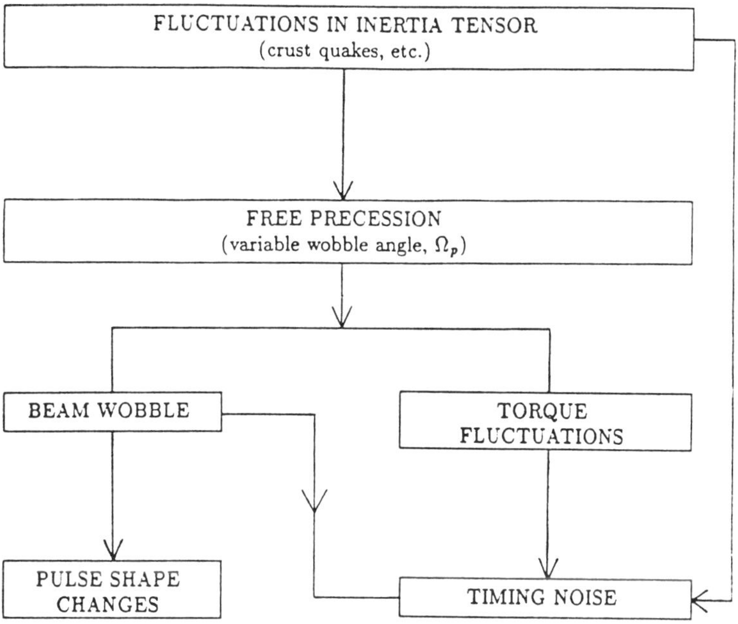

FIGURE IV Beam wobble and induced torque fluctuations in free precession.

pulse shape changes and timing variations by geometric wobble of the beam; however, timing variations also receive contributions from induced torque variations and from changes in the moment-of-inertia tensor that drive the precession in the first place. It may be shown that (1) the geometrical pulse shape and timing variations over the precessional cycle are out of phase; while (2) induced torque and pulse shape variations are in phase.

Crab Pulsar

Phase residuals from the Crab pulsar, after accounting for glitches and after fitting for a cubic polynomial, show cyclical (though non-periodic) phase residuals, as shown in Figure V. Interpretation in terms of free precession and assumption that phase residuals are dominated by induced-torque variations (eg. Jones 1988; Blaskiewicz 1991) implies a spin-orbit angle $\theta_p \approx 2.3$ arc sec and pulse shape changes at the level $\delta S/S \sim \theta_p/W \sim 10^{-4}$, where we have used $\dot{\nu} = 10^{-9.4}$ Hz s^{-1}, $W \sim 5°$, and $\delta\phi_{torque} \sim 0.3$ cycles. The Crab pulsar, therefore, represents a case where precession appears only in the timing residuals: pulse shape changes are too small, by a factor ~ 100, to be detectable.

The JPL Sample

The timing program at JPL extended from 1968 to 1983 using antennas of the Deep Space Network (Downs & Reichley 1983). Data were obtained on 32 pulsars, of which 24 were observed extensively. We have analyzed the pulse shape data on a subset of 14 pulsars (those with adequate S/N) to search for long-term pulse-shape changes. To do so, we applied a principal component analysis that expands fluctuations in pulse waveforms onto an orthogonal set of basis vectors (eg. Mittaz, Penston, & Snijders 1990). Of the 14 objects, we found that (1) 11 show variations in pulse shape consistent with the finite number of pulses

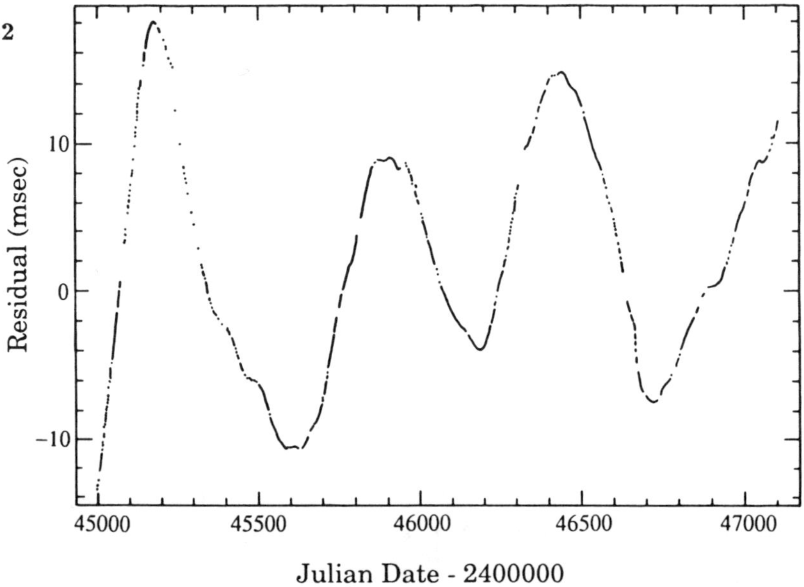

FIGURE V Phase residuals from the Crab pulsar (after Lyne, Pritchard, and Smith 1988).

summed; (2) one object, PSR 0329+54, shows three well-defined pulse shapes associated with 'mode changes' occurring on short time scales ($\sim$ minutes) (eg. Bartel *et al.* 1982); while (3) two objects, the Vela pulsar and PSR 1642−03, show long term secular effects in the pulse shape. For the Vela pulsar, pulse shape changes have a characteristic time scale $\sim$ 100 days, similar to that seen in timing variations. While it is difficult to estimate a wobble angle, given that a precession frequency cannot be identified, with $P_p \sim 100$ days, one obtains $\theta_p \approx 0.^\circ1$. For 1642−03, the pulse shape changes are cyclical with a period of about 10^3 days; PSR1642−03 also shows cyclical timing residuals (Figure VI). The precession model can account for shape changes and timing residuals with a wobble angle $\theta_p \sim 0.^\circ5$.

Torque vs. Wobble

The induced-torque effect 'amplifies' timing residuals to much larger amplitudes than would be obtained from geometrical effects alone. Objects with large $\dot{\nu}$ show the greatest amplification. It is instructive to compare torque and wobble variations by solving for $\dot{\nu}$ such that $\delta\phi_{torque} = \delta\phi_{wobble}$, yielding

$$\dot{\nu} = 10^{-9.1}\ \mathrm{Hz\ s^{-1}}\ P_p^{-2}(\mathrm{days}). \qquad (17)$$

This relation is shown in figure VII along with values of $\dot{\nu}$ and P_p for the Crab pulsar and other objects described above. Millisecond pulsars with $P_p \leq 10^3$ days and (intrinsic) $\dot{P} \leq 10^{-19}$ s s^{-1} will show precession only through geometrical wobble, as induced torque variations would be dominated by wobble. However, it seems unlikely that precession is self-excited in millisecond pulsars in the first place because rotation fluctuations driven by differential rotation also seem to be negligible for these objects.

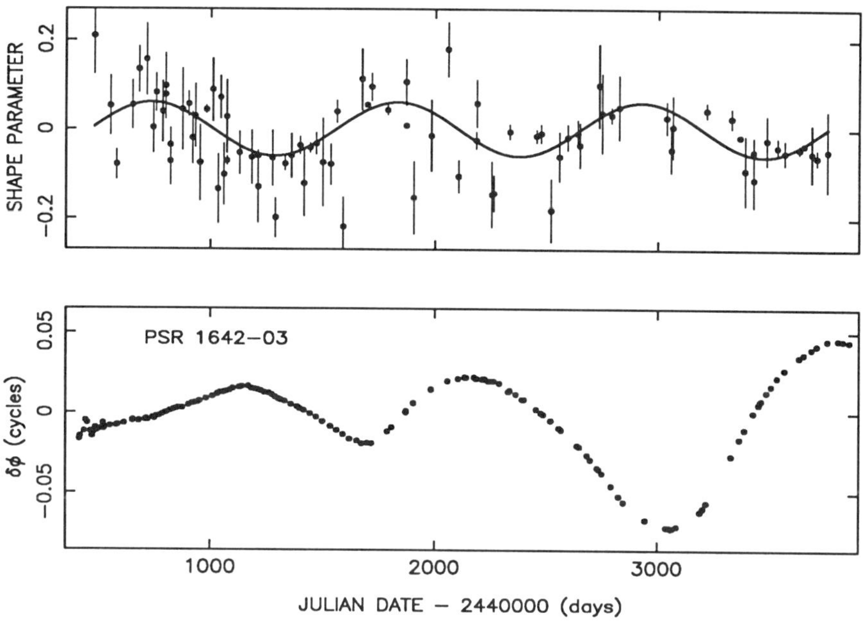

FIGURE VI Evidence for free precession from PSR1642−3. (Bottom) Phase residuals after second order fit; (Top) Pulse shape parameter vs. time. The solid curve is the best fit sinusoid implied by Fourier analysis of the time series.

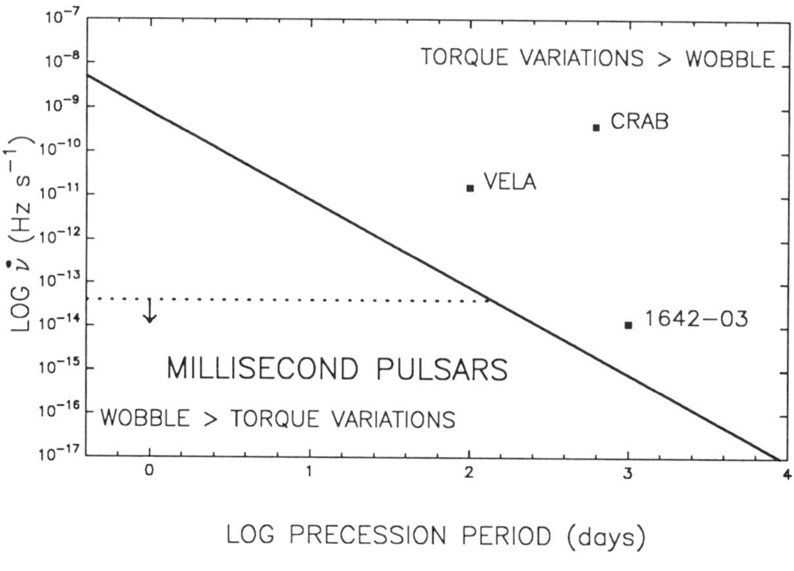

FIGURE VII Definition of regimes of torque-dominated timing residuals vs. wobble dominated residuals.

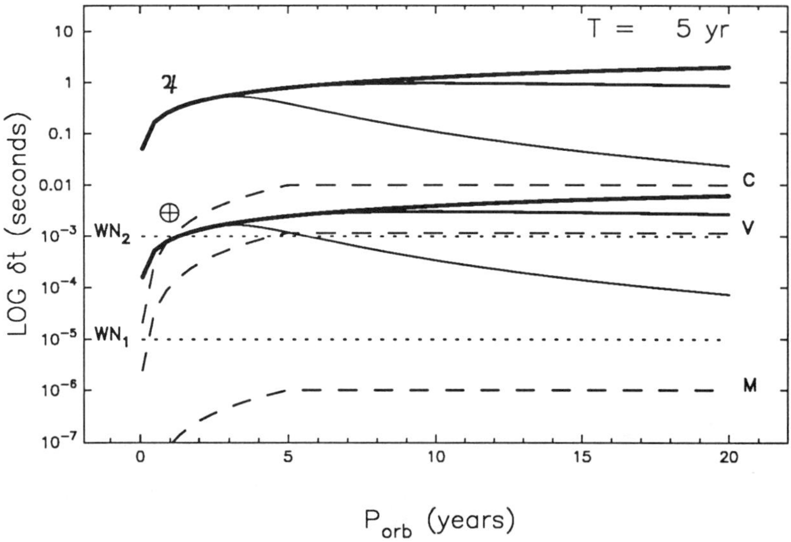

FIGURE VIII Timing perturbation δt from an Earth mass planet and a Jovian mass planet as a function of orbital period. The perturbations are those seen in a time span of length $T = 5$ yr. *(Heavy solid lines:)* Full amplitude of orbital perturbations, eqn (18). *(Medium solid lines:)* Amplitude of perturbation over the data span. *(Light solid lines:)* Perturbation after removal of 2nd order polynomial. *Dashed lines:* Timing noise assuming FN ($\sigma_{TN} \propto T_{effective}^{3/2}$, with $T_{effective} = P_{orb}$ for $P_{orb} \leq T$ and $T_{effective} = T$ otherwise). The 'c', 'v', and 'm' refer to TN for the Crab pulsar, 'vanilla' pulsars, and millisecond pulsars, respectively. *Dotted lines:* Constant white noise levels as designated by $WN_1 = 10\ \mu s$ and $WN_2 = 1$ ms.

DETECTABILITY OF PLANETS

Coherent oscillations in pulsar timing data generally will be intermixed with white-and-red noise, as discussed above. Here we consider the detectability of planetary perturbations in terms of these obfuscating effects.

In figure VIII we show planetary perturbations (for circular orbits with $\sin i = 1$ and a NS mass of $1.4 M_\odot$),

$$\delta t = 1.21\,\mathrm{ms}\ (M_p/M_\oplus) {P_{orb}}^{2/3}(\mathrm{yr}) \tag{18}$$

for Jupiter and Earth-massed planets plotted against P_{orb}. For data spans of fixed length (eg. $T = 5$ yr) only a fraction of the entire perturbation is seen for $P_{orb} > 4T$, as indicated. Also shown is the post-fit rms residual after removing a 2nd-order polynomial (to simulate removal of the pulsar spindown). For comparison, we show lines of constant rms white noise (at 1 ms and $10 \mu s$) and similar contours for timing noise that assume an FN type random walk, evaluation of equation (9) using equation (10), and choices of P and $\dot{P}$ that correspond to young, Crab-like pulsars, millisecond pulsars, and 'vanilla' pulsars ($P \sim 1$ s and $\dot{P} \sim 10^{-15}$ s s^{-1}). Figure VIII demonstrates that Earth-mass planets are easy to detect around millisecond pulsars (TN $\leq 1\mu s$, WN $\leq 10\mu s$) but are marginally detectable around some vanilla pulsars only for orbital periods $P_{orb} \sim 1-5$ yr, and are undetectable from Crab-type objects because of timing noise. Jupiter-mass objects are easily detectable from nearly all pulsars unless they have orbital periods ≥ 15 yr.

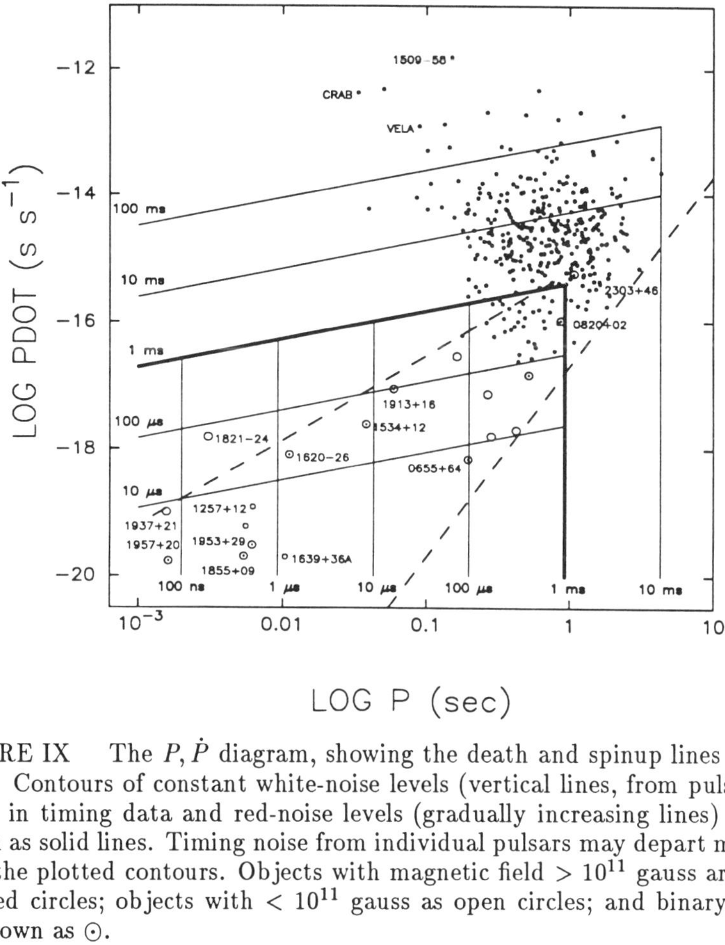

FIGURE IX The $P, \dot{P}$ diagram, showing the death and spinup lines (dashed lines). Contours of constant white-noise levels (vertical lines, from pulse phase jitter) in timing data and red-noise levels (gradually increasing lines) are also shown as solid lines. Timing noise from individual pulsars may depart markedly from the plotted contours. Objects with magnetic field $> 10^{11}$ gauss are shown as filled circles; objects with $< 10^{11}$ gauss as open circles; and binary pulsars are shown as $\odot$.

It is also useful to consider the masking effects of noise in the $P - \dot{P}$ plane, as shown in figure IX. Using equation (1) with $f = m_I = 1$ and constant, 5% duty cycle (W/P), we show vertical contours of constant white noise level due to pulse phase jitter and assuming that 6 observations yr^{-1} are made over a span of 5 yr. Also shown are contours of constant timing noise using equations (9) and (11) to give

$$\dot{P}_{-15} = \left[\frac{\sigma_{TN}(\mathrm{ms})}{0.2\,\mathrm{ms}}\right]^{1.11} P^{0.44} T_{yr}^{1.67}. \tag{19}$$

We have assumed $T = 5$ yr to construct the contours. The region in the diagram to the left and below the lines labelled '1 ms' is where Earth-mass planets are detectable for data spans of 5 yr. This excludes most of the known pulsar population.

MULTIPLE PLANETARY SYSTEMS

Multiple planetary systems around NS, if like the solar system, will display a large range of masses and orbital periods, yielding a composite signal in pulsar timing data that may be difficult to interpret. Finite and nonuniformly sampled data sets, combined with the need to remove at least a quadratic polynomial from timing data (not to mention yearly, periodic terms for fitting position and proper motion), require a careful analysis in the face of alternative interpretations involving precession or spin noise. If the solar system's planets were placed around a NS, the timing perturbation would appear as in Figure X, where only a 2nd order polynomial has been subtracted to yield post-fit residuals. These are not dissimilar from real timing residuals. Figure XI shows the same data with a fit for a single Keplerian orbit, showing that the spectral analysis is much improved if the complete orbit (rather than a circular orbit) of the dominant planet is removed from the data. Residuals from Jupiter, with a fit of a circular orbit, yield $\sim$ 25 ms residuals at the second harmonic.

Lessons learned from simulations of this kind include:

- *To find the terrestrial planets, the dominant effects from the more massive, outer planets must be removed.* With a 10-year data set, Saturn's contribution is removed by a polynomial while a Keplerian orbit removes Jupiter's contribution. The power spectrum in figure XI before fitting (solid line) shows no convincing spike associated with the Earth. The post-fit power spectrum shows three significant spikes associated with the Earth, Venus, and Mercury; note the power that has been removed at low frequencies due to removal of the fitting function. With white-noise levels modestly larger than the 10 μs used, none of these planets would be detectable in the pre-fit power spectrum.

- *Irregularly and less-frequently sampled data are more difficult to analyze.*

- *Inclusion of red noise to the timing perturbation produced by planets makes it even harder to detect planets.*

- *Predictability of future data from an orbital solutions is thwarted by additional, long-orbital period objects and by red noise.*

Regimes for Hypothesis Testing

With multiple planetary systems, three cases can be defined based on the relative values of the data span T and orbital periods P_{orb}. For a planetary system $[(M, P_{orb})_j \; j = 1, \ldots, N_p]$:

- $P_{orb,j} \leq T \; \forall \; j$: timing residuals will be periodic, with periods detectable in a power spectrum, given adequate S/N. As $N_p \rightarrow \infty$, the time series tends toward a noiselike process, as with a hypothetical asteroid belt.

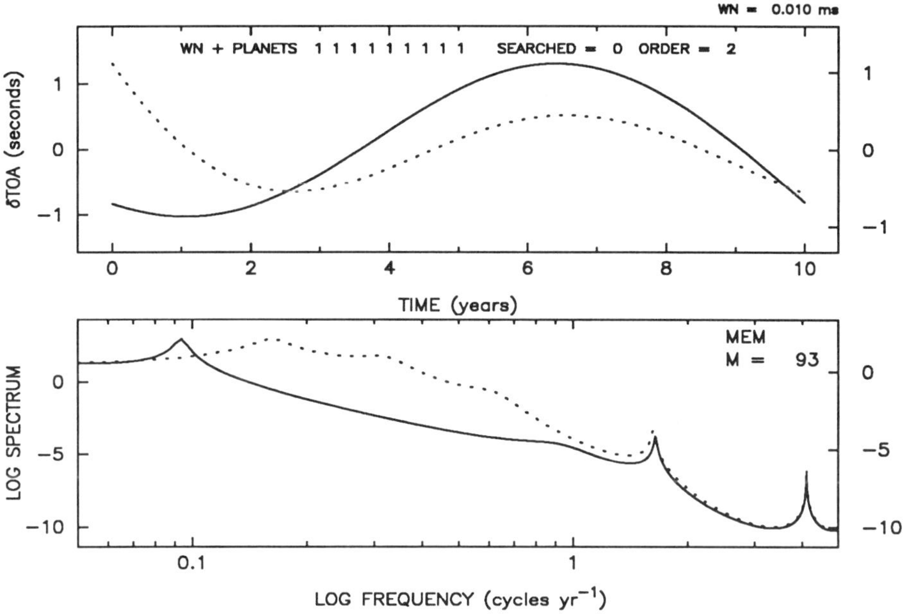

FIGURE X Simulated data consisting of orbital perturbations on a $1.4 M_{\odot}$ neutron star with the solar systems' planets viewed at an inclination angle of 45°. *(Top)* The solid curve shows the perturbation before fitting, with obvious dominance by the outer planets. The dashed curve results after fitting a 2nd order polynomial. The left-and-right-hand scales apply, respectively, to the pre- and-post fit perturbations. *(Bottom)* Maximum entropy power spectra of the pre-and-post fit perturbations in the top panel.

- $P_{orb,j} \ll T \ll P_{orb,k}$, having $N_<$ planets with $P_{orb} \leq T$ and $N_p - N_<$ planets with $P_{orb} > T$: $\sigma_{\mathcal{R}}$ is nonstationary in T. With adequate S/N, the short period planets are detectable in the power spectrum, though perturbations from long-period planets may need to be removed to see short-period planets in the power spectrum.

- $P_{orb,j} \gg T \; \forall j$: $\sigma_{\mathcal{R}}$ is nonstationary in T and the phase residuals are smooth. A polynomial fit will remove much of the variation. The power spectrum is also smooth.

These properties of planetary systems must be compared with effects from red, intrinsic timing noise:

- Individual events can be identified if the sample interval $\Delta t \ll R^{-1}$ (Δt = sample interval; R = rate of events). This seems to be the case with PSR 1929+10 and some other objects in figure I.

- Relatively fast changes in slope are also produced collectively from unresolved events; these may be quantified using structure functions (eg. Cordes &

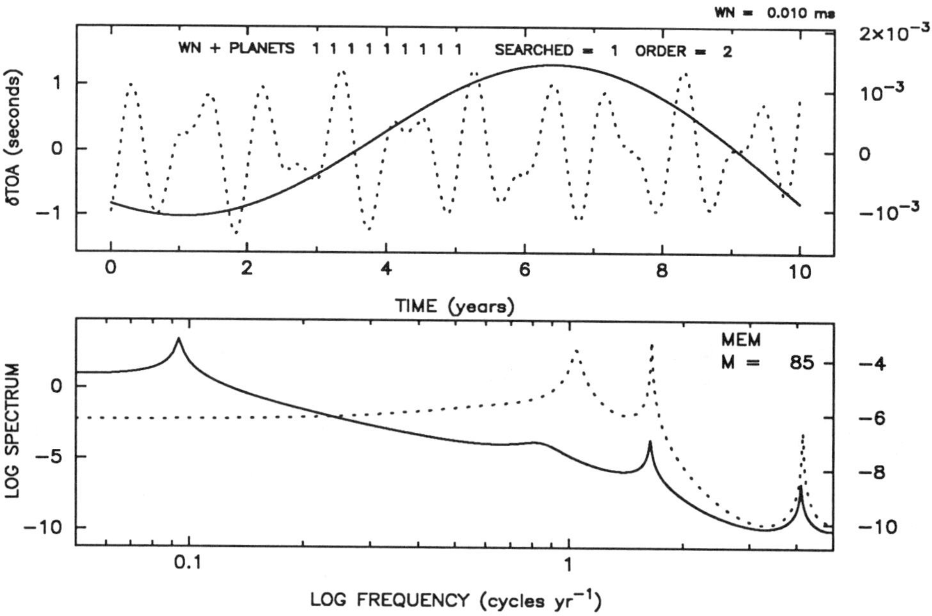

FIGURE XI Same as Figure X except that simulated data have been fitted for a Keplerian orbit as well as a polynomial.

Downs 1985).

- Self-similar, nonstationary noise is expected for $R^{-1} \ll \Delta t$. The power spectrum is a smooth power law that may contrast markedly from that seen from planetary systems in the first two regimes defined above.

Criteria for testing the presence of planetary signatures in timing data include the following, none of which are fail-safe:

- *Deep well(s) in χ^2 while fitting for planets.* These may be expected only where a single planet dominates the timing perturbation. Comparably massed planets will not show a deep minimum unless all planets are fitted simultaneously. The general case of elliptical orbits with N_p more than a few makes this extremely difficult because of the large parameter space ($5N_p$). Thorsett & Phillips (1992) have placed upper bounds on perturbing planets by searching for single planets with circular orbits.

- *Spikes in the power spectrum* will be seen only for $P_{orb} \leq T$. Small planets may be swamped by the effects of larger ones.

- *Peaks and valleys in the spectrum* may signify the presence of planets in contrast to the smooth spectra expected from random-walk processes in the large rate limit.

- *Predictability of time series* will hold only if all planets have $P_{orb} < T$ for the data used to predict ahead. Longer period planets cause departures from the prediction that must be assessed along with alternative hypotheses, such as the presence of red noise from rotation fluctuations.

Bayesian Methods

While individual realizations of timing noise may mimic (or be mimiced by) an ensemble of planets, the orbital-and-red-noise hypotheses may be compared on a statistical basis from a Bayesian point of view. Letting D be the data, I the background information regarding pulse timing and planetary systems, and H_R and H_P be, respectively, red-noise and planetary hypotheses, we may compare the odds favoring the red-noise model over the planetary model using ratios of probabilities $\mathcal{P}$ (eg. Loredo 1990, eqn [23])

$$O_{RP} = \left[\frac{\mathcal{P}(R|I)}{\mathcal{P}(P|I)}\right] \frac{\mathcal{P}(D|RI)}{\mathcal{P}(D|PI)}, \tag{20}$$

where the quantity in brackets is independent of the data, representing the prior odds, and the trailing ratio modifies the odds according to the data. Characterization of the data may include those properties of one process that are the most unlikely to occur from the alternative hypothesis. For example, red timing noise is always nonstationary and can show dramatic changes in the slope of the phase (or its derivatives) either due to resolution of individual events that make up the noise or, spuriously, through the collective effects of individual events. Planetary signatures may appear nonstationary over finite data spans. However, rapid changes in slope (relative to the dominant orbital periods) are unlikely unless some planets have large eccentricities. Thus, a Bayesian analysis could focus on the different characters of red noise and planetary signatures in the time and frequency domains.

ACKNOWLEDGMENTS

I thank Mike Blaskiewicz, David Chernoff, Tom Loredo, Phil Nicholson, Tony Phillips, Steve Thorsett, and Ira Wasserman for helpful discussions. This work was supported by the National Astronomy and Ionosphere Center, which operates the Arecibo Observatory by cooperative agreement with the National Science Foundation.

REFERENCES

Alpar, A., Nandkumar, R., & Pines, D. 1986, *ApJ*, **311**, 197.

Alpar, A. & Ögelman, H. 1987, *ApJ*, **185**, 196.

Backer, D., Sallmen, S., & Foster, R. 1992, *Nature*, in press; see also Backer, D. these proceedings.

Bailes, M., Lyne, A., & Shemar 1991, *Nature*, **352**, 311.

Bartel, N., Morris, D. Sieber, W. & Hankins, T 1982, *ApJ*, **258**, 776.

Bhattacharya, D. 1992, in *The Magnetospheric Structure and Emission Mechanisms of Radio Pulsars*, eds. T. Hankins, J. Rankin, & J. Gil, (Zielona Góra, Poland: Pedagogical Univ. Press), 27.

Blandford, R., Narayan, R. & Romani, R. 1984, *J. Ap. Ast.*, **5**, 369.

Blaskiewicz, M. 1991, Ph.D. Thesis, Cornell University.

Boynton, P. *et al.* 1972, *ApJ*, **175**, 217.

Cordes, J. 1980, *ApJ*, **237**, 216.

Cordes, J. & Downs, G. 1985, *ApJS*, **59**, 343.

Cordes, J., Downs, G. & Krause-Polstorff 1988, *ApJ*, **330**, 847.

Cordes, J. & Wasserman, I. 1984, *ApJ*, **279**, 798.

Deeter, J. 1984, *ApJ*, **281**, 482.

Demianski, M. & Prószyński 1979, *Nature*, **282**, 383.

Dewey, R. & Cordes, J. 1989 in *Timing Neutron Stars*, NATO ASI, eds. H. Ögelman & van den Heuvel, E., 119.

Downs, G. & Reichley, P. 1983, *ApJS*, **53**, 169.

Flanagan, C. 1989, IAU Circular 4695.

Goldreich, P. & Julian, W. 1969, *ApJ*, **157**, 869.

Groth, E. 1975, *ApJS*, **29**, 453.

Jones. P. 1988, *MNRAS*, **235**, 545.

Kristian, J. *et al.* 1989, *Nature*, **338**, 234.

Loredo, T. 1990, in *Maximum Entropy and Bayesian Methods*, ed. P. Fougère, (Kluwer), 81.

Lyne, A., Pritchard, R., & Smith, F. 1988, *MNRAS*, **233**, 667.

McCulloch, P., Hamilton, P., McConnell, & King, E. 1990, *Nature*, **346**, 822.

Mittaz, J., Penston, M., & Snijders, M. 1990, *MNRAS*, **242**, 370.

Nelson, R., Finn, S., & Wasserman, I. 1990, *ApJ*, **348**, 226.

Pines, D. & Shaham, J. 1972, *Comments. Ap.*, 37.

Rankin, J. 1990, *ApJ*, **352**, 247.

Richards, D., Pettengill, G., Counselman, C., & Rankin, J. 1970, *ApJ*, **160**, L1.

Ryba, M. 1991 Ph.D. Thesis, Princeton University.

Shaham, J. 1977, *ApJ*, **214**, 251.

Smarr, L. & Blandford, R. 1976, *ApJ*, **207**, 574.

Taylor, J. & Weisberg, J. 1989, *ApJ*, **345**, 434.

Thorsett, S. & Phillips, J. 1992, *ApJ*, **387**, L69.

Weisberg, J., Romani, R., & Taylor, J. 1989, *ApJ***347**, 1029.

Wolszczan, A. & Frail, D. 1992, *Nature*, **355**, 145.

Planets Around Pulsars
ASP Conference Series, Vol. 36, 1993
J.A. Phillips, J.E. Thorsett, and S.R. Kulkarni

ON QUASIPERIODIC VARIATIONS OF PULSARS' PERIODS: AN ALTERNATIVE TO THE PLANETARY INTERPRETATION OF PSR1257+12

A. Z. DOLGINOV[1] AND T.F. STEPINSKI
Lunar and Planetary Institute, 3600 Bay Area Blvd., Houston, Texas, 77058

ABSTRACT The quasi-periodic timing residuals of PSR1257+12 are considered to be a result of two large-scale internal motions of the neutron star: core-crust relative motion and precession. The model requires that the superfluid core of a neutron star is not pinned to the solid crust. A very good fit to the observations is obtained for the set of parameters that corresponds to a nearly orthogonal rotator, and the period of the core-crust relative rotation that is nearly twice the period of precession.

INTRODUCTION

Recently Wolszczan and Frail (1992) (hereafter W&F) reported observations of quasiperiodic changes in the period of millisecond pulsar PSR1257+12. The authors attribute those changes to motion of the pulsar as it and its two planetary-mass companions orbit around their common center of mass. This "planetary" interpretation of PSR1257+12 has been very quickly embraced by the astrophysical community, partly because of the remarkable agreement between the observations and behavior expected if this interpretation is correct, and partly because of the intellectual appeal of the first discovery of a planetary system outside the solar system. However, it has proven difficult to find a physically plausible scenario that would explain the formation of planets around pulsars. A large number of various scenarios has been proposed, but all of them are rather unsatisfactory (see papers by Podsiadlowski and Phinney in this proceedings). Thus, notwithstanding the attractiveness of the planetary model, it is important to examine its alternatives.

We believe that large-scale internal motions of neutron star provide an equally plausible and more readily understandable explanation of the timing residuals periodicity. In this paper we present preliminary results of our work on a model that seems to be capable of explaining the basic features observed in PSR1257+12. Our model relies on taking into account two physical processes: free precession of the neutron star and the relative rotation of the neutron star crust with respect to its core. The magnetic field of the neutron star is frozen in the charged component of the star and can be considered, in the first approx-

[1]On leave from A.F. Ioffe Physico-Technical Institute, Russian Academy of Sciences, 194021 St. Peterburg, ul. Politekhnicheskaya 26, Russia

imation, as fixed with respect to the crust. The crust rotates around the total angular momentum vector, as well as precessing around its axis of symmetry. In addition, the core-crust friction involves the crust in a slow rotation around the core angular velocity vector. The radiating region of the neutron star is connected with its magnetic field, and thus undergoes a complex motion that results in nonperiodic times of pulse emission.

The outline of the remainder of this paper is as follows. We first discuss the two-component model, which suggests that neutron stars consist of a solid crust and superfluid core that rotate independently but are loosely coupled to each other. In order for our model to explain W&F observations, a relative core-crust angular velocity of the order of $10^{-6} - 10^{-7}$ sec^{-1} must be physically plausible. We then discuss whether a millisecond pulsar may undergo a precessional movement with angular velocity of the order of $10^{-6} - 10^{-7}$ sec^{-1} (period of the order of months). Following this discussion we calculate general formulae for pulse times of arrival residuals as predicted by our model and apply our model to observations of PSR1257+12 as published by W&F. Finally, we summarize our results and discuss the strong and weak points of our model.

CORE-CRUST RELATIVE ROTATION

We postulate that one of two periodicities detected in the post-fit residuals of pulse arrival times from PSR1257+12 is due to the slow, systematic rotation between the crust and the core of the neutron star. The idea that the pulsar does not rotate as a single body, but has two components that are only loosely coupled together, was originally advocated to explain the form of the so-called "glitch function" in young pulsars like Crab and Vela (Baym et al., 1969). In its simplest version, the two-component model consists of a solid crust weakly coupled to the superfluid core. The crust is assumed to rotate with the observed pulsar frequency, and the core lags behind the crust by some relative amount ω_c. Furthermore, as it was first pointed out by Anderson and Itoh (1975), and later elaborated by Alpar and Ögelman (1987), angular velocities of the crust and the core should not necessarily be coaxial. In our model, we would specifically assume that those velocities are not coaxial, resulting in a systematic rotation of the crust around the core rotation axis. There are various possible causes that may initiate the relative core-crust motion. Examples include starquakes that restore a star's equilibrium shape, which becomes nonequilibrated because of the star spin down or spin up, neutron star crustal plate tectonics (Ruderman, 1991), and any form of mass accretion that took place not longer than a relaxation time ago.

The superfluid core of a neutron star takes part in the star's rotation by carrying a large number of quantized vortex lines. It is still not completely established whether the cores of the vortex lines are "pinned" to the crustal nuclei, thus providing the core-crust friction and the relatively short relaxation time for frictional dissipation τ, or instead thread the spaces between them. This last possibility is referred to in the literature as "unpinning", and would results in weak coupling and relatively long τ. Alpar et al. (1984a,b) argued that whether the vortices are pinned to the crust depends on the magnitude of the relative core-crust angular velocity ω_c. If ω_c is smaller than some critical value

ω_{crit}, then the vortices are pinned and τ is very short, about $10^2 - 10^3$ times the pulsar period (Alpar and Sauls, 1988). On the other hand, if $\omega_c > \omega_{crit}$, the vortices becomes unpinned and τ is large. The actual value of ω_{crit} is model dependent. Microscopic calculations of Alpar et al. (1984a,b) suggest large values of ω_{crit} in the range $10^{-2} - 10$ sec^{-1}. That would suggest that vortex lines remain pinned to the crust for any reasonable values of ω_c. However, as was noticed by Shaham (1986), this conclusion remains valid only if the angular velocities of the core and the crust are coaxial. That opens the possibility that if the core and the crust rotate around different axes, vortex lines may become unpinned.

As the quasiperiodic variation in time residuals from PSR1257+12 has now been observed for almost two years, our model required the relaxation time for frictional dissipation to be at least two years and preferably much longer. In the framework of the present theory of the neutron star interior, that can be only achieved if the vortex lines are not pinned to crustal nuclei. Furthermore, because both periodicities in time residuals from PSR1257+12 are of the order of months, unpinning must persist despite very slow relative core-crust rotation.

PRECESSION

We postulate that the second of two periodicities detected in the post-fit residuals of pulse arrival times from PSR1257+12 is due to free precession. Free precession of neutron stars has been first adduced as the mechanism behind some long-period variations in the fluxes from X-ray binaries. In particular, the observed 35-day cycle of Her X-1 has been interpreted as the result of free precession (Brecher, 1972). The strongest argument that a neutron star in Her X-1 undergoes precession comes from a sampling of the 35-day cycle in X-rays, which provides clear evidence for changes of the 1.24-sec pulse profile with the 35-day phase (Trümper et al., 1986). Single, radio pulsars, such as Crab pulsar, often show quasiperiodic time residuals. Lyne et al. (1988) found that the period of the residuals from the Crab pulsar is approximately 620 days, which has been interpreted by Jones (1988) as due to small-amplitude free precession of the neutron star. Despite many efforts, the changes in pulse profiles of radio pulsars, which could be attributed to free precession, have not yet been observed (see the paper by Cordes in this proceedings). Nevertheless, strictly from an observational point of view, the precession of neutron star is the most simple explanation of the quasiperiodic timing residuals.

Ruderman (1970) was the first to consider quantitatively the timing residuals due to free precession. He noted that if the magnetic dipole axis lies near the star's symmetry axis, the resulting timing residuals would have a period equal to the precessional period. If, in the other limiting case, the magnetic dipole axis is almost perpendicular to the star's symmetry axis, time residuals would oscillate with half of the precessional period. We calculated numerically the intermediate cases and have found that, in general, timing residuals due to precession show two strict periodicities of frequencies ω_{prec} and $2\omega_{prec}$. Thus, precession *alone* is the leading candidate to explain the quasiperiodic timing residuals exhibiting either one periodicity or two periodicities that are in the 2:1 ratio. However, because timing residuals from PSR1257+12 exhibit two periodicities that are in the 3:2 rather than 2:1 ratio, precession alone cannot, even in principle, explain those observations.

There has been one major objection to a precessing neutron star, and it comes from the two-component model of the neutron star interior (see previous section). Shaham (1977) has shown that, if the vortex lines remain pinned, then the precession frequency ω_{prec} must be large, of the order of $10^{-2}\omega_{puls}$. For PSR1257+12 with $\omega_{puls} \approx 10^3$ sec^{-1} that would mean a precession with the period of the order of seconds rather than months. On the other hand, if the vortex lines are not pinned to crustal nuclei, one can estimate a neutron star precession rate without considering any gyroscopic response of vortices. In such a case, the frequency of precession could be very small, especially for neutron stars with a low surface magnetic field like PSR1257+12 (see discussion in Shaham, 1986). Thus, an unpinned superfluid core is essential for the physical plausibility of both major components of our model: core- crust relative rotation and precession.

Finally, the causes of a precession are mostly the same internal or external perturbations that are responsible for the initiation of the relative core-crust motion. In addition, the equatorial mass shift within the crust during the cooling of a weakly magnetized neutron star (Schaaf, 1990) may lead to the changes in the moments of inertia and thus to the star's precession. The precession damping timescale τ_{prec} is again related to the vortex pinning and is given by $\tau_{prec} \approx \tau\omega_{puls}/\omega_{prec}$ (Alpar and Ögelman, 1987). For PSR1257+12, $\omega_{puls}/\omega_{prec} >> 1$, so that it is a damping timescale for core-crust relative rotation τ, not the precession damping time $\tau_{prec} >> \tau$, that provides the effective damping time in our model.

TIMING RESIDUALS

The kinematics of our model is the simple superposition of three rotations around three different axes (see Fig 1). We have chosen the inertial coordinate system $[x, y, z]$ so that the z-axis coincides with the direction of the total angular momentum vector $\mathbf{J}$. Further, we have chosen the x-axis so the direction toward the observer is in the $x - z$, $x > 0$ half-plane. For simplicity, we will assume that the pulsar emission is observed when some special vector, for example, magnetic dipole axis vector $\mathbf{M}$, is in the half-plane of the observer. Thus, once the trajectory of vector $\mathbf{M}$ in the inertial coordinate system $[x, y, z]$ is known, emission times t_{em} are given by conditions $\mathbf{M}_y(t_{em}) = 0$ and $\mathbf{M}_x(t_{em}) > 0$. Let us define the coordinate system $[x'', y'', z'']$, where z'' coincides with the symmetry axis of the crust and the magnetic dipole axis $\mathbf{M}$ is inclined by the angle λ with respect to z''. Because of the precession, vector $\mathbf{M}$ rotates around z'' with the angular velocity ω_{prec}. Assume that in the coordinate system $[x', y', z']$, where z' coincides with the rotation axis of the core, the angle between z' and z'' is μ. Thus the movement of vector $\mathbf{M}$ in the $[x', y', z']$ system is a superposition of two rotations, one due to the precession, as described above, and the second due to the relative rotation between the core and the crust, characterized by ω_c. Finally, the axis of rotation of the core is inclined by a constant angle θ with respect to $\mathbf{J}$. Thus, in the inertial coordinate system $[x, y, z]$, the movement of vector $\mathbf{M}$ is described as a superposition of three rotations, the first two as described above, and the third characterized by the angular velocity $\omega_{puls} = 2\pi/P_{puls}$, where P_{puls} is the core rotational period, often identified with the period of pulsation.

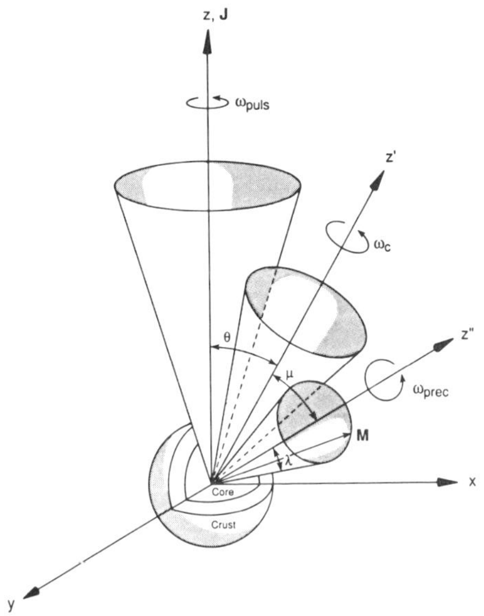

FIGURE I Illustration of the kinematics of our model.

In the inertial coordinate system the vector $\mathbf{m} = \mathbf{M}/|\mathbf{M}|$ has following components:

$$\begin{aligned} \mathbf{m}_x &= F\sin\phi + G\cos\phi \\ \mathbf{m}_y &= G\sin\phi - F\cos\phi \\ \mathbf{m}_z &= D\cot\theta - C\tan\theta - H \end{aligned} \tag{1}$$

in those equations $\phi = \omega_{puls}t$, where t is time, $F = -(A+B)$, $G = C+D+E$, and H are function of λ, μ, θ, $\psi = \omega_{prec}t$, and $\chi = \omega_c t$. Explicit forms of functions A to H are as follows

$$\begin{aligned} A &= \sin\lambda(\cos\mu\sin\chi\cos\psi + \cos\chi\sin\psi) \\ B &= \cos\lambda\sin\mu\sin\chi \\ C &= \sin\lambda\cos\theta(\cos\mu\cos\chi\cos\psi - \sin\chi\sin\psi) \\ D &= \cos\lambda\cos\mu\sin\theta \\ E &= \sin\mu(\cos\lambda\cos\theta\cos\chi - \sin\lambda\sin\theta\cos\psi) \\ H &= \sin\mu(\cos\lambda\sin\theta\cos\chi + \sin\lambda\cos\theta cos\psi) \end{aligned} \tag{2}$$

Using formulas (2) it can be checked that for the case when the axes of rotation of the core and the crust coincide ($\mu = 0$), equations (1) reduce to the description of simple precession as given by Nelson et al. (1990). The observer identifies pulse emission at times when $\mathbf{m}_y = 0$ or when $\tan\phi = F/G$. In the limit of vanishing precession and aligned axes of crust symmetry and core rotation, the observer identifies pulses at times $t_{em,n}$ given by the condition

$$2\pi n = \begin{cases} \phi_n + \psi_n + \chi_n = (\omega_{puls} + \omega_{prec} + \omega_c)t_{em,n} & \text{if } \lambda > \theta \\ \phi_n = \omega_{puls} t_{em,n} & \text{if } \theta > \lambda \end{cases} \tag{3}$$

However, if the precession and misalignment of core and crust axes are present, the times of emission are advanced or retarded from expected values $t_{em,n}$ by $\delta t(t, \theta, \lambda, \mu, \omega_{prec}, \omega_c)$. This *time residual* can be most conveniently expressed in terms of *phase residual* $\delta\phi$ - the angular distance between the vector $\mathbf{M}$ at the time equal to $t_{em,n}$ and the plane of the observer. The phase residual $\delta\phi$ can be expressed simply as

$$\tan\delta\phi(t) = \begin{cases} \frac{\tan\phi + \tan(\psi+\chi)}{1 - \tan\phi\tan(\psi+\chi)} & \text{if } \lambda > \theta \\ \tan\phi & \text{if } \theta > \lambda \end{cases} \tag{4}$$

Thus, the time residuals, which the observer would measure as the difference between the predicted and the actual times of arrival, are given by the expression

$$\delta t(t) = \frac{2\pi}{P_{puls}} \arctan\left(\begin{cases} \frac{\tan\phi + \tan(\psi+\chi)}{1 - \tan\phi\tan(\psi+\chi)} & \text{if } \lambda > \theta \\ \tan\phi & \text{if } \theta > \lambda \end{cases}\right) \tag{5}$$

Assuming particular values for θ, λ, μ, ω_{prec}, and ω_c, equation (5) gives time residuals as predicted by our model that can be directly compared to observations. Note that δt does not depend on ω_{puls}, because ϕ appearing formally in the equation (5) is expressed in terms of ψ and χ using condition $\mathbf{m}_y = 0$.

We have applied our model to observations of PSR1257+12 as published by W&F. Our goal was to establish for what parameters of θ, λ, μ, ω_{prec}, and ω_c our model would fit the residuals of pulse arrival times from PSR1257+12. We have found that theoretical curves of timing residuals, as predicted by our model, fall into two distinct families. For some sets of parameters, predicted residuals exhibit the continuous growth corresponding to steady advancement or retardation of pulse arrival times. Clearly, such configurations could not explain data from PSR1257+12. For most sets of parameters, predicted residuals exhibit periodic or quasiperiodic variations; however, the amplitude of those variations seems to be limited to $P_{puls}/4$. Thus, in order for our model to explain the magnitude of observed residuals we have to assume that pulses are emitted from both sides of the magnetic dipole axis and the real period of neutron star is 12.4 ms. Under this assumption we have been able to find a particular set of parameters, $\theta \approx 90^o$, $\lambda = 50^o$, $\mu = 30^o$, $P_{prec} \approx 205$ days, $P_c = 2\pi/\omega_c \approx 428$ days, that results in a good fit between observations and our model. Figures 2 and 3 show how timing residuals and period variations predicted by this model fit observations. It is important to note that our prediction fits observed timing residuals as well as does the planetary model. In fact, using the parameters of the planetary system as published by W&F we obtained a virtually identical fit. Period variations are slightly better fitted by the planetary model.

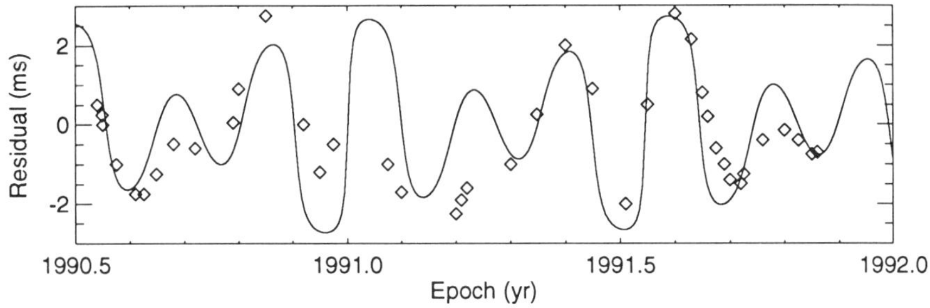

FIGURE II The residuals of pulse arrival times from PSR1257+12 are taken from W & F and are represented by diamond-shaped marks. The solid line denotes residuals predicted by our model.

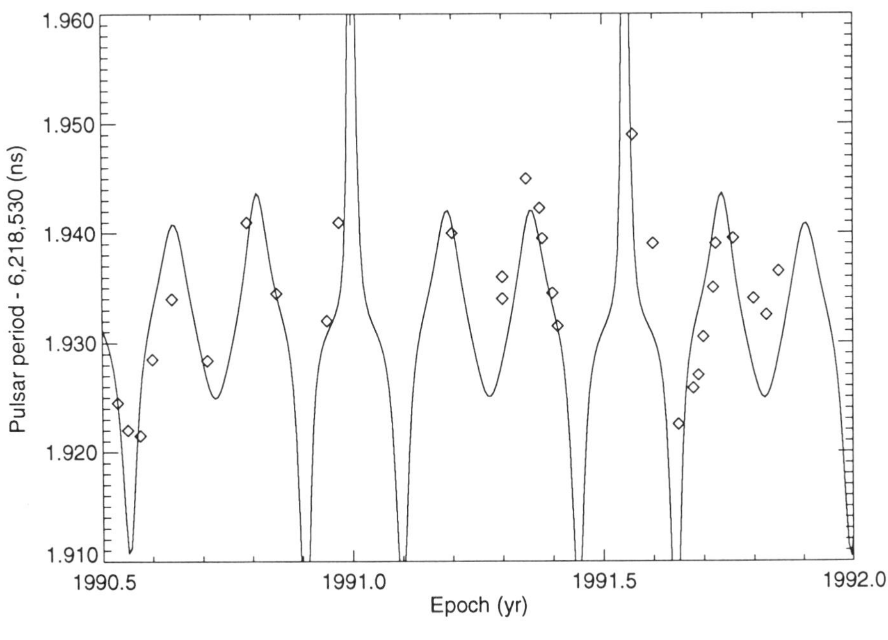

FIGURE III Period variations of PSR1257+12. Diamond-shaped marks are observations by W & F, the solid line denotes changes in period predicted by our model.

DISCUSSION

The purpose of this paper was to investigate whether the reported timing residuals from PSR1257+12 could be explained it terms of internal motions of the neutron star: core-crust relative rotation and free precession. We have first discussed the physical plausibility of such a model in the framework of currently accepted theories of the neutron star interior and concluded that for our model to work, the star's superfluid must be not pinned to the crustal nuclei. Further, unpinning must persist despite very slow relative core-crust rotation, and must be relatively long-lived. Those requirements are not consistent with existing models of pinning (Alpar et al., 1984a,b), but those models are unique and limited in their applicability, opening the principal possibility that the requirements of our model could be met by some neutron stars.

We constructed a simple geometric model to analyze quantitatively timing residuals resulting from postulated internal motions of neutron star. In the absence of definitive understanding of the pulsar signal (see review by Michel, 1982) such an approach is justified. We believe that even such a simple model is sufficient to explain the shape and the magnitude of timing residuals for a variety of pulsar configurations. One exception is the aligned rotator, where neutron star axes of rotation, symmetry, and magnetic field are almost aligned with each other. In such a configuration, the phenomenon of the pulsar can only be observed if the radiation region is described by the hollow cone and the observer is positioned inside that cone. This is a very particular, but important case that we shall consider in a future paper. Figures 2 and 3 show that our model can predict the observations of timing residuals from PSR1257+12. A noteworthy feature of the model is that it requires two internal motions to be periodic with the ratio of periodicities close to 2:1 to explain observed periodicities that are in the 3:2 ratio. In fact, due to the complicated form of predicted timing residuals (equation 5) one would not expect the ratio of physical periodicities to be equal to the ratio of observed periodicities.

In order to obtain the proper magnitude of timing residuals, we had to assume that pulses are emitted from both sides of the magnetic dipole axis. That is observationally feasible when the magnetic dipole axis and the direction to the observer are nearly orthogonal to the axis of rotation. The observer sees both poles, and the period of pulsation is half the period of rotation. Such a configuration is not necessarily contradictory to the observed average pulse profile of PSR1257+12. It is feasible that the emission from two poles may form an overlapping pattern that, upon averaging, may look like the average pulse profile of PSR1257+12. The fit of our model to the observations yields $\theta \approx 90^o$. As the crust, and with it the magnetic field, are slowly moving around the axis of core rotation, $\theta \approx 90^o$ corresponds to a magnetic field that is, on average, nearly orthogonal to the axis of rotation. Thus there is no internal contradiction in our model! In addition, calculations by Ruderman (1991) indicate that nearly orthogonal rotators should naturally occur in spun-down pulsars. Thus, if one consider that PSR1257+12 has formed by accretion induced collapse of white dwarfs (Michel, 1987; Narayan and Popham, 1989) the geometry considered in our model is, in fact, preferable and expected. On the other hand, if PSR1257+12 is an old neutron star spun- up by accretion, then an aligned rotator configuration is expected (Ruderman, 1991). As we have mentioned earlier, our present calculations cannot handle aligned rotator configurations. Nearly aligned rotators require future investigation, since they may offer the preferable explanation for PSR1257+12. In fact, it was suggested by Gil and Jessner (see their paper in this proceedings) that PSR1257+12 is in fact the nearly aligned rotator. They argue that in such a configuration the average pulse profile would not vary significantly with time, whereas, in non-aligned models the projection of the magnetic dipole axis along the observer's line of sight varies significantly with time, which may result in pulse shape variation. However, it is not clear from Gil and Jessner's qualitative discussion whether the aligned rotator geometry can result in periodical timing residuals or would lead to systematic time residuals.

With more observation of PSR1257+12 coming every month, it should be possible to distinguish between planetary and internal motions interpretations of timing residuals. If there really are planets around this pulsar, the interaction between them would lead to small changes in orbital parameters that could become deductible (Malhotra et al. 1992, Rasio et al. 1992). On the other hand, if such changes are not observed, alternative interpretations, such as ours, gain in significance. Finally, it should be stressed that our model offers the most natural explanation for other pulsars exhibiting quasiperiodic or even semichaotic timing residuals.

We wish to thank D.C. Black and R. Malhotra for valuable discussions during this work. This research was done while A.D. was a visiting scientist, and T.S. was a staff scientist at the Lunar and Planetary Institute operated by Universities Space Research Association under contract NASW-4574 with the National Aeronautic and Space Administration. This is Lunar and Planetary Institute Contribution No. 785.

REFERENCES

Alpar, M.A., Anderson, P.W., Pines, D., and Shaham, J. 1984a, *ApJ*, **276**, 325

Alpar, M.A., Anderson, P.W., Pines, D., and Shaham, J. 1984b, *ApJ*, **278**, 791

Alpar, A., and Ögelman, H. 1987, *Astro. Astrophys.*, **185**, 196

Alpar, M.A., and Sauls, J.A. 1988, *ApJ*, **327**, 723

Anderson, P.W., and Itoh, N. 1975, *Nature*, **256**, 25

Baym, G., Pethick, C., Pines, D., and Ruderman, M. 1969, *Nature*, **224**, 872

Brecher, K. 1972, *Nature*, **239**, 325

Jones, P.B. (1988), *MNRAS*, **235**, 545

Lyne, A.G., Pritchard, R.S., and Smith, F.G. 1988, *MNRAS*, **233**, 667

Malhotra, R., Black, D, Eck, A., and Jackson, A. 1992, *Nature*, **356**, 583

Michel, C.F. 1982, *Rev. Mod. Phys.*, **54**, 1

Michel, C.F. 1987, *Nature*, **329**, 310

Narayan, R., and Popham, R. 1989, *ApJ*, **346**, L25

Nelson, R.W., Finn, L.S., and Wasserman, I. 1990 *ApJ*, **348** 226

Rasio, F.A., Nicholson, D.D., Shapiro, S.L., and Teukolsky, S.A. 1992, *Nature*, **355**, 325

Ruderman, M. 1970, *Nature*, **225**, 838

Ruderman, M. 1991, *ApJ*, **366**, 261

Schaaf, M.E. 1990, *Astron. Astrophys.*, **235**, 499

Shaham, J. 1977, *ApJ*, **214**, 251

Shaham, J. 1986, *ApJ*, **310**, 780

Trümper, J., Kahabka, P., Ögelman, H., Pietsch, W., and Voges, W. 1986, *ApJ*, **300**, L63

Wolszczan, A., and Frail, D.A. 1992, *Nature*, **355**, 145

Planets Around Pulsars
ASP Conference Series, Vol. 36, 1993
J.A. Phillips, J.E. Thorsett, and S.R. Kulkarni

ARE THERE REALLY PLANETS AROUND PSR 1257+12?

J.A. GIL
Max-Planck Institute for Radioastronomy, Auf dem Hügel 69,
5300 Bonn 1, Germany
The Astronomical Centre, Lubuska 2, 65-001 Zielona Góra, Poland

A. JESSNER
Max-Planck Institute for Radioastronomy, Auf dem Hügel 69,
5300 Bonn 1, Germany

INTRODUCTION

Evidence of planet-like bodies orbiting pulsars, appearing from time to time in the literature (Richards et al. 1970, Demiański and Prószyński, 1979; Bailes et al. 1991) may indicate that planets are fairly common in the universe. Although the planet around the relatively young pulsar PSR 1829-10 announced last year (Bailes et al. 1991) has been retracted (Lyne and Bailes, 1992), Wolszczan and Frail (1992) have recently presented the case of an old, millisecond pulsar that might be orbited by two planetary companions. By precise timing observations of PSR 1257+12 they have found significant post-fit residuals (involving $P, \dot{P}$, VLA position and Solar System Barycenter corrections) of time-of-arrivals (TOA) in the form of a quasi-periodic 'wandering' over the entire period $P = 6.2$ ms. A closer examination indicated that they could be attributed to the presence of two strictly periodic processes with $\mathcal{P}_1 = 98.2$ days and $\mathcal{P}_2 = 66.6$ days, which they interpreted as two planets with masses of at least 2.8 $M_\oplus$ and 3.4 $M_\oplus$, circling around the pulsar at distances 0.47 AU and 0.36 AU, respectively. Although, one could in principle accept the existence of planetary companions around old, millisecond pulsars, their circular orbits with excentricities $e_1 \leq 0.020$ and $e_2 \leq 0.022$, respectively, seem rather unlikely. Another problem concerns with the amplitude of the observed TOA residuals. Within the planetary interpretation, the pulsar is gravitationally tugged back and forth, and the line-of-sight component of this motion can be measured as TOA residuals. Coincidentally, the combination of six keplerian parameters (masses, distances and excentricities) is such, that the maximum radial displacement of the pulsar is about 900 km, which corresponds to about 3 ms amplitude of TOA residuals. This is embarrasingly close to half of the pulsar period $P = 6.2$ ms. In this paper we present an alternative explanation of the TOA residuals observed in

PSR 1257+12, in which the above problems do not exist. On the contrary, the two strict periodicities contributing to the observed TOA variations, as well as the amplitude of TOA residuals close to $P/2$ are natural consequences of the proposed model.

The discovery article of Wolszczan and Frail (1992) contains the following short statement: 'The integrated pulse profiles of PSR 1257+12 do not show any morphological changes that might indicate the presence of a free precession of the pulsar spin axis'. Although, this does not seem like a strong statement, it indicates that the relative intensities and separations of the components remain unchanged over a period of at least 486 days. In general, if the pulsar spin axis performs a precession, both profile and TOA variations are expected. However, as indicated in Fig.1a, the observed TOA residuals should be much smaller than $P/2$ (the absolute limit for amplitude is $P/4$ (Nelson et al. 1990)). Below we discuss two geometrical configurations in which the above predictions do not apply.

ALTERNATIVE EXPLANATION

PSR 1257+12 has a wide pulse profile with a width $\Delta\varphi = 120$ degrees of rotation at low intensity level. The opening angle between the magnetic axis **m** and tangents to last open field lines is $\alpha \approx 15°(r/R)^{1/2}$ for this pulsar, where r is the altitude of radio-emission region and R is the neutron star radius. Thus, in order to atribute the observed 120° pulse width to the opening angle α of the last open field lines, one has to assume $r \approx 15R$, which would mean that the emission region resides at about 50% of the light cylinder radius $R_L = cP/2\pi$. But the radio emission region is believed to lie much closer to the neutron star (e.g. Phillips, 1992). This indicates that the inclination angle θ between the spin $\boldsymbol{\Omega}$ and the magnetic **m** axes is small in PSR 1257+12. The observed pulse width $\Delta\varphi = 2 \cdot \varphi$ (Fig.1), where the pulse longitude

$$\varphi = 2\sin^{-1}\left\{\frac{\sin\left(\frac{\alpha-\sigma}{2}\right)\sin\left(\frac{\alpha+\sigma}{2}\right)}{\sin\theta \cdot \sin\xi}\right\}^{1/2}. \qquad (1)$$

Here $\sigma = \xi - \theta$ is the impact angle and ξ is the radius of the line-of-sight trajectory (Gil, 1981). A 120° pulse width and 20° separation between the profile components can be obtained for $\theta = -5°$, $\sigma = 19°$ and $\alpha = 17°$ or close to these values (Fig.2a). We also present a second plausible geometrical configuration in Fig.3. The radio-emission altitude $r \approx 1.3R$ is then below 10% of the light cylinder.

In the case of a nearly aligned rotator, the line-of-sight trajectory can stay all or most of the time within the pulsar beam (Figs. 1b, 2a and 3). This completely changes the nature of possible effects caused by precession of the spin axis $\boldsymbol{\Omega}$ mentioned previously (Fig.1a). If both the inclination angle θ and the precession angle δ are small, then it is possible that the axis of the precession orbit almost coincides with the magnetic axis (Figs. 1b and 2a). In such a case, the TOA will systematically change due to a periodic motion of the line-of-sight

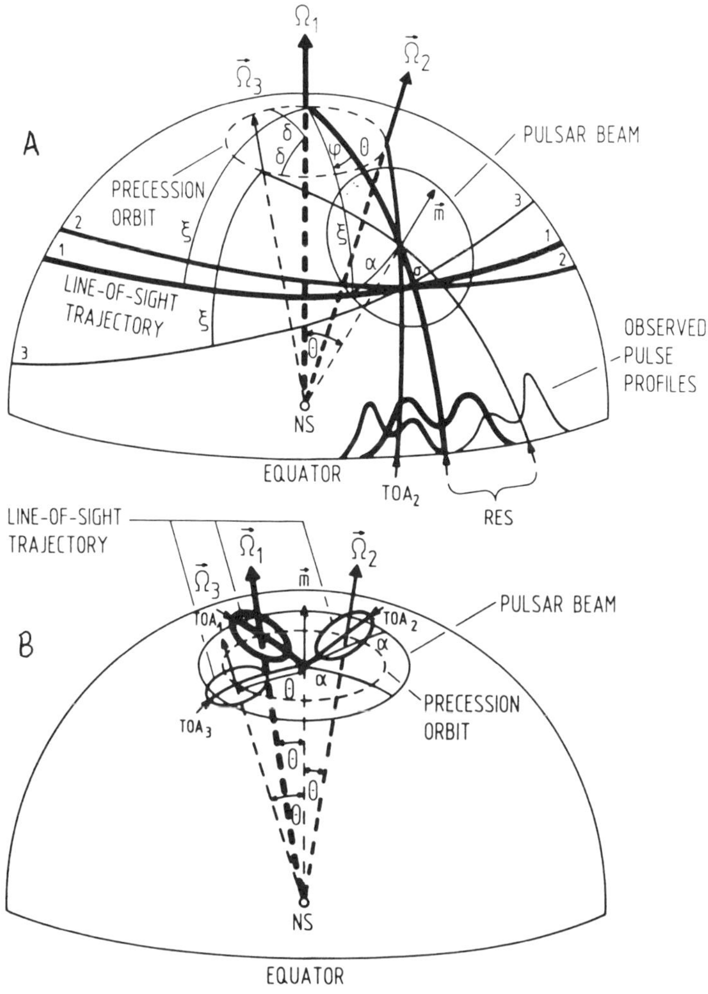

Fig. 1 The celestial sphere centered on a neutron star. The following angles are marked: θ - inclination angle between the spin axis $\mathbf{\Omega}$ and the magnetic axis $\mathbf{m}$, ξ - radius of the line-of-sight trajectory, α - the opening angle of the pulsar beam, σ - impact angle between the magnetic axis and the line-of-sight measured in the fiducial plane containing both the $\mathbf{\Omega}$ and the $\mathbf{m}$ axis and φ - pulse longitude. The spin axis performs precession along the precession orbit with radius δ. Three line-of-sight trajectories corresponding to three different orientations of the spin axis are presented. Three observed pulse profiles corresponding to these trajectories are marked with appriopriate line thickness. The fiducial TOA corresponds to $\mathbf{\Omega}_1$ and TOAs corresponding to other orientations of $\mathbf{\Omega}$ are measured as residuals (RES). Fig.1a corresponds to the general situation while Fig.1b corresponds to the case of small inclination angle θ.

trajectory along the conical pulsar beam. The corresponding TOA residuals are schematically shown in Fig.2b. As one can immediately see, the TOA will move over entire period, with the maximum possible amplitude close to half the pulsar period. The resulting period variation can easily be misinterpreted as a Doppler shift due to pulsar motion in the gravitational field of planetary companions. Within our interpretation, the modulation of the pulsar period is not due to radial motion of the neutron star, but instead, due to precessional drift of the pulse profile. The actual amplitude of TOAs depends on the secondary periodic procees which modifies the precessional effects.

As if follows from geometry sketched in Fig. 2a, the pulsar beam is systematically probed by the line-of-sight trajectory along a cone of average emission. As a consequence of the assumed symmetry of the precession orbit with respect to the pulsar beam axis, the pulse-width (component separation) should remain unchanged over the precession cycle. However the intensity of the two profile components is not equal, as in many other pulsars. The trailing component appears about 20% stronger than the leading one. To keep the relative intensities of the components at the same level during the precession cycle, one has to make assumptions about the energy structure of the pulsar beam. One possibility is that the mean radiation energy increases at a constant rate along the emission cone. This would imply a systematic increase of the total pulse energy, as indicated in Fig.2b. At some place, however, the mean intensity should drop back to the lower level. If the profile is observed at the precessional phase corresponding to this feature, one should observe a rare event in which one of the components is decreased. We do not expect any signifficant systematic changes in the component separation or the overall profile width. However, any small misalignment of the precession axis with respect to the magnetic axis, would produce minute profile variations (small changes in the ratio of intensity components), which could probably be detected by very sensitive profile measurements over an interval of $\mathcal{P}_1$. A hypothetical increase of the total pulse energy indicated schematically in Fig.2b could be verified, if one could correct the data for the effect of long term refractive scintillation.

We therefore propose the precession of the pulsar spin axis is one of the periodic processes contributing to the observed quasi-periodic variations of TOAs in PSR 1257+12. A tantalizing candidate for the second periodic process with frequency is nutation. This possibility arises naturally as a general solution of Euler's equations for a rotating, magnetized neutron star $\dot{\mathbf{L}} = \mathbf{\Gamma}$, where $\mathbf{L}$ is the angular momentum and $\mathbf{\Gamma}$ is a braking torque due to either magnetic dipole radiation, or currents flowing through the polar cap or both. It has been argued that isolated pulsars might show free precession (e.g. Shaham, 1977) but one would not expect nutation as the torques are small. One has to bear in mind however, that over a long time, even minute braking torques accumulate into a significant effect, as they do not change their sign and act on all three axes of the inertial ellipsoid.

The asymmetries of the inertial ellipsoid which lead to precession and nutation are simply the ratios of the involved periods $P = 0.00621$ s, $\mathcal{P}_1 = 8.64 \cdot 10^6$ s, and $\mathcal{P}_2 = 5.7 \cdot 10^6$ s. These asymmetries could be caused by a non-aligned magnetic field and its interaction with the iron crust of the neutron star. The energy flow at the planets' position (0.4 AU) is about $4 \cdot 10^7$ erg cm^{-2} s^{-1} (Wolszczan and Frail, 1992), which implies a luminosity $\mathcal{L} = 16 \cdot \pi(0.4AU)^2 10^7$ erg cm^{-2} s^{-1} $= 1.79 \cdot 10^{34}$ erg s^{-1}. Thus the moment of inertia derived from the

a

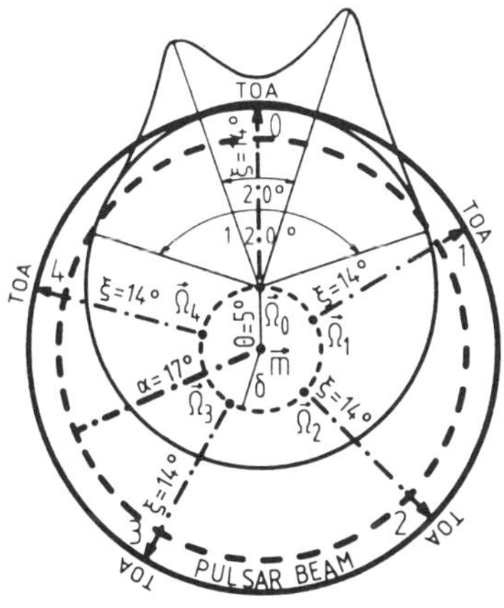

b

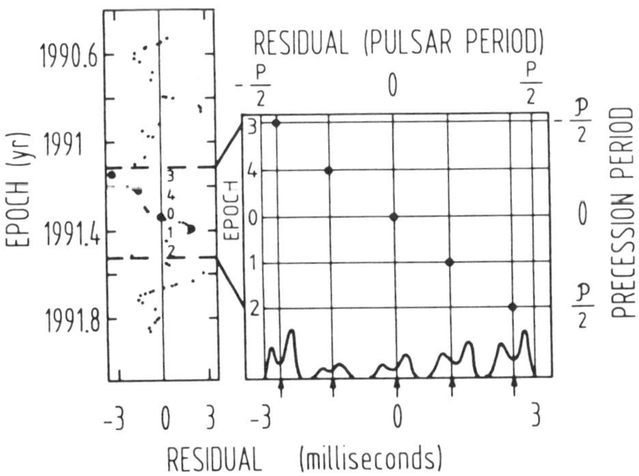

Fig. 2 **a.** Detailed pulsar geometry for PSR 1257+12 derived from Eq.1 applied to the almost aligned rotator with $\theta = -5°$ and $\xi = 14°$. The precession orbit (radius δ) of the spin axis is almost concentric with the pulsar beam. **b.** Schematic TOA residuals resulting from a periodic motion of the line-of-sight trajectory along a conical pulsar beam shown at different epochs numbered from 0 to 4. The residuals observed in the PSR 1257+12 are shown on insert for comparison. The zero residual corresponds to epoch 0 for which the line-of-sight trajectory (solid circle centered on $\mathbf{\Omega}_0$) is shown in Fig.2a. The conical pulsar beam is centered on the **m** axis. Notice, that in this case the opening angle α corresponding to full 120° pulse width is smaller than the opening angle corresponding to 20° separation of the profile components.

pulsar spindown $I = \mathcal{L}/\Omega\dot{\Omega} = 9.01 \cdot 10^{44}$ g cm^2, where $\Omega = 2\pi/P$ and $\dot{\Omega} = 2\pi\dot{P}/P^2$. This can be compared with $I_{ns} = \frac{2}{5}MR^2 = 2.5 \cdot 10^{45}$ g cm^2 computed for $M = 1.4M_{\odot}$ and $R = 15$ km. The radial deformation of the neutron star corresponding to asymmetries in the inertial tensor are $\delta R_1 = 2 \cdot R \cdot P/\mathcal{P}_1 = 0.002$ cm and $\delta R_2 = 2 \cdot R \cdot P/\mathcal{P}_2 = 0.003$ cm. The estimated magnetic field $B = 8.8 \cdot 10^8$ G does supply an energy density per Fe atom $B^2/(8\pi n_{Fe}) = 18.0$ eV (where $n_{Fe} = \rho_{Fe}/56m_p$, $\rho_{Fe} \approx 10^5$ g cm^{-3} is the average iron crust density, $m_p = 1.672 \cdot 10^{-24}$ g, which matches the potential energy required to lift each atom by δR from the neutron star's surface $GM \cdot 56m_p\delta R_1/R^2 = 10.4$ eV and $GM \cdot 56m_p\delta R_2/R^2 = 15.8$ eV, respectively. The one kilometer iron crust has a moment of inertia $I_{Fe} = (8/15)\pi\rho_{Fe}[R^5 - (R - 10^5\text{cm})^5] = 1.3 \cdot 10^{36}$ g cm^2, so $I_{Fe}/I = 1.4 \cdot 10^{-9}$, which is close to the ratios $P/\mathcal{P}_1 = 7.2 \cdot 10^{-10}$ and $P/\mathcal{P}_2 = 1.1 \cdot 10^{-9}$, which represent asymmetries ($\delta I/I \sim P/\mathcal{P}$) in the inertial tensor. A similar contributory effects of deeper layers can not be ruled out. Thus, as we see, all numbers are in suprisingly good agreement (they depend only weakly on the mass of the neutron star), supporting the idea that the two periodicities observed in PSR 1257+12 could be caused by precession and nutation. Obviously, a motion of the spin axis of triaxial ellipsoid might be very complicated, producing phase residuals which can mimic those caused by planets around the pulsar. Due to nearly alignment of spin and magnetic axes, the residuals can achieve large values, exceeding $P/4$ (Nelson et al. 1990).

Another periodic process capable of causing the observed periodicities in PSR 1257+12 is the migration of the magnetic moment of neutron star. The time scale for this effect on the Earth is of the order of 10^6 years. Interestingly, the ratios $24h/10^6y \approx 10^{-9}$ and $6.2ms/100d \approx 10^{-9}$ are almost the same. The pulsar spin axis is coupled to the superfluid core and the magnetic moment is probably frozen into the charged component of the neutron star. A relative motion of the crust against the core is certainly possible, due to imperfect frictional coupling of both components. This motion will manifest itself as the rotation of the magnetic axis around the spin axis, even if the latter performs precession. An interesting relevant geometry is presented in Fig. 3. In this case the impact angle σ is negative, i.e. an observer passes between the spin and the magnetic axes. Since the migration orbit is concentric with the precessing spin axis, the cut of the line-of-sight trajectory through the pulsar beam occurs always in the same region. Therefore, no changes of the pulse profile should be observed. Let us recall that the geometry discussed in Fig. 2a predicts systematic increase of the total pulse energy, as indicated in Fig. 2b. These two geometrical configurations differ also by means of the position angle variations. In the first case, the maximum sweep rate is $\sin 5°/\sin 19° = 0.25°$ and the position angle curve should be that flat over the entire 120° of the pulse profile. In the second case with $\theta = 20°$ and $\sigma = -5°$, the maximum sweep rate is $\sin 20°/\sin(-5°) = -4$. Thus, the position angle should perform about 80° swing (with negative slope) over about 20° between component peaks and remain flat through rest of the profile.

The magnetic axis $\mathbf{m}(t)$ is expected to migrate around an instantenous spin axis $\mathbf{\Omega}(t)$, which is allowed to precess slightly. The radius δ of the precession orbit can be estimated from excentricities $e \approx 0.02$ reported by Wolszczan and Frail (1992). Since the radius of the migration orbit is equal to the inclination angle $\theta \approx 20°$, then $\delta/\theta \approx 0.02$ and $\delta = 0.4°$. Thus, the only important effect of precession in this case is one extra rotation of the pulsar in about every 10^9

basic rotations. It is clear that residuals following from $\mathbf{\Omega}(t)$ and $\mathbf{m}(t)$ can explain observed variations of TOAs in PSR 1257+12, provided that directions of precession of $\mathbf{\Omega}$ and migration of $\mathbf{m}$ are opposite to each other

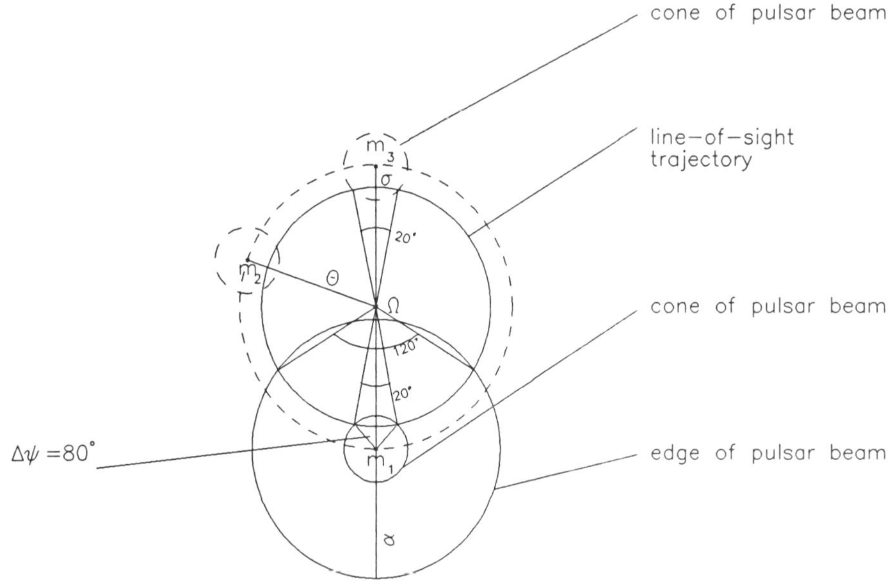

Fig. 3 An alternative geometry for PSR 1257+12 derived from Eq.1. The inclination angle $\theta = 20°$ and the impact angle $\sigma = -5°$. The magnetic axis $\mathbf{m}$ migrates around the spin axis $\mathbf{\Omega}$, which is assumed to perform small precession.

CONCLUSIONS

We propose that the combination of effects involving precession and/or nutation of the spin axis and/or migration of the magnetic axis around the spin axis is responsible for the observed residuals in PSR 1257+12. To account for the observed TOA variations, the actual processes should work in opposite directions and have time scale equal or longer than periodicities identified in PSR 1257+12. Our paper does not rule out the possibility that the observed effect is due to planets orbiting the pulsar, we only propose an alternative mechanism. The planetary hypothesis can be tested observationally as proposed by Rasio et al. (1992). If this test fails to confirm planets, one has to explain the observed residuals differently. The obvious periodic processes that should be seriously considered are precession and nutation of the pulsar spin axis and migration of the magnetic axis around the spin axis. The two relevant geometrical configurations proposed to account for the observed profile of PSR 1259+25 can probably position angle should be flat over the entire 120° pulse profile. Occasional, rapid changes of intensity of one of the components, reflecting the scanning of pulsar beam structure by precessing line-of-sight trajectory, could be observed. In the second case corresponding to Fig. 3, the position angle should undergo about 80° swing over 20° between the peaks of the profile components and remain constant through the rest of the profile. If the migration of the magnetic moment around the spin axis is the major periodic process determining TOA residuals in PSR 1257+12, no significant profile changes should be observed. A minute profile variations, on the level of about few percent, could be produced by slight misaligment of the corresponding symmetry axes.

Finally, if one proposes the precessional origin of the TOA variations observed in PSR 1257+12, then the obvious question to ask is why free precession has not been detected in other pulsars so far. If radial deformations $\delta R/R \sim 10^{-9}$ are of the same order for all neutron stars, then in longer period pulsars with $P \sim 1$ s the time scales would probably be prohibitively long. In millisecond pulsars the time scales are of the order of 1 year but the amplitude of TOA residuals will be much less than 1 ms, unless the pulsar is an almost aligned rotator, in which case (as in PSR 1257+12) the amplitude is half of the pulsar period. We believe that free precession can only be observed in short period pulsars with nearly aligned magnetic and spin axes. Interestingly, Cordes (1992) has recently reported evidence for precession in the Crab pulsar. It is worth mentioning that TOA residuals of this pulsar are large, approaching 17 ms $= P/2$ and that the apparent quasiperiodicity is about 1 year (Lyne and Smith, 1990), consistent with a 10^{-9} ratio in PSR 1257+12. Also the position angle at ratio wavelenghts seems flat over the entire pulse profile (Lyne and Smith, 1990, p. 87), being consistent with almost aligned rotator. An evidence of bridge of emission between the main pulse and the interpulse is visible at about 200 MHz (Lyne and Smith, 1990, Fig. 7.1). At this frequency the main pulse of the Crab pulsar looks almost identical to the profile of PSR 1257+12 at 400 MHz (Wolszczan and Frail, 1992).

ACKNOWLEDGEMENTS.

We thank J. Kijak and J. Seiradakis for helpful discussions. One of us (JAG) wishes to thank the Alexander von Humboldt Foundation and the Polish KBN Grant 2 0385 9101 for partial support and E. Gil for technical assistance.

REFERENCES

Bailes, M., Lyne, A.G. & Shemar, S.L. 1991, *Nature* **352**, 311-313
Cordes, J.M. 1992, this volume
Demiański, M. & Prószyński, M. 1979, *Nature* **282**, 383-385
Gil, J. 1981, *Acta Phys.Pol.* **B12**, 1081
Gil, J. 1983, *Astron.Astrophys.* **127**, 267
Lyne, A.G. & Bailes, M. 1992, *Nature* **355**, 213

Narayan, R. & Vivekanand, M. 1982, *Astron.Astrophys.* **113**, L3
Nelson, R.W., Finn, L.S. & Wasserman, I. 1990, *Astrophys.J.* **348**, 226-231
Phillips, J.A. 1992, *Astrophys.J* **385**, 282
Rasio, F.A., Nicholson, D.D., Shapiro, S.L. & Teukolsky, S.A. 1992, *Nature* **355**, 325
Richards, D.W., Pettengill, G.H., Counselman, C.C. III & Rankin, J.M. 1970, *Astrophys J.*, **160**, L1-6
Shaham, J. 1977, *Astrophys.J.* **214**, 251
Wolszczan, A. & Frail, D.A. 1992, *Nature* **355**, 145

Planets Around Pulsars
ASP Conference Series, Vol. 36, 1993
J.A. Phillips, J.E. Thorsett, and S.R. Kulkarni

A PRELIMINARY ANALYSIS OF PULSE PROFILE STABILITY IN PSR 1257+12

V. M. KASPI AND A. WOLSZCZAN[1]
Joseph Henry Laboratories and Physics Department, Princeton University, Princeton, New Jersey 08544

ABSTRACT We have examined data from the millisecond pulsar PSR 1257+12 in order to search for systematic changes in the pulse profile morphology. Such changes are expected in models that invoke precession of the neutron star spin axis to explain the observed periodicities in timing residuals. We have calculated difference profiles for each observing session by subtracting the observed profiles from a high signal-to-noise ratio standard profile. The average noise-corrected modulation index of all difference profiles for PSR 1257+12 is 0.010 (σ=0.007). We compare our results for PSR 1257+12 with those of PSR 1534+12, and find similar levels of difference profile modulation. The most straightforward interpretation of our results is that low level structure in the difference profiles of PSR 1257+12 is due to noise, interference, or instrumental effects.

INTRODUCTION

Precession of the spin axes of neutron stars, if observable, could have important implications for the study of neutron star interiors (Pines and Shaham 1974). In addition to periodic modulation of timing residuals, precession may result in measurable changes in pulsar profile morphology (see papers by Cordes, by Gil and by Stepinski and Dolginov, this volume). The absence of changes large enough to be seen by eye in the integrated profiles of PSR 1257+12 led Wolszczan and Frail (1992) to conclude that precession was an unlikely explanation for the observed periodicities in timing residuals.

In order to investigate further whether any precession-based models could be applied to PSR 1257+12 in lieu of the "standard" model in which two planet-sized bodies orbit the neutron star, we have undertaken a search for low-level pulse profile changes in existing data. We report here the preliminary results of this analysis.

SUMMARY OF OBSERVATIONS

We have used data for PSR 1257+12 from a total of 60 observing sessions in the period from September 7, 1990 through April 5, 1992. All of the observations

---

[1]ON LEAVE FROM ARECIBO OBSERVATORY, ARECIBO, PR 00613

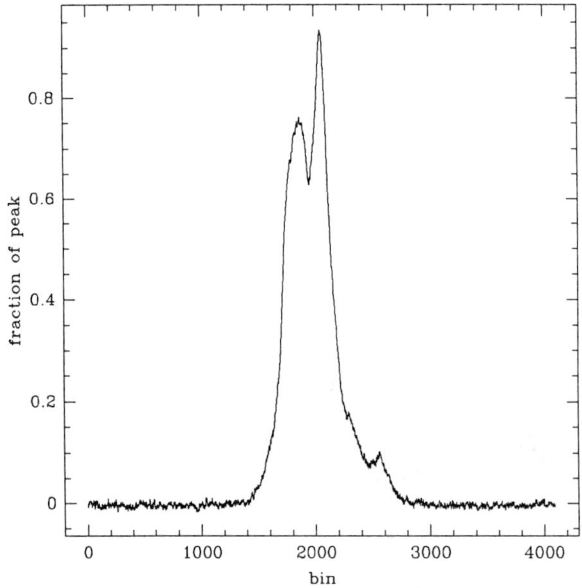

FIGURE I Standard profile of PSR 1257+12 at 430 MHz.

were made with the Princeton Mark III data acquisition system at the 305-m telescope in Arecibo, Puerto Rico, at an observing frequency of 430 MHz. The 60 data profiles in this study were obtained by summing two coherently dedispersed and detected circular polarizations. Real-time coherent dispersion removal was used to prevent smearing of the details of the pulse profile morphology (Hankins and Rajkowski 1987). Figure I shows the high signal-to-noise, 4096 time-bin standard profile derived by aligning and adding 19 of the 60 pulse profiles. This alignment was done by cross-correlation, so that the standard profile was created independently of any pulse time-of-arrival ephemeris. All our analysis was done with the standard profile shown in Figure I.

ANALYSIS AND RESULTS

To search for possible changes in pulse profiles, we have used the following procedure. For each of the 60 observing sessions, we obtained a difference profile by subtracting the appropriately scaled data profile from the standard profile. The alignment of the data profile with the standard profile was done in the frequency domain by assuming the data profile, $p(j)$, was of the form $p(j) = a + b \times s(j - \tau) + n(j)$, where s is the standard profile and n is a noise background. The parameters a, b, and τ were determined using a chi-squared minimization fitting procedure, and the resulting time delay τ was used to make the alignment (Taylor 1990).

Figure IIa shows a typical example of approximately 90% of the difference

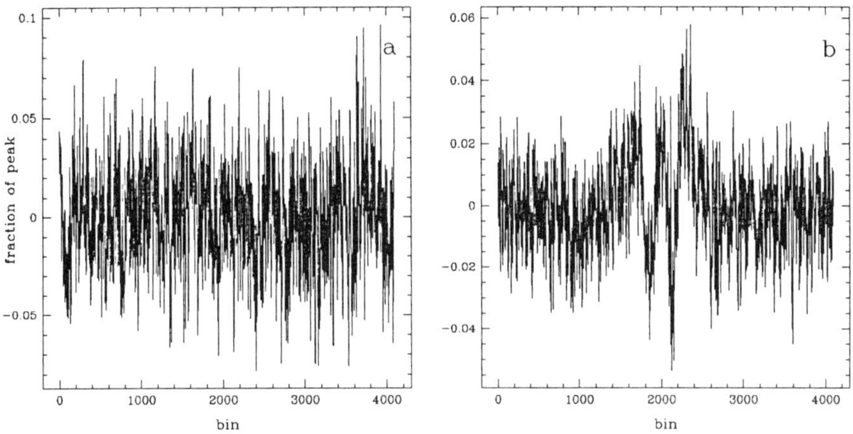

FIGURE II Two examples of difference profiles for PSR 1257+12. Figure (a) is a typical example of the 90% of our difference profiles that contained no structure. Figure (b) is representative of the few difference profiles that contain some structure.

profiles, while Figure IIb is representative of the few difference profiles that appear to contain some structure.

We divided each difference profile into on- and an off-pulse regions, with each having N=2048 bins. We then calculated noise-corrected autocorrelation functions (ACFs) by subtracting off- from on-pulse ACFs for each difference profile. Figure IIIa shows the on- and off-pulse ACFs, as well as their difference, for the difference profile shown in Figure IIa. Similarly, Figure IIIb corresponds to Figure IIb. Of the 60 data profiles, 9 contain low-level interference that results in periodicities in both the on- and off- pulse ACFs. We did not include that subset in the subsequent analysis.

We used the noise-corrected ACFs to calculate a modulation index ($m = \sqrt{\mathrm{ACF}(0)/N - 1}$), and a characteristic ACF width at half maximum for each observing session. Approximately one third of the ACFs had ACF(0) $<$ 0, indicating that baseline effects dominated the difference profile. For those, no modulation indices could be calculated. ACF widths could not be measured for approximately half of the ACFs, since they did not show any structure (see Figure 3a for example).

We obtained time series of modulation indices and ACF widths, and folded them modulo 66.54 and 98.20 days to search for low-level structure that might be associated with the observed periodicities in the timing residuals. We found no significant periodicities in the time series of the either quantity.

The average modulation index for those difference profiles that had positive ACFs at lag 0 is $< m >= 0.010$ ($\sigma = 0.007$). For comparison, the identical analysis was done on data for the pulsar PSR 1534+12. This pulsar is in a relativistic binary orbit with a second neutron star, and is not known to have any other unusual timing characteristics. The data for PSR 1534+12 were taken

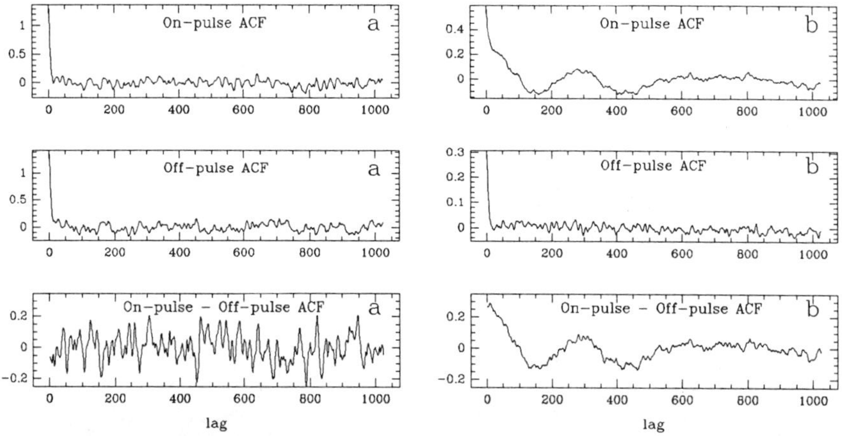

FIGURE III On- and off-pulse ACFs and their difference. The ACFs in (a) correspond to the difference profile shown in Figure IIa, while the ACFs in (b) correspond to Figure IIb.

at Arecibo using the Princeton Mark III system at the same epochs as the data taken for PSR 1257+12, and at the same observing frequency. For PSR 1534+12, the average modulation index from noise-corrected autocorrelation functions calculated in the manner described above is $< m > = 0.007$ ($\sigma = 0.005$).

SUMMARY AND CONCLUSIONS

The data from the 60 observing sessions do not show any evidence for large pulse profile changes; the largest feature in all the difference profiles has height of approximately 9% of the peak. The average modulation index of difference profiles is at the 1% level, and is comparable to that obtained from difference profiles of PSR 1534+12, which is not known to have any unusual timing characteristics. The most straightforward interpretation of our results is that any low level structure in the difference profiles of PSR 1257+12 is due to noise, interference, or instrumental effects.

We plan to continue this study, and will report on a more detailed analysis elsewhere.

ACKNOWLEDGMENTS

VMK wishes to thank the Natural Sciences and Engineering Research Council of Canada for fellowship support. The Arecibo Observatory, part of the National Astronomy and Ionosphere Center, is operated by Cornell University under contract with the NSF.

REFERENCES

Hankins T. H., and Rajkowski, J. M. 1987, *Rev. Sci. Instrum.*, **58**, 674.

Pines, D., and Shaham, J., 1974, *Nature*, **248**, 483.

Taylor J. H. 1990, in *Impact of Pulsar Timing on Relativity and Cosmology*, ed. D. C. Backer, (Berkeley: Center for Particle Astrophysics), m1.

Wolszczan, A., and Frail, D. A. 1992, *Nature*, **355**, 145.

Section 3:
Orbital Dynamics

Planets Around Pulsars
ASP Conference Series, Vol. 36, 1993
J.A. Phillips, J.E. Thorsett, and S.R. Kulkarni

ORBITAL DYNAMICS OF PSR1257+12 AND ITS TWO PLANETARY COMPANIONS

RENU MALHOTRA
Lunar and Planetary Institute, Houston, TX 77058

ABSTRACT I present a review of the orbital dynamics of the three body problem of PSR1257+12 and its two putative planetary companions. The orbital dynamics provides two "reality checks" on the planetary interpretation of the pulse arrival time residuals observed by Wolszczan and Frail. Firstly, by examining the dynamical stability of the system for the deduced orbital periods, one can place upper limits on the companion masses of about two times the mass of Jupiter. Since the lower limits of the companion masses deduced from the observations are well below this upper limit, the planetary interpretation passes this first "reality check". Secondly, the 3:2 near-commensurability of the orbital periods effects periodic variations of the keplerian orbital parameters. The amplitude of these variations provides a direct measure of the masses of the companions relative to the mass of the pulsar. The signature of these variations in pulse time-of-arrival data is not straightforward, but, if the planetary interpretation is correct, it should become detectable as an additional quasiperiodic phase residual over a period of a few years.

1. INTRODUCTION

The quasi-periodic variations in the arrival times of pulses from the millisecond radio pulsar PSR1257+12 reported by Wolszczan and Frail (1992) (hereafter referred to as W&F) are most straightforwardly interpreted as arising from the gravitational perturbations of two planetary mass companions orbiting the star. W&F's multiparameter fit to the time-of-arrival (TOA) residuals has yielded nearly circular orbits (eccentricities of about 0.02) for the two planets with the orbital periods close to a 3:2 commensurability. The "projected masses", $m \sin i$, of the companions were found to be $3.4 M_\oplus$ and $2.8 M_\oplus$, where i is the inclination of the orbit to the plane of the sky.

Here I review the orbital dynamics of the three body problem in the context of the putative planetary system of PSR1257+12. The importance of this discussion lies in the dynamical test it provides for the planetary interpretation of the observations (Rasio *et al.* 1992, Malhotra *et al.* 1992).

The paper is organised as follows. In Section 2, I describe the relationship of the pulse arrival time residuals to the kinematics of the pulsar and its two companions. The requirement of dynamical stability of the system and the upper limits it provides on the companion masses is discussed in Section 3. Section 4 describes the theory for the first order secular and resonant perturbations to the orbital parameters that arise due to the mutual gravitational interactions of

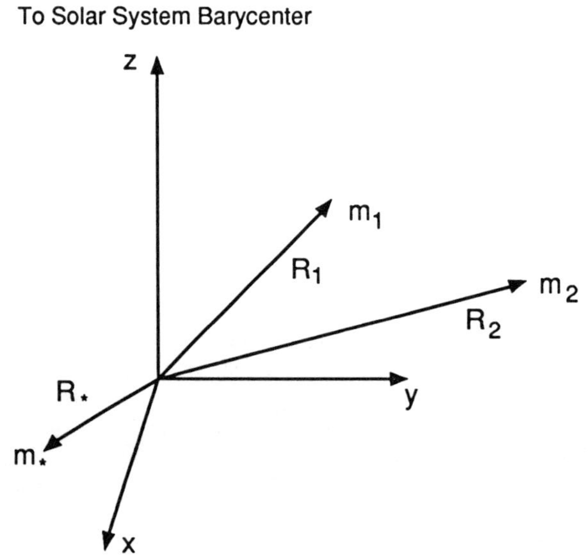

FIGURE I

the planetary companions. However, as pointed out in Malhotra *et al.* 1992, first order perturbation theory breaks down for mass ratios m_j/m_* (planet mass to star mass) greater than about 6×10^{-5} because the 3:2 near-resonance becomes an exact resonance. Section 5 descibes orbital evolution in this case. And, finally, in section 6, I summarise the results and discuss the observational consequences of the three-body orbital dynamics.

2. KINEMATICS

Figure 1 shows a schematic sketch of the barycentric reference frame, where the (x, y) plane represents the plane of the sky and the z axis points from the pulsar+planets system barycenter to the solar system barycenter. $\mathbf{R}_*, \mathbf{R}_1, \mathbf{R}_2$ are the barycentric position vectors of the three bodies (the pulsar, planet 1 and planet 2) with masses m_*, m_1, m_2, respectively:

$$\mathbf{R}_* = -\frac{m_1}{m_*}\mathbf{R}_1 - \frac{m_2}{m_*}\mathbf{R}_2 \,. \tag{1}$$

The solar system barycentric pulse time of arrival residual, Δt, due to the motion of the pulsar about the pulsar+planets barycenter is given by

$$c\Delta t = -\hat{\mathbf{z}} \cdot \mathbf{R}_* \,. \tag{2}$$

For the perturbative analysis of the nearly keplerian motion of the planets around the pulsar, it is more convenient to use Jacobi coordinates (see section

4) which are defined as follows:

$$\mathbf{r}_1 = \mathbf{R}_1 - \mathbf{R}_*\,,$$
$$\mathbf{r}_2 = \mathbf{R}_2 - \frac{m_*\mathbf{R}_* + m_1\mathbf{R}_1}{m_* + m_1}\,. \tag{3}$$

$\mathbf{r}_1$ is the position of m_1 relative to m_* and $\mathbf{r}_2$ is the position of m_2 relative to the center of mass of m_* and m_1. In terms of the Jacobi coordinates, we have

$$\mathbf{R}_* = -\kappa_1\mathbf{r}_1 - \kappa_2\mathbf{r}_2\,, \tag{4}$$

where

$$\kappa_1 = \frac{m_1}{m_* + m_1}\,, \qquad \text{and} \qquad \kappa_2 = \frac{m_2}{m_* + m_1 + m_2}\,. \tag{5}$$

To first approximation, the solution $\mathbf{r}_j(t)$ is an elliptical trajectory (see Figure II):

$$\begin{aligned} \mathbf{r} = r\,\{&[\cos(f+\omega)\cos\Omega - \sin(f+\omega)\sin\Omega\cos i]\,\hat{\mathbf{x}} \\ &+[\cos(f+\omega)\sin\Omega + \sin(f+\omega)\cos\Omega\cos i]\,\hat{\mathbf{y}} \\ &+\sin(f+\omega)\sin i\,\hat{\mathbf{z}}\,\}\,, \end{aligned} \tag{6}$$

where

$$\begin{aligned} r &= \frac{a(1-e^2)}{1+e\cos f}\,, \\ f &= \mathcal{M} + 2e\sin\mathcal{M} + \frac{5}{4}e^2\sin 2\mathcal{M} + \mathcal{O}(e^2)\,, \\ \mathcal{M} &= nt + \sigma\,, \\ n^2a^3 &= Gm_*/(1-\kappa)\,. \end{aligned} \tag{7}$$

In celestial mechanics terminology, n is the *mean motion* and $\mathcal{M}$ is called the *mean anomaly*; σ is related to the time of periastron. In equations (6) and (7), the subscript $j = 1, 2$ is implicit for the orbital elements as well as κ.

Not surprisingly, in the first approximation, the motion of the pulsar, $\mathbf{R}_*(t)$, about the system barycenter is a superposition of two elliptical motions. However, note that it differs from the superposition of the independent solutions to the two 2-body problems of (m_*, m_1) and (m_*, m_2) by terms of $\mathcal{O}(m_j/m_*)^2$. It is important not to overlook this when including the effects of the mutual perturbations of the planets which are also of the same order.

Using equations (4)–(6) in equation (2), we have

$$c\Delta t = c(\Delta t_1 + \Delta t_2)\,, \tag{8}$$

where

$$c\Delta t_j = \kappa_j a_j \sin i_j (1 - e_j^2)\frac{\sin(f_j + \omega_j)}{1 + e_j\cos f_j}\,, \qquad j = 1, 2. \tag{9}$$

For circular orbits, $e_1 = e_2 = 0$, $f_j = \mathcal{M}_j = n_jt + \sigma_j$, so that

$$c\Delta t_j = \kappa_j a_j \sin i_j \sin(n_jt + \sigma_j)\,.$$

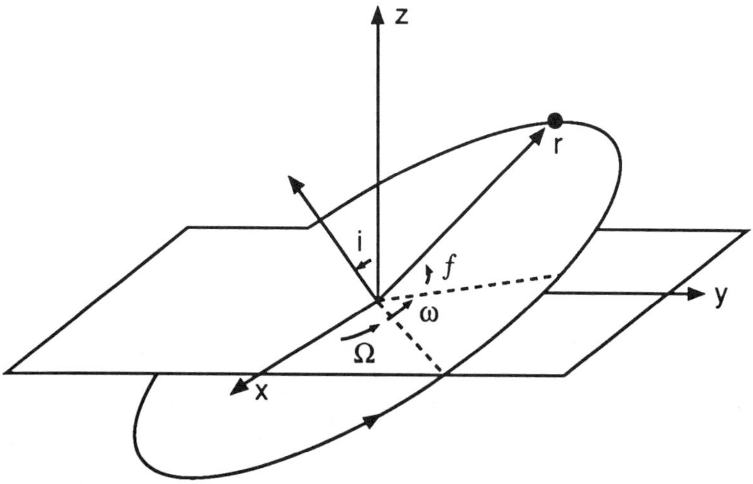

FIGURE II

Thus, in this case, the TOA residuals would be precisely the superposition of two sines with frequencies equal to the mean motions, n_1 and n_2, and amplitudes $\mathcal{A}_j = \kappa_j a_j \sin i_j$. Using the keplerian relationship between n and a (equation 7), we get

$$\left(\frac{Gm_*}{1-\kappa_j}\right)^{1/3} \kappa_j \sin i_j = \mathcal{A}_j n_j^{2/3}. \tag{10}$$

The above equation is valid in the case of non-circular orbits as well, except that $\mathcal{A}_j$ would then be identified with the amplitude of that component in the TOA residuals which has frequency n_j. As reported by W&F, the TOA residuals of PSR1257+12 yield

$\mathcal{A}_1 = 1.31$ light-ms, $2\pi/n_1 = 5751011\,\mathrm{s}$,

$\mathcal{A}_2 = 1.41$ light-ms, $2\pi/n_2 = 8487388\,\mathrm{s}$,

so that equation (10) yields the following estimates for the planet masses:

$m_1 \sin i_1 = 3.4 M_\oplus (\frac{m_*+m_1}{1.4M_\odot})^{2/3}$,

$m_2 \sin i_2 = 2.8 M_\oplus (\frac{m_*+m_1}{m_*})(\frac{m_*+m_1+m_2}{1.4M_\odot})^{2/3}$.

For the general case of non-circular orbits, to first order in the orbital eccentricities, we have

$$\begin{aligned} c\Delta t_j &= \mathcal{A}_j \sin(n_j t + \sigma_j + \omega_j) \\ &\quad + \frac{1}{2}\mathcal{A}_j e_j [\sin 2(n_j t + \sigma_j)\cos\omega_j + (\cos 2(n_j t + \sigma_j) - 3)\sin\omega_j] \\ &\quad + \mathcal{O}(e^2). \end{aligned} \tag{11}$$

From equations (10) and (11), we see that (i) from the n_j component in the TOA residuals, the planetary companion masses, m_j, may be determined only in combination with $\sin i_j$ and m_* (*i.e.* the left hand side of equation (10)), and (ii) the eccentricities and periastron longitudes can be determined from the amplitudes of the $2n_j$ frequency components in the TOA residuals. The orbital inclinations, i_j, and the longitudes of the ascending nodes, Ω_j, remain indeterminate from the above kinematical considerations. As discussed later, dynamical considerations can, in principle, yield further information on the companion masses, the orbital inclinations and the longitudes of ascending nodes, while simultaneously providing a test for the planetary hypothesis.

3. DYNAMICAL STABILITY

There is no general formula for the dynamical stability of the gravitational three-body problem. For the nearly circular orbits deduced from the observations of PSR1257+12, there exist two stability criteria in the literature that can be applied to this system: the empirical criterion found by Graziani and Black (1981), and the semi-analytical resonance overlap criterion (Wisdom 1980). Strictly-speaking, both of these criteria apply only to co-planar systems. (It can be stated also that other studies indicate that the dynamical stability of planetary-type three-body systems is a weak function of the relative inclinations of the orbits, as long as the relative inclinations are less than 90°.) Another caveat for the applicability of these criteria is that they define the boundary of "gross" instabilities: the system either becomes unbound or two of the bodies collide on a timescale of a few to a few hundreds of orbits if these criteria are not satisfied. Weaker instabilities that may make the system unstable over much longer timescales cannot ruled out very simply.

Graziani and Black (1981) conducted direct numerical integrations of the co-planar three body problem consisting of three masses, m_*, m_1 and m_2, where m_1 and m_2 were in orbit about the "central" body, m_*, in initially circular orbits of radii a_1 and a_2. They defined two dimensionless parameters, μ and Δ,

$$\mu = \frac{m_1 + m_2}{2m_*} \qquad \text{and} \qquad \Delta = \frac{a_2 - a_1}{(a_2 + a_1)/2}\,. \tag{12}$$

For μ in the range of 10^{-3} to 1.0 and Δ in the range of 0.2 to 1.4, their numerical integrations yielded the following empirical criterion for the stability of the system:

$$0.175\Delta^3(2 - \Delta)^{-3/2} > \mu\,. \tag{13}$$

Wisdom (1980) applied the "resonance overlap criterion" of Chirikov (1979) to the circular restricted three body problem (*i.e.* the three body problem in which $m_* \gg m_1$, and $m_2 = 0$, and m_* and m_1 move in circular orbits about their center of mass). The idea is quite simple: if m_2 were initially placed in a circular orbit of radius a_2 such that its unperturbed period of revolution about the center of mass of m_* and m_1 would be in a $j : j+1$ commensurability with the orbital period of the two primaries, then the resonant perturbations would produce variations of amplitude $\propto (m_1/m_*)^{1/2}$ in its semimajor axis. If the distance between neighboring $j : j+1$ resonances is smaller than the sum of the

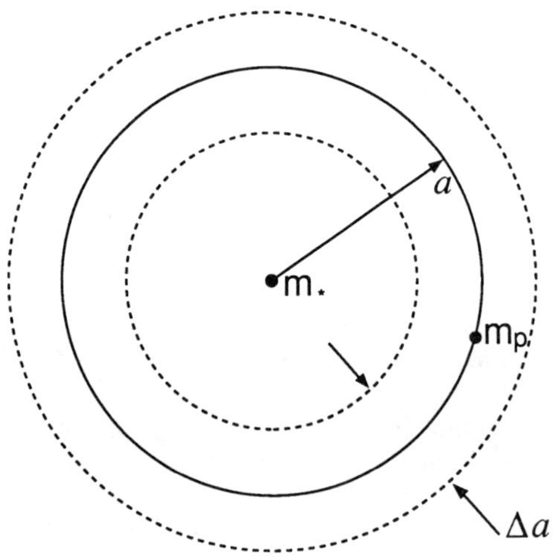

FIGURE III

half-widths, $(\delta a)^{(j)}_{\rm res} + (\delta a)^{(j+1)}_{\rm res}$, of those resonances, then a gross instability of the motion sets in. This instability due to the overlap of all $j : j+1$ resonances with $j > j_{\rm min}(\frac{m_1}{m_*})$ occurs in the range of orbital radii $[a_1 - \frac{1}{2}\Delta a, a_1 + \frac{1}{2}\Delta a]$ (see Figure III), where

$$\frac{\Delta a}{a_1} \simeq C(\frac{m_1}{m_*})^{2/7}. \tag{14}$$

In Wisdom's analysis, the numerical constant C was about 1.3, but later numerical results have found it to be about 1.5 (Duncan *et al.* 1989). By an extrapolation of this result, if we require that the two putative planets, m_1 and m_2, orbiting PSR1257+12 are not to be in each other's zones of instability, then we must have

$$a_2 - a_1 \gtrsim \frac{1}{2}C[(\frac{m_1}{m_*})^{2/7}a_1 + (\frac{m_2}{m_*})^{2/7}a_2]. \tag{15}$$

Then, with the approximation $\frac{m_1}{m_*} \approx \frac{m_2}{m_*} \approx \mu$, and noting that the left hand side is equal to Δ, the Graziani-Black parameter, we find that the resonance overlap criterion is simply $\Delta \gtrsim C\mu^{2/7}$.

Using the nominal values, $a_1 \sin i_1 = 0.36\text{AU}, a_2 \sin i_2 = 0.47\text{AU}$, reported in W&F, and assuming coplanar orbits, the Graziani-Black criterion gives $\mu < 1.4 \times 10^{-3}$ while the resonance overlap criterion gives $\mu < 2.3 \times 10^{-3}$. Assuming that the mass of the pulsar has the canonical value of $1.4M_\odot$, the requirement of dynamical stability of the planetary system provides an upper limit to the planet masses of about *two times the mass of Jupiter.* (Direct numerical integrations of the three-body equations of motion have confirmed this estimate of the upper limit.)

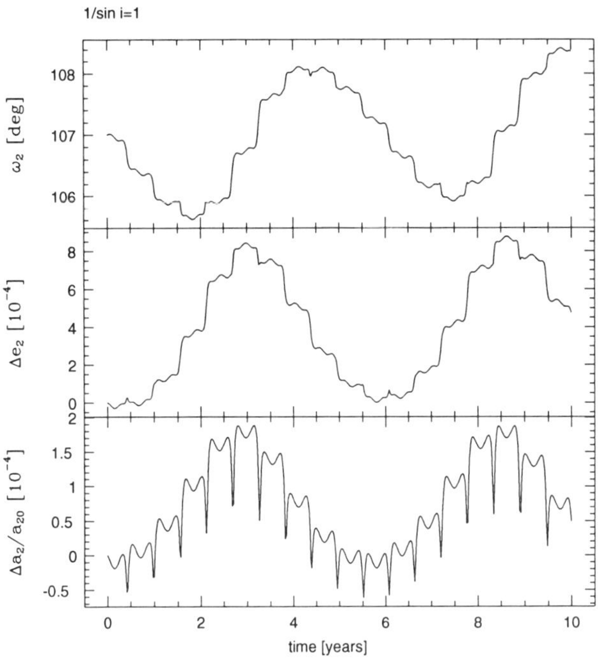

FIGURE IV

4. PERTURBATION THEORY

Figure IV shows the time variation of a_2, e_2 and ω_2, three of the ten orbital elements that determine the TOA residuals arising from the motion of the pulsar about the planetary system barycenter for the case of coplanar, edge-on orbits. The other elements exhibit similar variations. The variations are due to the mutual gravitational perturbations of the two planets and their signature can, in principle, be detected in the TOA residuals. The amplitudes and periods of these variations are determined by the masses of the planets relative to the mass of the pulsar and also by their relative separations. In order to take account of these perturbations in the timing model, we need an analytical form for the characteristic orbital evolution indicated in Figure IV.

In order to proceed with the analysis, recall that in section 2, I introduced the Jacobi coordinates $\mathbf{r}_1, \mathbf{r}_2$ for the 3-body problem and asserted that, in the first approximation, the solutions $\mathbf{r}_j(t)$ were keplerian ellipses. So, the first order of business here is to motivate this choice of coordinates. We can do this most directly by writing down the total center-of-mass energy of the three bodies:

$$\begin{aligned} H &= \frac{1}{2}(m_*\dot{\mathbf{R}}_*^2 + m_1\dot{\mathbf{R}}_1^2 + m_2\dot{\mathbf{R}}_2^2) \\ &\quad -\frac{Gm_*m_1}{|\mathbf{R}_1 - \mathbf{R}_*|} - \frac{Gm_*m_2}{|\mathbf{R}_2 - \mathbf{R}_*|} - \frac{Gm_1m_2}{|\mathbf{R}_2 - \mathbf{R}_1|}\,. \end{aligned} \tag{16}$$

The center-of-mass position vectors $\mathbf{R}_*, \mathbf{R}_1, \mathbf{R}_2$ are related to the Jacobi coordinates as follows:

$$\begin{aligned} \mathbf{R}_* &= -\kappa_1\mathbf{r}_1 - \kappa_2\mathbf{r}_2\,, \\ \mathbf{R}_1 &= (1-\kappa_1)\mathbf{r}_1 - \kappa_2\mathbf{r}_2\,, \\ \mathbf{R}_2 &= (1-\kappa_2)\mathbf{r}_2\,, \end{aligned} \tag{17}$$

where κ_j were defined in equation (5). Using these identities in equation (16) above, we get

$$H = \frac{1}{2}\mu_1\dot{\mathbf{r}}_1^2 + \frac{1}{2}\mu_2\dot{\mathbf{r}}_2^2 - \frac{Gm_*m_1}{r_1} - \frac{Gm_*m_2}{|\mathbf{r}_2+\kappa_1\mathbf{r}_1|} - \frac{Gm_1m_2}{|\mathbf{r}_2-(1-\kappa_1)\mathbf{r}_1|}\,. \tag{18}$$

Adding and subtracting the term $-\frac{Gm_*m_2}{r_2}$ and re-arranging terms, we have

$$\begin{aligned} H &= \frac{1}{2}\mu_1\dot{\mathbf{r}}_1^2 - \frac{Gm_*m_1}{r_1} + \frac{1}{2}\mu_2\dot{\mathbf{r}}_2^2 - \frac{Gm_*m_2}{r_2} \\ &\quad -\frac{Gm_1m_2}{|\mathbf{r}_2-(1-\kappa)\mathbf{r}_1|} + Gm_*m_2\left[\frac{1}{r_2} - \frac{1}{|\mathbf{r}_2+\kappa\mathbf{r}_1|}\right], \end{aligned} \tag{19}$$

where

$$\mu_1 = \frac{m_*m_1}{m_*+m_1}, \qquad \text{and} \qquad \mu_2 = \frac{(m_*+m_1)m_2}{m_*+m_1+m_2}\,. \tag{20}$$

The first two terms in equation (19) can be identified as the energy of a particle of mass μ_1 moving in a central potential

$$V_1 = -\frac{Gm_*(m_1/\mu_1)}{r_1}\,.$$

This is the familiar reduced 2-body kepler problem. The next two terms can be similarly identified. The remaining terms are smaller by $\mathcal{O}(m_j/m_*)$. Therefore, the natural choice is to consider the 3-body Hamiltonian, H, to be the sum of two unperturbed 2-body Hamiltonians, $H_0 = H_{10} + H_{20}$,

$$H_{j0} = \frac{1}{2}\mu_j\dot{\mathbf{r}}_j^2 - \frac{Gm_*m_j}{r_j} = -\frac{Gm_*m_j}{2a_j}, \qquad (j=1,2) \tag{21}$$

and a perturbation:

$$H_{\rm int} = -\frac{Gm_1m_2}{|\mathbf{r}_2-(1-\kappa)\mathbf{r}_1|} + Gm_*m_2\left[\frac{1}{r_2} - \frac{1}{|\mathbf{r}_2+\kappa\mathbf{r}_1|}\right]. \tag{22}$$

This "natural" choice is not apparent in the center-of-mass coordinates (equation (16)). Hence the choice of Jacobi coordinates. (Jacobi coordinates are easily generalised to the case of N bodies.)

First order perturbation theory

To first order in the product, $m_1 m_2$, of the planetary masses, the perturbation Hamiltonian is given by

$$H_{\mathrm{int}} = -Gm_1m_2\left[\frac{1}{|\mathbf{r}_2 - \mathbf{r}_1|} - \frac{\mathbf{r}_1 \cdot \mathbf{r}_2}{r_2^3}\right]. \tag{23}$$

The expression in square brackets is called the disturbing function in celestial mechanics terminology and is usually denoted as R:

$$R = \frac{1}{|\mathbf{r}_2 - \mathbf{r}_1|} - \frac{\mathbf{r}_1 \cdot \mathbf{r}_2}{r_2^3}. \tag{24}$$

Perturbation analysis for the departure from keplerian motion is more naturally and conveniently handled in terms of the orbital elements rather that the position and velocity vectors. We can take account of the perturbations induced by H_{int} by using a variation-of-parameters method: *i.e.* we allow that the solution $\mathbf{r}_j(t)$ is an elliptical trajectory (equation (6)), but an ellipse whose parameters vary with time. The classical form of Lagrange's perturbation equations for the orbital elements is as follows (see, for example, Brouwer and Clemence 1961).

$$\begin{aligned}
\frac{da}{dt} &= \frac{-2}{\mu n a}\frac{\partial H_{\mathrm{int}}}{\partial \sigma}, \\
\frac{di}{dt} &= \frac{1}{\mu nab\sin i}\frac{\partial H_{\mathrm{int}}}{\partial \Omega} - \frac{\cos i}{\mu nab\sin i}\frac{\partial H_{\mathrm{int}}}{\partial \omega}, \\
\frac{d\Omega}{dt} &= \frac{-1}{\mu nab\sin i}\frac{\partial H_{\mathrm{int}}}{\partial i}, \\
\frac{de}{dt} &= \frac{b}{\mu na^3 e}\frac{\partial H_{\mathrm{int}}}{\partial \omega} - \frac{b^2}{\mu na^4 e}\frac{\partial H_{\mathrm{int}}}{\partial \sigma}, \\
\frac{d\omega}{dt} &= \frac{-b}{\mu na^3 e}\frac{\partial H_{\mathrm{int}}}{\partial e} + \frac{\cos i}{nab\sin i}\frac{\partial H_{\mathrm{int}}}{\partial i}, \\
\frac{d\sigma}{dt} &= \frac{2}{\mu n a}\frac{\partial H_{\mathrm{int}}}{\partial a} + \frac{b^2}{\mu na^4 e}\frac{\partial H_{\mathrm{int}}}{\partial e},
\end{aligned} \tag{25}$$

where $b = a\sqrt{1-e^2}$ is the semiminor axis, and n is related to a as in equation (7). In order to determine the time variation of the orbital elements, we need to express H_{int} in terms of the orbital elements.

In the context of the solar system, the disturbing function, R, is usually developed as a Fourier series in the phase angles Ω, ω and $\mathcal{M}$. The coefficients in the Fourier series are then given as a power series in orbital eccentricities and inclinations and are also functions of the ratio $\alpha = a_1/a_2 < 1$ of the semimajor axes.

In the present context, recall that for a random orientation of orbital planes, it is more probable that the planetary orbits are observed nearly edge-on rather than face-on. Therefore, if the reference (x, y) plane is the plane of the sky, then we are especially interested in the case of orbital inclinations close to 90°. Using equations (6)–(7), and doing some algebra, the following expansion of R is obtained for the case of coplanar, edge-on orbits. The expansion is given to first order in the orbital eccentricities for the periodic terms, and to second order for the secular terms.

$$\begin{aligned} a_2 R = \; & P(\psi,\alpha) - \alpha\cos\psi \\ & -e_1\cos(\ell_1-\omega_1)[\partial_\alpha P(\psi,\alpha) + \alpha\cos\psi] \\ & +e_1\sin(\ell_1-\omega_1)[2\partial_\psi P(\psi,\alpha) + 2\alpha\sin\psi] \\ & +e_2\cos(\ell_2-\omega_2)[(1+\partial_\alpha)P(\psi,\alpha) - 2\alpha\cos\psi] \\ & -e_2\sin(\ell_2-\omega_1)[2\partial_\psi P(\psi,\alpha) + 2\alpha\sin\psi] \\ & +\frac{1}{8}\alpha[b_{3/2}^{(1)}(\alpha)(e_1^2+e_2^2) - 2b_{3/2}^{(2)}(\alpha)e_1e_2\cos(\omega_1-\omega_2)], \end{aligned} \tag{26}$$

where

$$\begin{aligned} \ell_j &= \mathcal{M}_j + \omega_j, \\ \psi &= \ell_1 - \ell_2, \\ P(\psi,\alpha) &= \frac{1}{(1-2\alpha\cos\psi+\alpha^2)^{1/2}}, \end{aligned} \tag{27}$$

and $\partial_\alpha \equiv \alpha\frac{\partial}{\partial\alpha}$, and $\partial_\psi \equiv \frac{\partial}{\partial\psi}$. The variable ℓ is the mean longitude measured from the x-axis, and ψ measures the relative mean orbital phases of the two planets. The functions $b_s^{(j)}(\alpha)$ are the so-called Laplace coefficients defined as follows:

$$b_s^{(j)}(\alpha) = \frac{1}{2\pi}\int_0^{2\pi} d\theta\,\cos j\theta[1-2\alpha\cos\theta+\alpha^2]^{-s}. \tag{28}$$

It is also useful to know the Fourier series expansion of $P(\psi,\alpha)$:

$$P(\psi,\alpha) = \frac{1}{2}\sum_{j=-\infty}^{+\infty} b_s^{(j)}(\alpha)\cos j\psi. \tag{29}$$

If we insert equation (29) into the expression for R in equation (26) above, we can identify the two first order 3:2 resonance terms as those with the arguments $\phi_1 = 3\ell_2 - 2\ell_1 - \omega_1$ and $\phi_2 = 3\ell_2 - 2\ell_1 - \omega_2$. The dominant perturbations to the keplerian orbital elements are described by the following truncated disturbing function:

$$\begin{aligned} a_2 R = \; & P(\psi,\alpha) - \alpha\cos\psi \\ & +C_1(\alpha)e_1\cos(3\ell_2-2\ell_1-\omega_1) \\ & +C_2(\alpha)e_2\cos(3\ell_2-2\ell_1-\omega_2) \\ & +\frac{1}{8}\alpha[b_{3/2}^{(1)}(\alpha)(e_1^2+e_2^2) - 2b_{3/2}^{(2)}(\alpha)e_1e_2\cos(\omega_1-\omega_2)], \end{aligned} \tag{30}$$

where

$$C_1(\alpha) = -(3+\frac{1}{2}\alpha\frac{d}{d\alpha})b_{1/2}^{(3)}(\alpha), \tag{31}$$

$$C_2(\alpha) = +(\frac{5}{2}+\frac{1}{2}\alpha\frac{d}{d\alpha})b_{1/2}^{(2)}(\alpha). \tag{32}$$

The lowest order perturbation equations obtained from the truncated disturbing function of equation (30) are as follows.

The semimajor axes:

$$\begin{aligned}\frac{da_1}{dt} &= 2\frac{m_2}{m_*}n_1 a_1 \alpha\{(\partial_\psi P(\psi,\alpha) + \alpha\sin\psi) + 2C_1(\alpha)e_1\sin(3\ell_2 - 2\ell_1 - \omega_1)\\ &\qquad +2C_2(\alpha)e_2\sin(3\ell_2 - 2\ell_1 - \omega_2)\}, \qquad (33)\\ \frac{da_2}{dt} &= -2\frac{m_1}{m_*}n_2 a_2\{(\partial_\psi P(\psi,\alpha) + \alpha\sin\psi) + 3C_1(\alpha)e_1\sin(3\ell_2 - 2\ell_1 - \omega_1)\\ &\qquad +3C_2(\alpha)e_2\sin(3\ell_2 - 2\ell_1 - \omega_2)\};\end{aligned}$$

the eccentricities:

$$\begin{aligned}\frac{de_1}{dt} &= \frac{m_2}{m_*}n_1\alpha\{\frac{1}{4}\alpha b_{3/2}^{(2)}(\alpha)e_2\sin(\omega_1-\omega_2) + C_1(\alpha)\sin(3\ell_2 - 2\ell_1 - \omega_1)\},\\ &\qquad (34)\\ \frac{de_2}{dt} &= -\frac{m_1}{m_*}n_2\{\frac{1}{4}\alpha b_{3/2}^{(2)}(\alpha)e_1\sin(\omega_1-\omega_2) - C_2(\alpha)\sin(3\ell_2 - 2\ell_1 - \omega_2)\};\end{aligned}$$

and the periastrons:

$$\begin{aligned}\frac{d\omega_1}{dt} &= \frac{m_2}{m_*}n_1\alpha\{\frac{1}{4}\alpha b_{3/2}^{(1)}(\alpha) - \frac{1}{4}\alpha b_{3/2}^{(2)}(\alpha)\frac{e_2}{e_1}\cos(\omega_1-\omega_2)\\ &\qquad +\frac{1}{e_1}C_1(\alpha)\cos(3\ell_2 - 2\ell_1 - \omega_1)\}, \qquad (35)\\ \frac{d\omega_2}{dt} &= \frac{m_1}{m_*}n_2\{\frac{1}{4}\alpha b_{3/2}^{(1)}(\alpha) - \frac{1}{4}\alpha b_{3/2}^{(2)}(\alpha)\frac{e_1}{e_2}\cos(\omega_1-\omega_2)\\ &\qquad +\frac{1}{e_2}C_2(\alpha)\cos(3\ell_2 - 2\ell_1 - \omega_2)\}.\end{aligned}$$

Recall that in the unperturbed solution, the only time variation is that of the mean anomaly, $\mathcal{M}_j = n_j t + \sigma_j$; all other orbital elements are fixed. Using the unperturbed solution in the right hand side of equations (33), the first order solution to the semimajor axes variations are given by $a_j(t) = a_{j,0} + \Delta a_j(t)$, where

$$\begin{aligned}\frac{\Delta a_1(t)}{a_{1,0}} &= 2\frac{m_2}{m_*}\alpha\left\{\frac{n_1}{n_1-n_2}[P(\psi(t),\alpha) - P(\psi_0,\alpha) - \alpha(\cos\psi(t) - \cos\psi_0)]\right.\\ &\qquad -\frac{2n_1}{3n_2-2n_1}[C_1 e_1(\cos\phi_1(t) - \cos\phi_{1,0})\\ &\qquad \left. +C_2 e_2(\cos\phi_2(t) - \cos\phi_{2,0})]\right\}, \qquad (36)\\ \frac{\Delta a_2(t)}{a_{2,0}} &= -2\frac{m_1}{m_*}\left\{\frac{n_2}{n_1-n_2}[P(\psi(t),\alpha) - P(\psi_0,\alpha) - \alpha(\cos\psi(t) - \cos\psi_0)]\right.\\ &\qquad -\frac{3n_2}{3n_2-2n_1}[C_1 e_1(\cos\phi_1(t) - \cos\phi_{1,0}))\\ &\qquad \left. +C_2 e_2(\cos\phi_2(t) - \cos\phi_{2,0}))]\right\},\end{aligned}$$

where

$$\begin{aligned} \psi(t) &= (n_1 - n_2)t + (\sigma_1 + \omega_1) - (\sigma_2 + \omega_2)\,, \\ \phi_j(t) &= (3n_2 - 2n_1)t + 3(\sigma_2 + \omega_2) - 2(\sigma_1 + \omega_1) - \omega_j\,, \end{aligned} \tag{37}$$

and a_j, n_j, e_j, ω_j and σ_j have their unperturbed values in the right hand side of equations (37). This solution shows that there is no secular evolution of the semimajor axes; there are relatively long period variations (of period $2\pi/|3n_2 - 2n_1|$) due to the 3:2 near-commensurability of the mean motions, and there are also perturbations of frequency equal to the synodic frequency, $n_1 - n_2$, which can be seen to be the short spikes in the numerical solution (see Figure IV).

The first order solution for the eccentricities and periastrons requires a somewhat more involved analysis, since it is clear from equations (34)–(35) that there is secular evolution of these variables. The apparent singularities at zero eccentricity in these equations can be removed by using the following variables:

$$h_j = e_j \sin\omega_j, \qquad \text{and} \qquad k_j = e_j \cos\omega_j. \tag{38}$$

Equations (32)–(33) can then be replaced by the following equations for the time variation of h_j, k_j (Malhotra *et al.* 1989):

$$\begin{aligned} \dot{h}_j &= \sum_{l=1,2} A_{jl}k_l + B_j \cos(3\ell_2 - 2\ell_1)\,, \\ \dot{k}_j &= -\sum_{l=1,2} A_{jl}h_l - B_j \sin(3\ell_2 - 2\ell_1)\,, \end{aligned} \tag{39}$$

where the coefficient matrix $\mathbf{A}$ is

$$\mathbf{A} = \begin{bmatrix} \frac{1}{4}\frac{m_2}{m_*} n_1 \alpha^2 b_{3/2}^{(1)}(\alpha) & -\frac{1}{4}\frac{m_2}{m_*} n_1 \alpha^2 b_{3/2}^{(2)}(\alpha) \\ -\frac{1}{4}\frac{m_1}{m_*} n_2 \alpha b_{3/2}^{(2)}(\alpha) & \frac{1}{4}\frac{m_1}{m_*} n_2 \alpha b_{3/2}^{(1)}(\alpha) \end{bmatrix}, \tag{40}$$

and

$$B_1 = \frac{m_2}{m_*} n_1 \alpha C_1(\alpha), \qquad B_2 = \frac{m_1}{m_*} n_2 C_2(\alpha)\,. \tag{41}$$

Equations (39) can be recognised as the equations for a pair of forced linear oscillators. If we use the unperturbed solution for the arguments of the forcing terms,

$$3\ell_2 - 2\ell_1 = (3n_2 - 2n_1)t + 3(\sigma_2 + \omega_2) - 2(\sigma_1 + \omega_1)\,, \tag{42}$$

then, to a good approximation, the evolution of $\{h_j, k_j\}$ is a superposition of free oscillations and forced oscillations:

$$\begin{aligned} h_j(t) &= \sum_{i=1,2} E_j^{(i)} \sin(g_i t + \beta_i) + F_j \sin(3\ell_2 - 2\ell_1)\,, \\ k_j(t) &= \sum_{i=1,2} E_j^{(i)} \cos(g_i t + \beta_i) + F_j \cos(3\ell_2 - 2\ell_1)\,, \end{aligned} \tag{43}$$

where g_i and $\mathbf{E}^{(i)}$ are eigenvalues and eigenvectors, respectively, of the coefficient matrix $\mathbf{A}$. The phases, β_i, and the normalisation of the eigenvectors, $\mathbf{E}^{(i)}$, are determined by the initial conditions, $\{e_{i,0}, \omega_{i,0}\}$. The amplitude of the forcing is given by

$$\mathbf{F} = -[\mathbf{A} - (3n_2 - 2n_1)\mathbf{I}]^{-1} \cdot \mathbf{B}\,, \tag{44}$$

where $\mathbf{I}$ is the 2×2 identity matrix.

Let us consider the magnitudes of some of these perturbations for the case of nearly edge-on orbits ($\sin i \approx 1$) for which the above analysis is valid (see Figure IV).

- In the semimajor axes variations, we see from equations (36) that there are two components: one of period equal to the synodic period of the planets, $(2\pi/|n_1 - n_2|) \approx 206$ days, and fractional amplitude $\frac{m_j}{m_*}\frac{n_1}{n_1-n_2} \approx 1 \times 10^{-4}$; and the other of period equal to $2\pi/(3n_2 - 2n_1) \approx 5.5$ years and fractional amplitude $\frac{m_j}{m_*}\frac{2n_1}{3n_2-2n_1} \approx 2 \times 10^{-4}$.
- In the coupled eccentricity-periastron perturbations, Rasio *et al.* found that for the initial conditions deduced from W&F's observations, the eigenmode corresponding to g_1, the larger of the two eigenfrequencies, dominates the variations of h_j, k_j. This means that the "free" secular precessions of the two planets' periastrons are locked to each other at the frequency g_1 with relative phase difference π. This may be of significance for the long term evolution of the planetary system. However, these secular variations have periods of $2\pi/g_j$ which are on the order of 10^5 years, and hence the secular precession rates of the orbits are exceedingly small and undetectable in the near future. The 3:2 near-resonance induced sinusoidal forced precession rate, on the other hand, has an amplitude of about half a degree per year and a period of 5.5 years; the corresponding fractional amplitudes for the eccentricity variations are on the order of 10^{-3}.

As a final comment on the first order perturbation analysis described above, we note that it is valid only for

$$\frac{3n_2 - 2n_1}{n_2} \gg C\left(\frac{m_j}{m_*}\right)^{2/3}, \tag{45}$$

where the numerical factor C is $\mathcal{O}(10)$. If this inequality is not satisfied, then the forced precession of the periastrons cannot be neglected and keeping ω_j fixed in the right hand sides of equations (37) and (42) is a poor approximation.

5. RESONANCE DYNAMICS

For the best-fit orbital periods reported in W&F, the inequality (45) is violated if $(\sin i)^{-1} \gtrsim 10$ (Malhotra *et al.* 1992). This corresponds to mass ratios $(m_j/m_*) \gtrsim 6 \times 10^{-5}$, or orbital inclinations less than about 6 degrees (for coplanar orbits). (See Figure V which shows a comparison of the numerical integration results and the perturbative solution for time variation of the semimajor axes when $(\sin i)^{-1} = 10$.) If this were the case, the planets would be in an exact

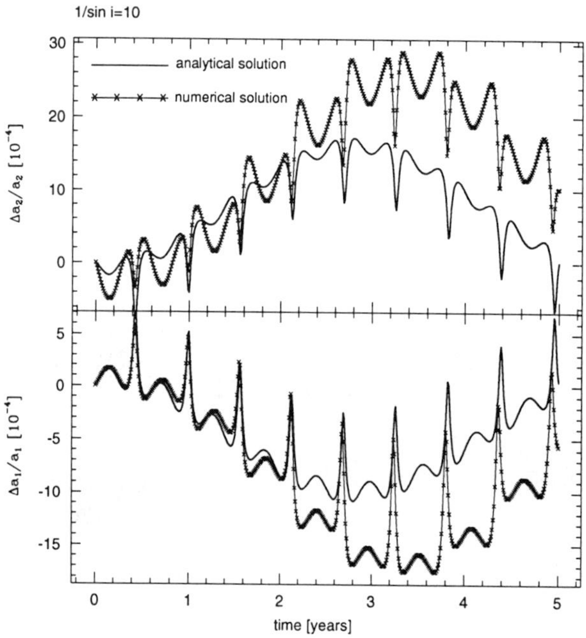

FIGURE V

resonance configuration, and the perturbative solutions of the previous section would not be valid.

The most elegant and powerful method to handle the resonance problem is canonical perturbation theory. Again confining the discussion to coplanar orbits, we may start with the canonical action-angle variables of the unperturbed two body problem (Brouwer and Clemence 1961),

$$\begin{aligned} &\mathcal{M}_j, \qquad L_j = \mu_j\sqrt{Gm_*(m_j/\mu_j)a_j}\,, \\ &-\omega_j, \qquad G_j = L_j\sqrt{(1-e_j^2)}\,. \end{aligned} \tag{46}$$

L is related to the total 2-body energy, and G is the total angular momentum in the 2-body problem.

By means of a simple canonical transformation, we can define the following pairs of canonical resonance variables as linear combinations of the above:

$$\begin{aligned} &\phi_j = 3\ell_2 - 2\ell_1 - \omega_j, && \Phi_j = L_j - G_j \approx \tfrac{1}{2}L_j e_j^2, \qquad (j=1,2) \\ &\psi = \ell_1 - \ell_2, && \Psi = L_1 + 2(L_1 - G_1 + L_2 - G_2), \\ &\psi' = \ell_2, && \Psi' = G_1 + G_2. \end{aligned} \tag{47}$$

where l_j, ψ, and ϕ_j have the same definitions as before. Note that Ψ' is the total 3-body angular momentum and is therefore a constant of the motion (the

3-body Hamiltonian is cyclic in the corresponding canonical variable ψ'). Note also that the canonical momenta Φ_1, Φ_2 are small, of $\mathcal{O}(e^2)$.

Making use of the relations

$$\begin{aligned} L_1 &= \Psi - 2(\Phi_1 + \Phi_2)\,, \\ L_2 &= \Psi' - \Psi + 3(\Phi_1 + \Phi_2)\,, \end{aligned}$$

we have the unperturbed Hamiltonian in terms of the new canonical variables (cf. equation (21)):

$$H_0 = -\frac{(Gm_*)^2 m_1^3}{[\Psi - 2(\Phi_1 + \Phi_2)]^2} - \frac{(Gm_*)^2 m_2^3}{[\Psi' - \Psi + 3(\Phi_1 + \Phi_2)]^2}\,. \tag{48}$$

This looks rather unwieldy, but if we define the constants $\tilde{a}_j$ as follows,

$$\begin{aligned} \Psi &= \mu_1\sqrt{Gm_*(m_1/\mu_1)\tilde{a}_1}\,, \\ \Psi' - \Psi &= \mu_2\sqrt{Gm_*(m_2/\mu_2)\tilde{a}_2}\,, \end{aligned} \tag{49}$$

and expand H_0 in a Taylor series to second order in the small quantities Φ_1, Φ_2, we get

$$H_0 = -\frac{Gm_*m_1}{2\tilde{a}_1} - \frac{Gm_*m_2}{2\tilde{a}_2} + \nu(\Phi_1 + \Phi_2) - \beta(\Phi_1 + \Phi_2)^2\,, \tag{50}$$

where

$$\nu = 3\tilde{n}_2 - 2\tilde{n}_1 \qquad \text{and} \qquad \beta = \left(\frac{6}{\mu_1\tilde{a}_1^2} + \frac{27}{4\mu_2\tilde{a}_2^2}\right)\,, \tag{51}$$

and $\tilde{n}_j^2\tilde{a}_j^3 = Gm_*(m_j/\mu_j)$.

The frequencies of the resonance variables obtained from H_0 are as follows

$$\dot{\phi}_j = \frac{\partial H_0}{\partial \Phi_j} = \nu - 2\beta(\Phi_1 + \Phi_2)\,. \tag{52}$$

The degeneracy of the two resonances is removed by the perturbation, $H_{\rm int}$. Since the resonances are only weakly coupled, they induce relatively independent perturbations in the orbits. Therefore, we can analyse the two resonances separately. The first order 3:2 resonance terms in $H_{\rm int}$ have the form $\varepsilon\sqrt{2\Phi}\cos\phi$ (cf. equation (30)), so that the resonance Hamiltonian for either the ϕ_1 or the ϕ_2 resonance is given by

$$H_{\rm res} = \nu\Phi - \beta\Phi^2 + \varepsilon\sqrt{2\Phi}\cos\phi\,. \tag{53}$$

A detailed analysis of this resonance Hamiltonian may be found in Henrard and Lemaitre (1983), and Malhotra (1988). Here we discuss some of its general properties. Figure VI shows the phase space structure of this Hamiltonian. In exact resonance, the mean value of $\dot{\phi}$ vanishes and ϕ librates about 0 or π

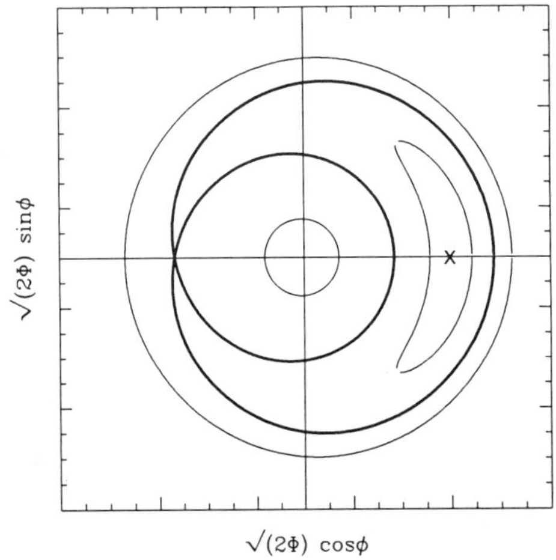

FIGURE VI

depending upon the sign of ε. Neglecting the $\mathcal{O}(\varepsilon)$ term, the exact resonance condition, $\langle\dot{\phi}\rangle = 0$, implies $\Phi_{\rm res} = \nu/2\beta$, or, in terms of the eccentricities,

$$e_{1,\rm res} \approx \left(\frac{\nu/n_1}{6(1+\frac{\mu_1/\mu_2}{\alpha})}\right)^{1/2}, \quad \text{and} \quad e_{2,\rm res} \approx \left(\frac{\nu/n_1}{9(1+\frac{\alpha\mu_2}{\mu_1})}\right)^{1/2}. \tag{54}$$

Note that the sign of ν is important: exact resonance can exist only for positive ν. For the best-fit orbital periods reported in W&F, the numerical values are $e_{1,\rm res} \approx 0.046$ and $e_{2,\rm res} \approx 0.047$.

The width of the resonance is approximately given by

$$\Delta\Phi \approx \left(\frac{4\varepsilon^2\nu}{\beta^3}\right)^{1/4},$$

or, in terms of the eccentricities,

$$\Delta e_1^2 \approx \left[\frac{4m_2}{3m_*}\frac{\alpha C_1(\alpha)}{(1+\frac{\mu_1/\mu_2}{\alpha})}e_{1,\rm res}\right]^{1/2}, \quad \text{and} \quad \Delta e_2^2 \approx \frac{2}{3}\left[\frac{4m_1}{3m_*}\frac{C_2(\alpha)}{(1+\frac{\alpha\mu_2}{\mu_1})}e_{2,\rm res}\right]^{1/2}. \tag{55}$$

From these expressions, we see that the amplitude of the resonant perturbations is dramatically larger that the $\mathcal{O}(m_j/m_*)$ amplitude of the near-resonant perturbations discussed in the previous section. Also, the mean eccentricities at exact resonance are a factor of two larger that the best-fit values reported in W&F. This means that the planets can be in resonance only if their masses are bigger than the lower limits of about $3M_\oplus$ by a factor $(\sin i)^{-1}$ of 10 or more (see Figure 2 in Malhotra *et al.* 1992).

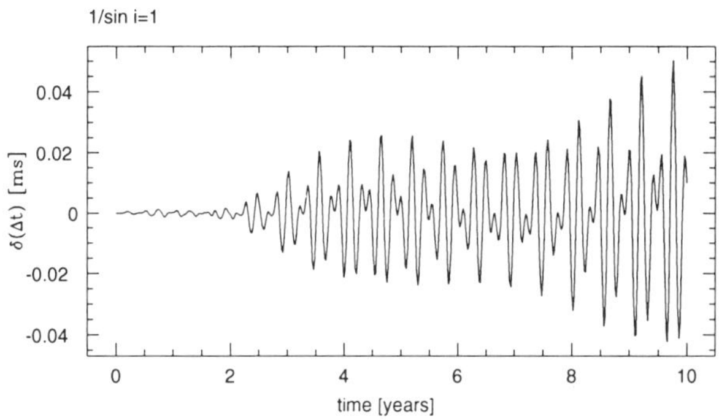

FIGURE VII

6. DISCUSSION

In this paper, I have discussed the constraints placed by the three-body dynamics upon the putative planetary system around PSR1257+12. It is important to realise that several quantitative as well as qualitative aspects of the mutual perturbations depend upon the viewing geometry that the planetary system presents to the observer on the earth. From the observations of W&F and from discussion of dynamical stability in section 3, we see that the unknown orbital inclinations allow for planetary masses ranging from about $3M_{\oplus}$ to two Jovian masses. As shown in sections 4 and 5, the time variation of the orbital elements can have dramatically different analytical forms over this range of masses. The TOA residuals Δt predicted from the *true* motion of the pulsar about the system barycenter differ by amount $\delta(\Delta t)$ from those predicted by assuming fixed orbital elements. These differences will accumulate over time and become detectable as a systematic variation of the rms residuals of the pulsar timing analysis used in W&F. In Figure VII, I show the growth in $\delta(\Delta t)$ for the case of coplanar, edge-on orbits. Detection of exactly such variations will provide confirmation of the existence of the first extra-solar planetary system, and will also yield more information about the system (in particular, the masses of the planets relative to the mass of the pulsar).

ACKNOWLEDGMENTS

Part of the work presented here is based on a paper that I co-authored with D.C. Black, A. Eck and A. Jackson. I am grateful to A. Wolszczan for several discussions about the observations and data analysis, and also to T. Stepinski and D. Black for general discussions. This research was done while the author was a Staff Scientist at the Lunar and Planetary Institute which is operated by the Universities Space Research Association under contract no. NASW-4574 with the National Aeronautics and Space Administration. This paper is Lunar and Planetary Institute Contribution no. 786.

REFERENCES

Brouwer, D. and Clemence, G.M. 1961, *Methods of Celestial Mechanics* Academic Press.

Duncan, M., Quinn, T., and Tremaine, S. 1989, *Icarus* **82** 402-418

Graziani, F. and Black, D. 1981,*ApJ*, **251** 337-341.

Henrard, J. and Lemaitre, A. 1983, *Cel. Mech.* **30** 197-218.

Malhotra, R. 1988, Ph.D. thesis, Cornell Univ.

Malhotra, R., Fox, K., Murray, C.D., and Nicholson, P.D. 1989, *A&A* **221** 348-358

Malhotra, R., D. Black, Eck, A., and Jackson, A. 1992, *Nature* **355** 583-585.

Rasio, F.A., Nicholson, P.D., Shapiro, S.L. and Teukolsky, S.A. 1992, *Nature* **355** 325-326.

Wisdom, J. 1980, *AJ* **85** 1122-1133.

Wolszczan, A. and Frail, D.A. 1992, *Nature* **355** 145-147.

Planets Around Pulsars
ASP Conference Series, Vol. 36, 1993
J.A. Phillips, J.E. Thorsett, and S.R. Kulkarni

ORBITAL EVOLUTION OF THE PSR1257+12 PLANETARY SYSTEM

FREDERIC A. RASIO, PHILIP D. NICHOLSON,
STUART L. SHAPIRO, AND SAUL A. TEUKOLSKY
Center for Radiophysics and Space Research, Cornell University, Ithaca, NY 14853

ABSTRACT The detection of orbital perturbation effects in the PSR 1257 +12 timing data would provide irrefutable confirmation that planets are indeed orbiting the pulsar. Here we give an overview of how perturbation effects are expected to affect the orbital elements of the two planets over the next few years. In particular, we give a simple calculation of resonant perturbations, including the nonlinear effects which could be important if $\sin i \lesssim 0.1$. We also present a new analysis of the effects of close encounters between the two planets and we discuss their detectability.

RESONANT PERTURBATIONS

One of the most intriguing, and potentially important, features of the putative planetary system around PSR1257+12 is the near-commensurability between the mean angular velocities (n_i) of the two planets. Wolszczan and Frail (1992) report orbital periods of $P_1 = 66.563 \pm 0.009$ days and $P_2 = 98.234 \pm 0.021$ days, or $n_1/n_2 = 1.4758 \pm 0.0004 \simeq 3/2$. Rasio *et al.* (1992) showed that, as a consequence of this near-commensurability, the orbital periods, eccentricities (e_i), and longitudes of pericenter (ω_i) should exhibit relatively large periodic perturbations. For masses near the minimum values of $m_1 = 3.4M_\oplus$ and $m_2 = 2.8M_\oplus$ (which correspond to $\sin i = 1$, i.e., orbits viewed edge-on), the period of these perturbations is $2\pi/(2n_1 - 3n_2) = 5.56\,\mathrm{yr}$, while the amplitudes are proportional to the planet masses.

If the masses of the planets are appreciably larger than the minimum values (i.e., if $\sin i \ll 1$), then it is possible that the planets are actually "locked" in the 3:2 resonance, and the dynamical evolution of the system is qualitatively different from that described by Rasio *et al.* (1992). This was first pointed out by Malhotra *et al.* (1992), who found that resonant locking was likely for values of $(\sin i)^{-1} \gtrsim 15$, or masses $m_i \gtrsim 45M_\oplus$. In this situation, the period and amplitude of the perturbations increase, and the perturbations themselves cease to show a simple sinusoidal character. This is illustrated in Fig. I, which shows the results of numerical integrations of the PSR1257+12 system for decreasing values of $\sin i$.

We present here a simplified dynamical model of the resonance which preserves the principal features of the full solution, while avoiding some of its complexity. The principal simplification is to treat the inner planet as a massless

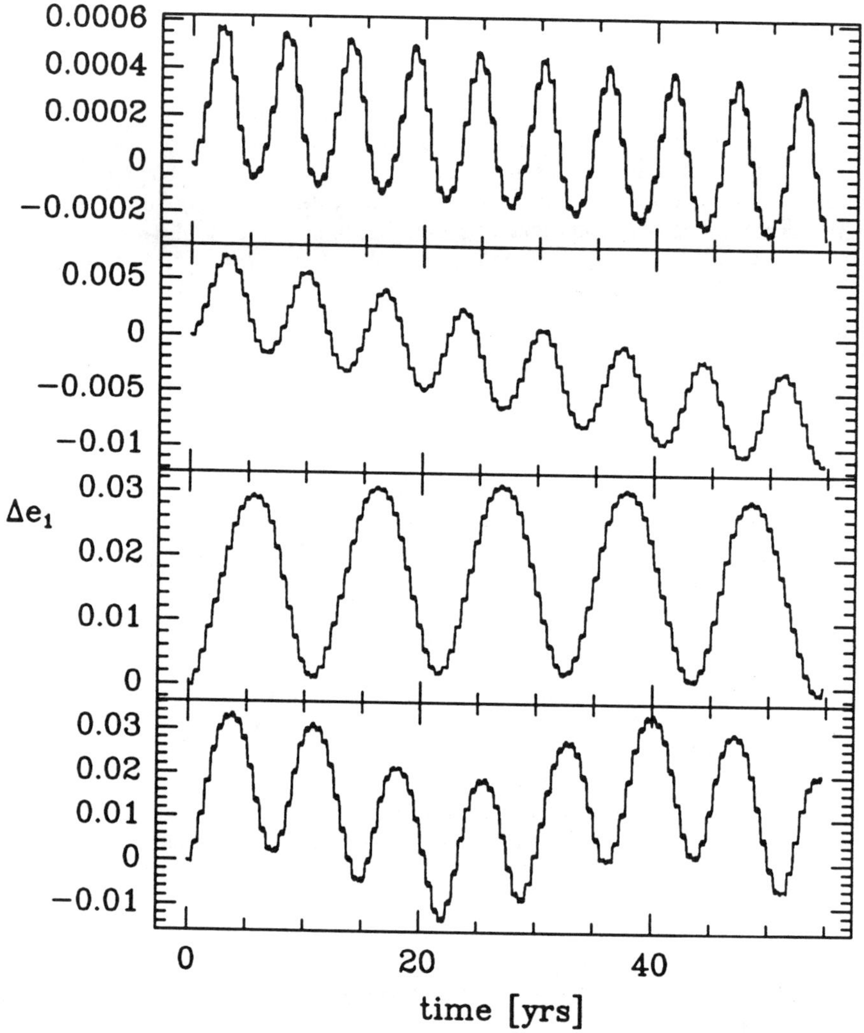

FIGURE I Numerical integration of the PSR1257+12 system, showing the changes in the eccentricity of the inner planet for (top to bottom) $(\sin i)^{-1} = 1$, 10, 20, and 40.

"test particle" perturbed by the outer planet, which is assumed to move on an unperturbed Keplerian orbit about the neutron star. We further assume that $e_i \ll 1$ and $m_2/M \ll 1$, where M is the mass of the neutron star, and that the planetary orbits are coplanar. Under these assumptions, and to lowest order in e_i and m_2/M, the Lagrange perturbation equations governing the variations in the orbital elements of the inner planet in the vicinity of the $m : m-1$ resonance may be written (Danby 1988)

$$\begin{aligned} \dot{n}_1 &= -3\,(m-1)\,e_1 n_1^2\,\beta \sin\phi_1, & (1)\\ \dot{e}_1 &= -n_1\,\beta \sin\phi_1, & (2)\\ \dot{\omega}_1 &= -n_1\,\beta \cos\phi_1\,/e_1, & (3) \end{aligned}$$

where the "resonance variable" ϕ_1 is defined by

$$\phi_1 \equiv (m-1)\,\lambda_1 - m\,\lambda_2 + \omega_1, \qquad (4)$$

λ_i are the mean longitudes of the two planets, and the dimensionless parameter β which governs the amplitude of the perturbations is given by (Brouwer and Clemence 1961)

$$\beta \equiv \frac{m_2}{M}\alpha \left[m + \frac{1}{2}\alpha\frac{d}{d\alpha}\right] b_{1/2}^{m}(\alpha). \qquad (5)$$

The function $b_{1/2}^{m}(\alpha)$ is a Laplace coefficient, which depends only on the ratio of semi-major axes $\alpha = a_1/a_2$. For the 3:2 resonance, $m = 3$, $\alpha \simeq (2/3)^{2/3} = 0.763$ and we find that $\beta = 1.546\,(m_2/M)$. For the PSR1257+12 system, $\beta = 0.937 \times 10^{-5} M_{1.4}^{-1/3}(\sin i)^{-1}$, where $M_{1.4} \equiv M/(1.4M_\odot)$.

In writing the above equations, we have implicitly averaged over many short period, small amplitude perturbations, to single out the principal, slowly-varying resonant term. Since the averaged equations depend on the planet longitudes and pericenters only through ϕ_1, we may simplify the problem by replacing eqn. (3) by

$$\begin{aligned} \dot{\phi}_1 &= (m-1)\,n_1 - m\,n_2 + \dot{\omega}_1 & (6)\\ &= \xi - n_1\,\beta \cos\phi_1\,/e_1, & (7) \end{aligned}$$

where ξ is the "slow" frequency $(m-1)\,n_1 - m\,n_2$. Henceforth, we will drop the subscript 1, except where clarity dictates otherwise. For the general problem, an analogous set of equations to (1)–(3) and (6) governs the evolution of n_2, e_2, ω_2 and the second resonance variable $\phi_2 = (m-1)\,\lambda_1 - m\,\lambda_2 + \omega_2$.

For sufficiently small perturbations, we see that ϕ circulates at a mean angular frequency ξ ($\simeq 2\pi/5.56$ yr for the PSR1257+12 system) and suffers small periodic perturbations, while both e and n show similar periodic perturbations about their mean Keplerian values. In the limit where we may treat e and n on the right-hand sides of eqns. (1-3) as approximately constant, and assume that $|\dot{\omega}| < |\xi|$, we obtain the approximate solutions

$$n(t) \simeq \bar{n} + \frac{3\,(m-1)\,\bar{n}^2\bar{e}}{\xi}\,\beta \cos[\xi(t-t_0)], \qquad (8)$$

$$e(t) \simeq \bar{e} + \frac{\bar{n}}{\xi} \beta \cos[\xi(t - t_0)], \qquad (9)$$

and

$$\omega(t) \simeq \omega_0 - \frac{\bar{n}}{\xi \bar{e}} \beta \sin[\xi(t - t_0)], \qquad (10)$$

where $\bar{n}$ and $\bar{e}$ are the mean angular velocity and eccentricity, and ω_0 is the pericenter longitude at time $t = t_0$. These are equivalent to the expressions derived by Rasio *et al.* (1992) and compared with numerical integrations for $\sin i = 1$.

For larger masses, or for sufficiently small ξ (i.e., n_1/n_2 very close to 1.5), these approximations are no longer valid and a full solution of eqns. (1), (2) and (6) is necessary. This is simplified by noting that

$$\frac{dn}{de} = 3(m - 1)\, ne, \qquad (11)$$

which may be integrated immediately to give the relation

$$\ln(n/n_*) = \frac{3}{2}(m - 1)\, e^2, \qquad (12)$$

where n_* is a constant of the motion, equal to the angular velocity when $e = 0$. As the original differential equations were accurate only for small e, we may replace this by the approximate relation

$$n \simeq n_* \left[1 + \frac{3}{2}(m - 1)\, e^2\right]. \qquad (13)$$

The quantity n_* is related to the conserved action $J_2 \equiv n^{-1/3}\left[1 + \frac{1}{2}(m - 1)e^2\right]$ which arises in a Hamiltonian treatment of the resonance dynamics (Malhotra 1988). Using eqn. (11), and introducing the *constant* quantity $\nu \equiv (m - 1)\, n_* - m\, n_2$, eqn. (6) becomes

$$\dot{\phi} = \nu + \frac{3}{2}(m - 1)^2 e^2 n_* - \frac{n}{e} \beta \cos \phi, \qquad (14)$$

which may be solved simultaneously with eqn. (2) for $\dot{e}$. The quantity ν may be thought of as the "distance from exact resonance" (Malhotra 1988).

Eqns. (2) and (14) admit of equilibrium solutions with constant n, e and ϕ for $\phi = 0$ or $\phi = \pi$ only, and with e given by the roots of

$$\nu + \frac{3}{2}(m - 1)^2 e^2 n_* - \frac{n}{e} \beta \cos \phi = 0. \qquad (15)$$

For $e \ll \beta^{1/3}$, we may ignore the second term and the fixed point is given approximately by

$$e_F \simeq \frac{n\beta}{\nu} \cos \phi_F, \qquad (16)$$

so that $\phi_F = 0$ for $\nu > 0$ and $\phi_F = \pi$ for $\nu < 0$. For $e \gg \beta^{1/3}$, fixed points exist only for $\nu < 0$, with

$$e_F \simeq \left[\frac{-2\nu}{3(m - 1)^2 n_*}\right]^{1/2} \qquad (17)$$

and $\phi_F = 0$ or π. Note that the large-e fixed point is almost independent of β, depending only on the distance from resonance, ν.

In general, eqn. (15) has a single root for $\nu > 0$, and either one or three roots for $\nu < 0$, depending on whether ν is greater or less than a critical value, ν_c given by

$$\nu_c \simeq -\frac{3}{2} n_* \left[3(m-1)^2\beta^2\right]^{1/3}. \tag{18}$$

The forced eccentricity corresponding to ν_c is

$$e_c = \left[\frac{\beta}{3(m-1)^2}\right]^{1/3}. \tag{19}$$

In Fig. II, we plot the equilibrium solutions $(e\cos\phi)_F$ vs. ν for $m = 3$ and $\beta = 10^{-5}$, appropriate to PSR1257+12 with $\sin i = 1$.

For small displacements from the fixed points, we find that

$$\frac{d^2\delta\phi}{dt^2} \simeq -3(m-1)^2\left[1 + (e_c/e_F)^3\cos\phi_F\right] n_*^2\beta e_F\cos\phi_F\,\delta\phi. \tag{20}$$

For $\phi_F = 0$ the fixed points are stable for all values of e_F, while for $\phi_F = \pi$ the fixed points are stable only for $e_F < e_c$. The stable and unstable fixed points merge at $\nu = \nu_c$ and $e = e_c$. The angular frequency of librations about the stable fixed points is given in the limiting cases of eqns. (16) and (17) by

$$\omega_L \simeq \begin{cases} \nu & \text{for } e \ll e_c \\ \left[-6(m-1)^2\nu\beta^2 n_*^3\right]^{1/4} & \text{for } e \gg e_c. \end{cases} \tag{21}$$

The nature of the solutions in the neighborhood of the resonance is best visualized in the form of a series of polar plots of (e, ϕ) for fixed values of n_*. Fig. III shows such a set of plots for $\beta = 10^{-5}$ and increasing values of n_*/n_2. The critical frequency $\nu_c \simeq -0.0024$, corresponding to $n_*/n_2 = \frac{3}{2} + \frac{1}{2}\nu = 1.4988$. For $n_*/n_2 < 1.4988$, two stable and one unstable fixed points are found, as predicted above and shown in Fig. II. For $n_*/n_2 > 1.4988$, only a single stable fixed point is present, at $\phi = 0$.

Three types of motion can be distinguished in these plots for $\nu < \nu_c$, or $n_*/n_2 < 1.4988$:

1. clockwise circulation about the small-e fixed point at $\phi = \pi$;
2. anticlockwise circulation at large-e encompassing all fixed points;
3. anticlockwise libration on bean-shaped trajectories about the fixed point at $\phi = 0$.

The trajectory passing through the unstable fixed point at $\phi = \pi$, or separatrix, (shown best in Fig. IIIb, for $n_*/n_2 = 1.497$) separates the type 3 orbits from those of types 1 and 2. Only type 3 trajectories are strictly classified as "resonant". For $\nu > \nu_c$, or $n_*/n_2 > 1.4988$, the type 1 trajectories and their enclosing separatrix cease to exist, and the motions of type 2 (large-e anticlockwise circulation) and type 3 (anticlockwise libration about the single fixed point)

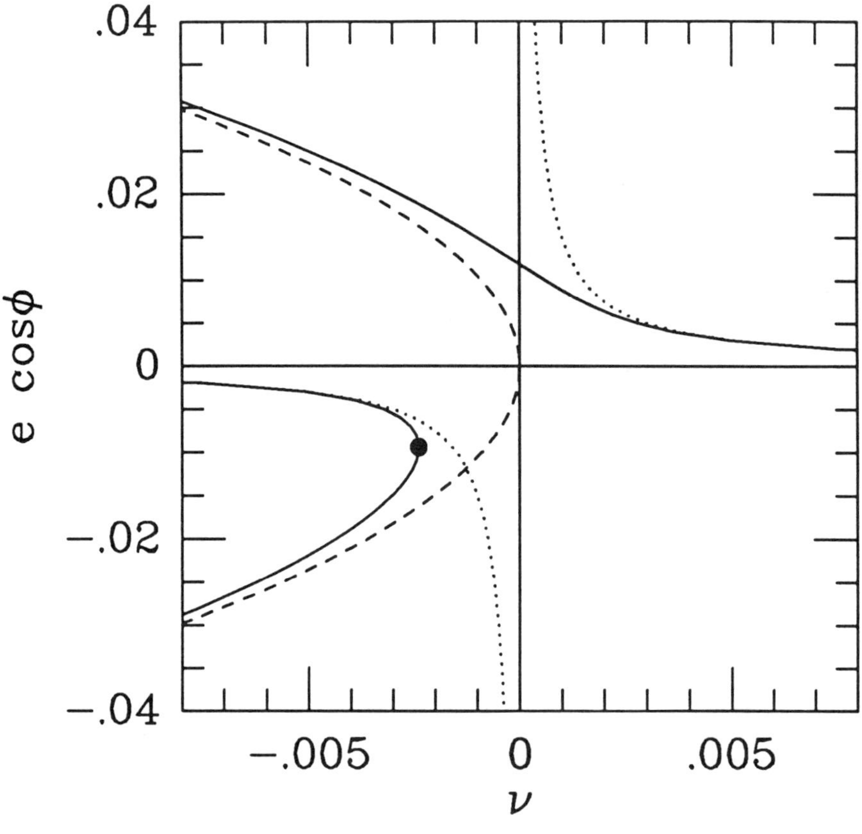

FIGURE II Equilibrium solutions $(e\cos\phi)_F$ vs. ν for $m = 3$ and $\beta = 10^{-5}$ (solid lines). Dotted and dashed curves show the asymptotic expressions of eqns. (16) and (17) for the small-e and large-e fixed points, respectively. The critical point, (ν_c, e_c) is indicated by the filled circle.

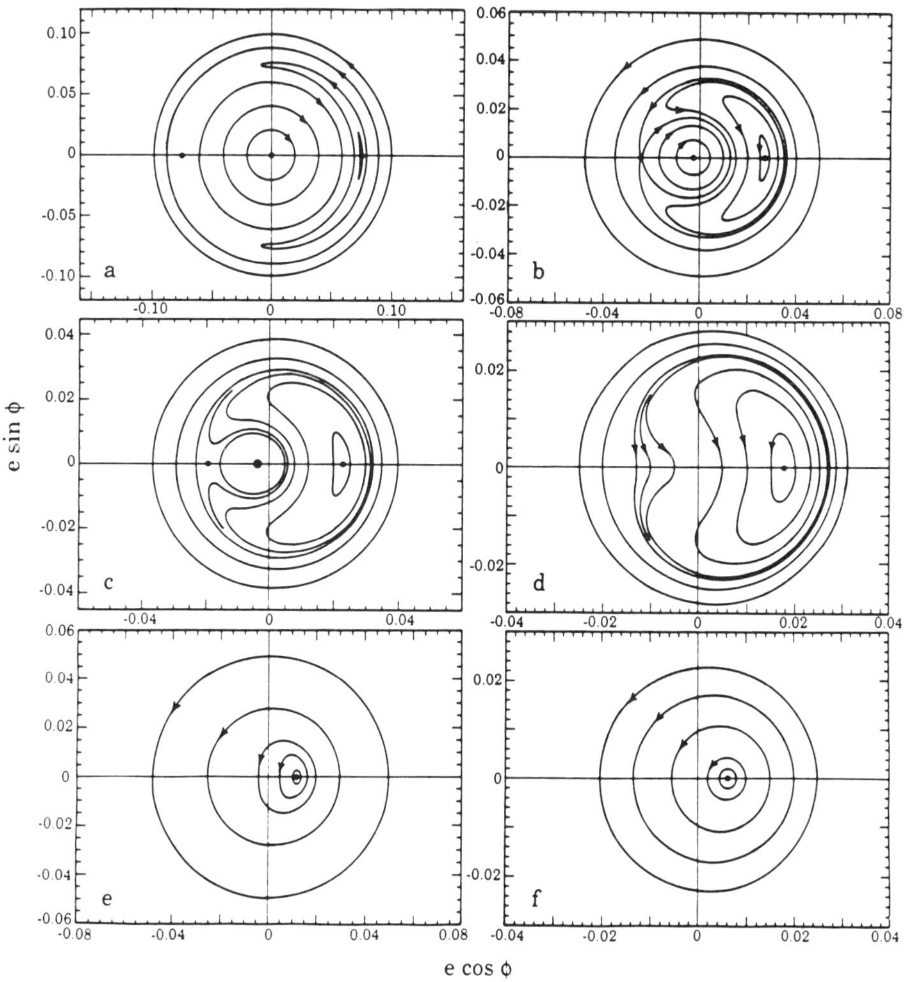

FIGURE III Phase space trajectories for $m = 3$, $\beta = 10^{-5}$ and $n_*/n_2 = 1.475$ (a), 1.497 (b), 1.498 (c), 1.499 (d), 1.500 (e), and 1.501 (f). The fixed points calculated from eqn. (15) are indicated by filled circles. Arrows indicate the direction of evolution.

merge into a single family of orbits. Note that strictly resonant motion, or resonance "locking", can occur for all values of $\nu < \nu_c$, but for $\nu \ll \nu_c$ is restricted to a very narrow range of eccentricities close to the stable equilibrium value of e_F given by eqn. (17), as shown in Fig. IIIa.

The principal observational significance of these solutions lies in the large variations in eccentricity that accompany resonant librations for small values of $|\nu|$, and the oscillatory variations in the pericenter, ω, that result from the non-uniform motion in ϕ. The above simplified model, which neglects the perturbations of the inner planet on the outer, does not, of course, provide a good quantitative description of PSR1257+12 (A more accurate model may be obtained by a straightforward extension of the above equations to include simultaneous perturbations of both planets). Nevertheless, the qualitative predictions of this model are in reasonable agreement with numerical integrations of the full three-body problem. From the observed values of n_1, n_2 and $e_1 = 0.022$, we find that $n_* = 1.4737\, n_2$, so that $\nu = 2n_* - 3n_2 = -0.0527\, n_2$. The critical point is at $\nu_c = -0.0024\, \beta_5^{2/3} n_2$ and $e_c = 0.0094\, \beta_5^{1/3}$, where $\beta_5 \equiv 10^5 \beta$.

For $\sin i = 1$, $\beta_5 \simeq 1$ and we find that $\nu \ll \nu_c$. The corresponding phase space trajectories are shown in Fig. IVa, and are similar to those of Fig. IIIa. There are two possible stable fixed points: the small-e point at $(e \cos \phi)_F \simeq -0.0003$ and the large-e solution at $(e \cos \phi)_F \simeq +0.0765$. The region of resonant librations is restricted to a narrow range of eccentricity about the latter value ($0.070 \lesssim e \lesssim 0.085$), and well above the observed eccentricity of 0.022. The motion for $\sin i \simeq 1$ is thus of type 1 above: small-e circulation about the fixed point at $e_F \simeq 0.0003$. The resulting fractional eccentricity variations are $\sim e_F/e \simeq 1.4\%$, comparable to that seen in the numerical integrations in Fig. I and predicted by Rasio *et al.* (1992).

For larger values of $(\sin i)^{-1} \simeq \beta_5$, $|\nu_c|$ increases and the phase space portrait more closely resembles that of Fig. IIIb or IIIc. Fig. IVb shows the situation for $(\sin i)^{-1} = \beta_5 = 40$, corresponding to $\nu_c = -0.028\, n_2$ and $e_c = 0.0322$. Since $\nu < \nu_c$, there are again two stable fixed points, at $(e \cos \phi)_F \simeq -0.0114$ and $+0.0765$, as well as an unstable point and a separatrix. The domain of resonant motion is now much larger than for $\sin i = 1$, as indicated in the figure. In this simple test-particle model, the nominal elements of the inner planet ($e_1 = 0.022$ and $\phi_1 = 354°$ at JD 244 8040) would place the system just outside the resonant region. Numerical integrations of the full system, however, show that both ϕ_1 and the corresponding argument for the outer planet, $\phi_2 = 2\lambda_1 - 3\lambda_2 + \omega_2$ are in resonant libration for $(\sin i)^{-1} = 40$, as illustrated in Fig. V. The critical value of $(\sin i)^{-1}$ necessary for the system to be in resonance is $\simeq 15$ (Malhotra *et al.* 1992), the exact value depending on the still uncertain eccentricities and pericenters of the two orbits.

Unlike the situation for $(\sin i)^{-1} \simeq 1$, the variations in e and the departures of $\dot{\phi}$ and $\dot{\omega}$ from their average values will become large and nonsinusoidal for $(\sin i)^{-1} \gtrsim 10$, as illustrated in Fig. I. As pointed out by Malhotra *et al.* (1992), this would make the perturbations much easier to detect. However, to observe the system with its orbital plane so nearly face-on seems rather unlikely. Indeed, for random orientations, the probability $P(i < i_o) = \frac{1}{2}(1 - \cos i_o)$, so that $P(\sin i < 0.1) \simeq 0.25\%$.

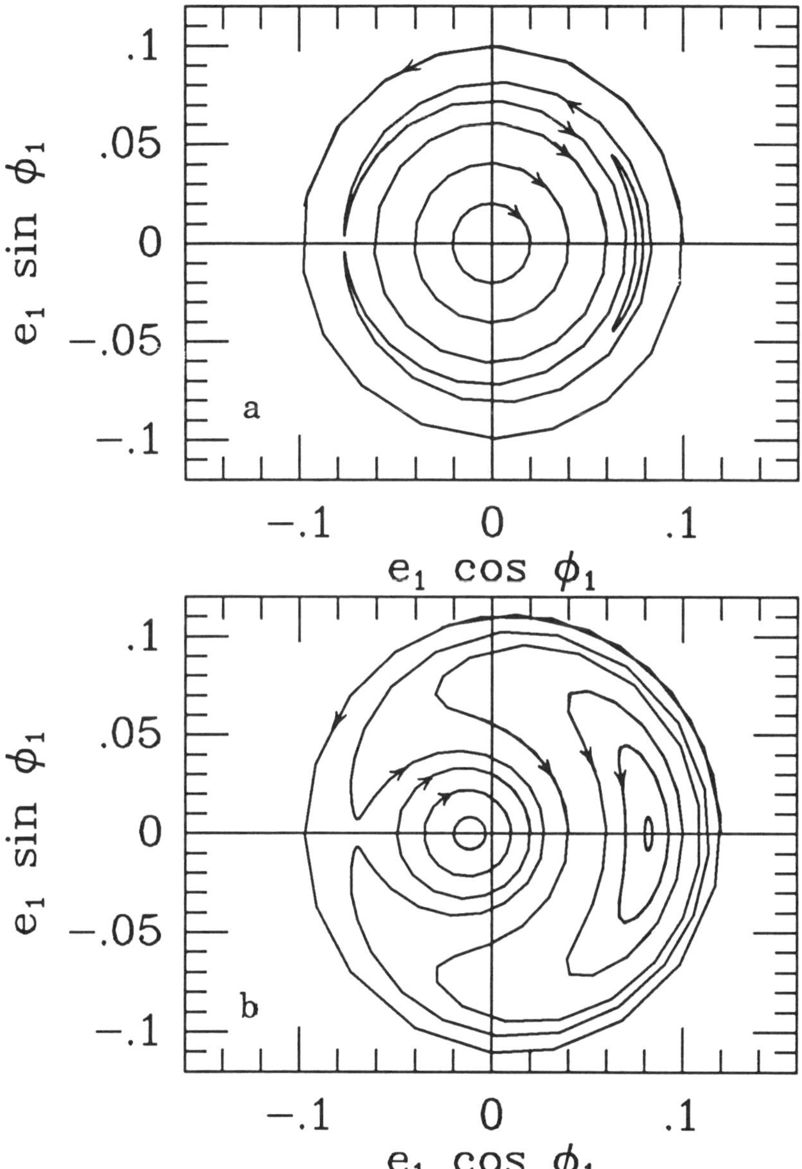

FIGURE IV Phase space trajectories for a test particle model of the inner planet in the PSR1257+12 system, for (a) $\beta = 10^{-5}$ ($\sin i = 1$) and (b) $\beta = 4 \times 10^{-4}$ ($\sin i = 1/40$).

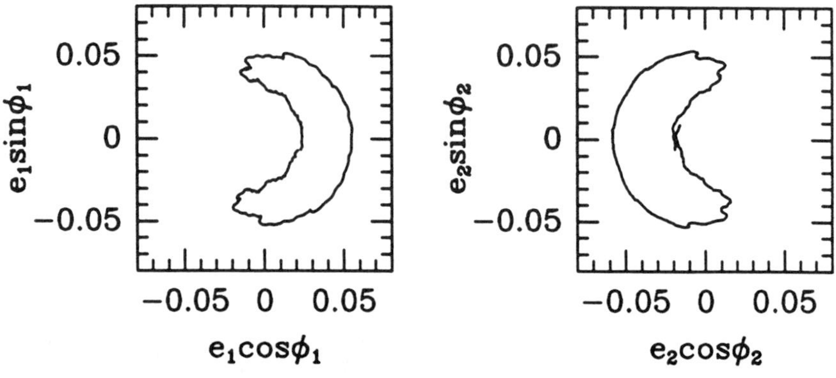

FIGURE V Phase-space trajectories for $(\sin i)^{-1} = 40$ obtained from a numerical integration as shown in he bottom panel of Fig. I. Both resonant variables ϕ_1 and ϕ_2 are clearly librating, with consequent large variations in e_1 and e_2 as seen in Fig. I. Compare the (e_1, ϕ_1) trajectory with the analytical model trajectories in Fig. IVb.

EFFECTS OF INDIVIDUAL CLOSE ENCOUNTERS

Quite independent from the presence of a near-resonance in the system, the fact that $(a_2 - a_1)/a_2 \ll 1$ implies that the two planets can perturb each other quite significantly during close encounters (near conjunctions). These close encounters occur every $2\pi/(n_1 - n_2) \simeq 200$ days in the PSR1257+12 system. Fig. VI shows the result of a numerical integration of the system for $M = 1.4 M_\odot$ and $\sin i = 1$. During close encounters, the instantaneous (osculating) orbital period of the inner planet increases briefly, while that of the outer planet decreases. These changes have amplitudes $|\Delta P_i/P_i| \sim 10^{-4}$, and take place over a timescale of about a month. In contrast, the long-term variations over a period $\simeq 5.5$ ¡yr are caused by the near 3:2 resonance (cf. previous section).

The orbital perturbations corresponding to close encounters are easy to calculate analytically. Here, for simplicity, we present this calculation for the case of two planets in coplanar circular orbits. As in the previous section, we treat the inner planet as a test particle perturbed by its interaction with the outer planet. We assume that a conjunction takes place at $t = 0$. Following Danby (1988), we write the perturbation equation for a_1 as

$$\dot{a}_1 = \frac{2n_1 a_1^3}{GM} B_{12}, \tag{22}$$

where B_{12} is the tangential component of the perturbing force on m_1,

$$B_{12} = \frac{-Gm_2 a_2 \sin(n_1 - n_2)t}{[a_1^2 + a_2^2 - 2a_1 a_2 \cos(n_1 - n_2)t]^{3/2}}. \tag{23}$$

Integrating eqn. (22) from $-t_o$ to t, where $t_o \equiv \pi/(n_1 - n_2)$ is the time of

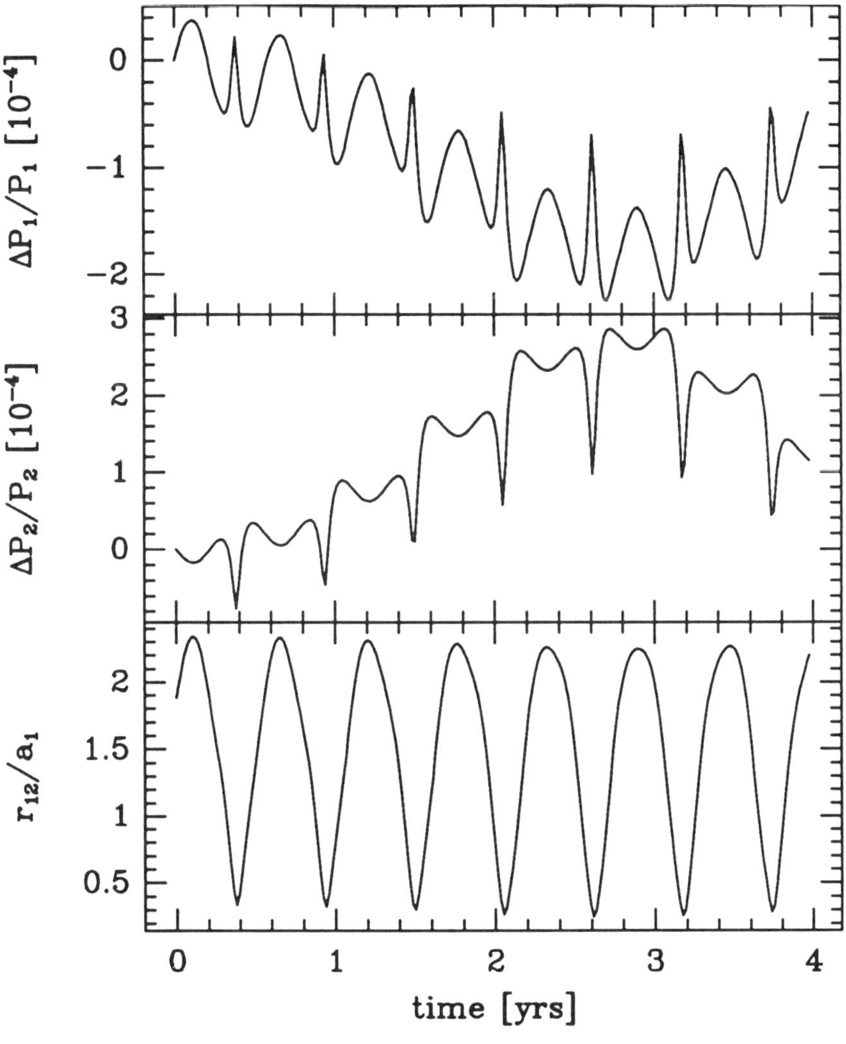

FIGURE VI Effects of close encounters (conjunctions) between the two planets in PSR1257+12. Changes in the osculating orbital periods of the two planets are shown as a function of time. Also shown is the separation r_{12} between the two planets. The large spikes of amplitude $\Delta P_i/P_i \sim 10^{-4}$ and width $\sim 1\,$month occur at conjunctions.

opposition, we find

$$\frac{\Delta a_1(t)}{a_1} = -2\alpha\left(\frac{1}{1+\alpha} - \frac{1}{[1+\alpha^2 - 2\alpha\cos(n_1-n_2)t]^{1/2}}\right)\left(\frac{n_1}{n_1-n_2}\right)\left(\frac{m_2}{M}\right). \tag{24}$$

The maximum amplitude of the change is obtained by setting $t = 0$ in this expression, giving

$$\left|\frac{\Delta a_1}{a_1}\right|_{max} = 4\,\frac{\alpha^2}{1-\alpha^2}\left(\frac{n_1}{n_1-n_2}\right)\left(\frac{m_2}{M}\right), \tag{25}$$

and the FWHM of the peak is

$$\Delta t_{\rm FWHM} = \left(\frac{2}{n_1-n_2}\right)\cos^{-1}\left[\frac{\alpha}{2}(3-\alpha^2)\right]. \tag{26}$$

Note that the semimajor axis $a_1(t)$ returns to its unperturbed value after the encounter ($\Delta a_1(t_o) = 0$ and expression (24) is symmetric in t). The same is true for the osculating orbital period P_1 and mean angular frequency $n_1 = 2\pi/P_1$, since $\dot{n}_1 = -(3/2)(n_1/a_1)\dot{a}_1$. However, the mean longitude of the planet, which is given by $\lambda_1 = const + \int n_1\,dt$ does *not* return to its unperturbed value. The total net shift in mean longitude from one opposition to the next is

$$\Delta\lambda_1 = \int_{-t_o}^{+t_o}\Delta n_1(t)\,dt = -6\alpha\left[2K(\alpha) - \frac{\pi}{1+\alpha}\right]\left(\frac{n_1}{n_1-n_2}\right)^2\left(\frac{m_2}{M}\right) \tag{27}$$

where $K(\alpha) \equiv \int_0^{\pi/2}(1-\alpha^2\sin^2 x)^{-1/2}\,dx$ is the complete elliptic integral of the first kind.

The net shift in mean longitude is negative for the inner planet (and positive for the outer one). When $\alpha \ll 1$ (and $n_2 \ll n_1$), we have $\Delta\bar{n}_1 \equiv \Delta\lambda_1/(2t_o) \simeq -3\alpha^2 n_1(m_2/M)$. This is the $\mathcal{O}(\alpha^2)$ correction to the mean angular velocity of the inner planet corresponding to the *tangential* component of the perturbing force. Note that, in this limit, the leading, $\mathcal{O}(\alpha)$ correction corresponds to the *radial* force of a distant ring of radius $a_2 \gg a_1$ and mass $m_2 \gg m_1$. In the opposite limit where $|\alpha - 1| \ll 1$, i.e., when the two orbits are very close to each other, the tangential component of the perturbing force becomes dominant, causing step-like shifts in the planet's longitudes occurring near conjunctions over a time short compared to the time between conjunctions.

For the PSR1257+12 system we have $\alpha \simeq (P_1/P_2)^{2/3} \simeq 0.77$ (giving $K(\alpha) \simeq 1.95$) and $n_1/(n_1 - n_2) \simeq 3.1$, independent of M and $\sin i$. From eqns. (25) and (26) we get an amplitude $|\Delta P_1/P_1|_{max} = \frac{3}{2}|\Delta a_1/a_1|_{max} \simeq 1.7\times 10^{-4}\,M_{1.4}^{-1/3}(\sin i)^{-1}$ and width $\Delta t_{\rm FWHM} \simeq 25\,$days, in good agreement with the results of our numerical integrations (Fig. VI). In reality, the orbital eccentricities produce slight variations in the amplitude $|\Delta P_i/P_i|_{max}$ from one encounter to the next. The predicted net angular shift $\Delta\lambda_1 = -2.6'\,M_{1.4}^{-1/3}(\sin i)^{-1}$. This corresponds to a shift in the time of zero longitude of $(-\Delta\lambda_1/2\pi)P_1 = +678\,{\rm s}\,M_{1.4}^{-1/3}(\sin i)^{-1}$ after each conjunction. While this is probably too small to be detectable at present (assuming $\sin i \simeq 1$), the effects of a sufficiently large number of these close encounters could become measurable after several years. One approach would be to compare orbital fits to timing data sets obtained during individual ~ 200 day segments between successive conjunctions (see Table I), but excluding data from the ~ 3 week periods around conjunctions. By combining fits to many such data sets, the signal-to-noise ratio might be increased to the point where the long-term average mean motions of the planets can be distinguished from the mean motion between conjunctions.

TABLE I Predicted dates of planet conjunctions.

JD 244 8040.31	=	05/28.81/1990
JD 244 8245.48	=	12/19.98/1990
JD 244 8450.77	=	07/13.27/1991
JD 244 8656.85	=	02/04.35/1992
JD 244 8863.79	=	08/29.29/1992
JD 244 9071.77	=	03/25.27/1993
JD 244 9280.21	=	10/19.71/1993
JD 244 9587.53	=	05/15.03/1994
JD 244 9693.83	=	12/07.33/1994

ACKNOWLEDGMENTS

We thank Doug Hamilton, Renu Malhotra, Scott Tremaine, and Alex Wolszczan for useful discussions. This work was supported in part by NSF Grants AST 91-19475 and AST 89-13112 and NASA Grants NAGW-2364 and NAGW-544.

REFERENCES

Brouwer, D., and Clemence, G. M. 1961, *Methods of Celestial Mechanics* (Academic, New York).

Danby, J. M. A. 1988, *Fundamentals of Celestial Mechanics* (Willmann-Bell, Richmond).

Malhotra, R. 1988, *Ph.D. Thesis*, Cornell University.

Malhotra, R. *et al.* 1992, *Nature*, **356**, 583.

Rasio, F. A. *et al.* 1992, *Nature*, **355**, 325.

Wolszczan, A., and Frail, D. A. 1992, *Nature*, **355**, 145.

Section 4:

Evolutionary Scenarios: Neutron Star and Planet Formation

Planets Around Pulsars
ASP Conference Series, Vol. 36, 1993
J.A. Phillips, J.E. Thorsett, and S.R. Kulkarni

FORMATION AND EVOLUTION OF PULSARS

E.P.J. VAN DEN HEUVEL
Astronomical Institute and Center for High Energy Astrophysics, University of Amsterdam
Kruislaan 403, 1098 SJ Amsterdam, The Netherlands

ABSTRACT Most radio pulsars are newborn single neutron stars with magnetic fields of dipole surface strength $10^{11.5}$ to $10^{13.5}$ G; they spin down and turn off as pulsars on a timescale of $\sim (0.5 - 1)10^7$ yrs. Some 10^9 such pulsars have formed and died since the birth of the Galaxy. The statistical properties of these pulsars are probably consistent with the absence of spontaneous decay of magnetic fields of single neutron stars.

Binary and millisecond pulsars -a few per cent of the total pulsar population- are characterized by much faster spin and much weaker magnetic fields (B $\sim 10^8 - 10^{11.5}$ G) than ordinary single neutron stars. All available evidence indicates that they have been spun-up by accretion ("recycled") in binaries, which in some way has weakened their magnetic fields. Systems like PSR 1913+16 are descendants of high-mass X-ray binaries, whereas systems like PSR 1953+29 descend from wide low-mass X-ray binaries.

Single millisecond pulsars are expected to be -in most cases- the remnants of close low-mass X-ray binaries, in which after a long accretion phase the companion star was disrupted, by tidal forces or "evaporation", leaving a single millisecond pulsar, surrounded by a disk of gas and/or debris.

1. INTRODUCTION: THE MAGNETIC FIELD STRENGTH VS. PERIOD DIAGRAM FOR PULSARS.

Within a year after the discovery of pulsars in 1967 (Hewish et al., 1968) it had become clear that these objects are rapidly rotating strongly magnetized neutron stars, which are formed in a supernova event (Gold, 1968; see the review by Shrinivasan, 1991). The pulsed emission is believed to arise because they emit a highly directional beam of radiation above their magnetic poles, which sweeps around the sky and illuminates a favourably located observer once per rotation period.

The rate of increase of the 0.033-second pulse period of the Crab pulsar $\dot{P} = 38 nsec/day$ was measured within a day after its discovery and showed that this "flywheel in the sky" loses its rotational energy at the huge rate:

$$\frac{dE_{rot}}{dt} = I\Omega\dot{\Omega} \simeq 10^{38} ergs/s (\sim 2.10^4 L_{\odot}) \qquad (1)$$

where I is the moment of inertia of the neutron star ($\sim 10^{45} g.cm^2$ for a standard neutron star with a mass of $\sim 1.4\ M_\odot$ and a radius of 10 km) and Ω its angular velocity of rotation. With this $\dot{P}$ the pulse period of the Crab pulsar doubles on a timescale of $\sim 10^3$ yrs.

It was almost immediately realized that the braking torque on this flywheel is an electromagnetic one (Gold, 1968, 1969). Both the presence of a strong -in first approximation dipolar- magnetic field in a neutron star and its expected rapid rotation had been predicted before pulsars were discovered, by Woltjer (1964) and Pacini (1967), respectively. The latter author had already predicted that neutron stars should be strong sources of magnetic dipole radiation and might be observable through this emission.

The rate of emission of magnetic dipole radiation for the case where the magnetic axis is perpendicular to the rotation axis, and the neutron star is in vacuum is:

$$\frac{dE_{em}}{dt} = \left(\frac{2R^6}{3c^3}\right) B_s^2 \Omega^4 \tag{2}$$

where R is the radius of the neutron star and B_s the strength of the dipole component of the magnetic field at its surface.

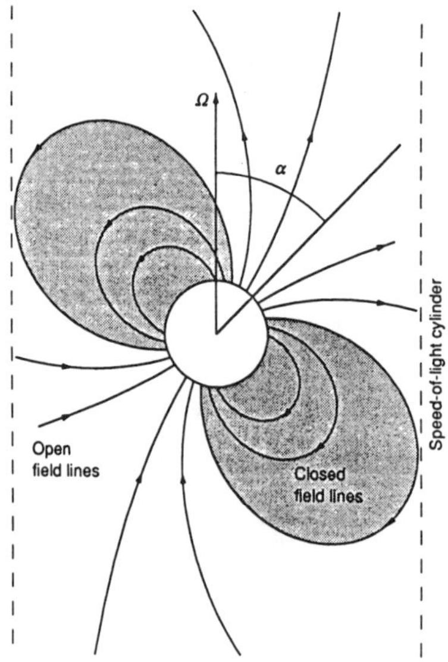

Figure 1. The basic geometry of a radio pulsar -an inclined dipole rotator. α is the angle between the spin axis and the axis of the magnetic dipole. The speed-of-light cylinder corresponds to a distance at which a particle co-rotating with the neutron star will move at the speed of light. Magnetic field lines that are closed within the speed of light cylinder are known as the "closed field lines" (the shaded region) and may contain plasma co-rotating with the star. Ultra-relativistic plasma flows out of the "open" field lines and radiates, generating beamed emission confined to cones around the magnetic poles.

Combination of equations (1) and (2), (using $\Omega = 2\pi/P$) yields

$$B_s = \left(\frac{3c^3 I}{8\pi^2 R^6} P\dot{P}\right)^{1/2} = 3.2 \times 10^{19} (P\dot{P})^{1/2} [G] \qquad (3)$$

where the numerical coefficient results from using the standard values for the neutron star radius $R \simeq 10^6 cm$ and moment of inertia $I \simeq 10^{45} gcm^2$ with P expressed in seconds ($\dot{P}$ is dimensionless).

Equation (3) shows that the surface magnetic field strength of a neutron star can be estimated from the two observed quantities P and $\dot{P}$ (Ostriker and Gunn, 1969; Gunn and Ostriker, 1970). In deriving eq.(3) it was assumed that the magnetic dipole axis is perpendicular to the rotation axis. The spindown torque is in fact proportional to $B_s^2 \sin^2 \alpha$ where α, called "obliqueness", is the angle between the spin and the dipole axis (figure 1). This may suggest that eq. (3) underestimates the field strength when α is small, but this appears not to be the case. The reason for this almost certainly lies in the fact that pulsars also lose a substantial amount of energy in a steady wind of relativistic particles.

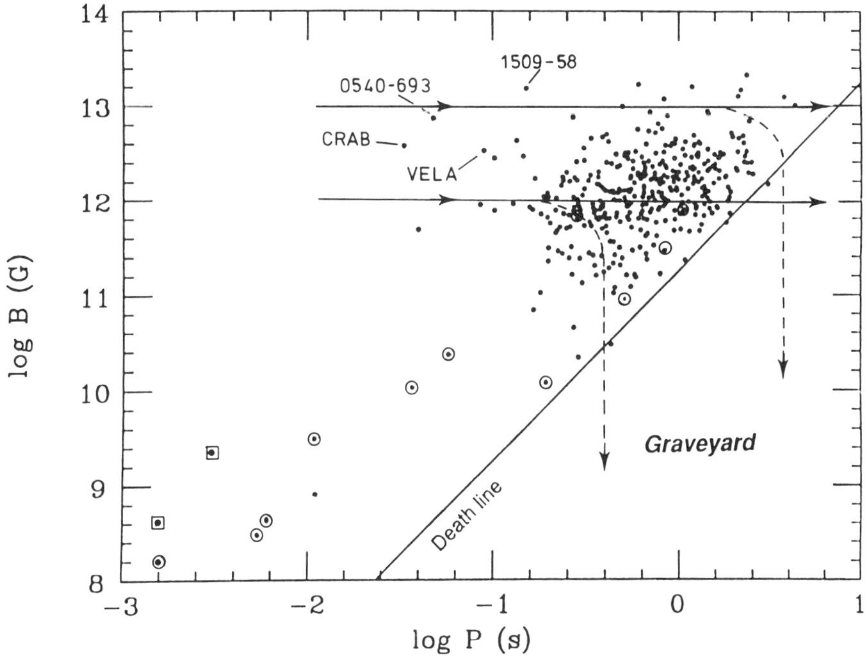

Figure 2. The derived magnetic fields ($\propto \sqrt{P}\,\dot{P}$) of 433 radio pulsars plotted against their spin periods. Pulsars in binaries are shown encircled and single millisecond pulsars are shown as squares. The "deathline" corresponds to a polar cap voltage below which pulsar activity is likely to switch off. Four young pulsars in supernova remnants are indicated. Horizontal lines indicate evolutionary tracks of pulsars without magnetic field decay. The dashed curves show evolutionary tracks with exponential field decay on a 5 Myr timescale.

The magnetospheric current associated with this wind gives rise to a spin-down torque that is non-zero even for an aligned rotator (Goldreich and Julian 1969; Beskin et al. 1984). The relative importance of the torques due to magnetospheric currents and vacuum dipole radiation is very difficult to predict theoretically because it has not yet been possible to obtain a completely self-consistent and stable model for a pulsar magnetosphere. Nevertheless it was shown by Goldreich and Julian (1969) that the spin-down torque on an aligned rotator due to currents can be nearly equal to that on an orthogonal rotator due to vacuum dipole radiation, and this seems to be consistent with recent observations for cases where α is known (for a review and discussion: see Bhattacharya and van den Heuvel, 1991). Therefore equation (3)is expected to yield a good estimate of B_s for any value of the obliqueness α.

The spin-down rates of over 400 pulsars are now known. Figure 2 shows the distribution of their spin periods and derived magnetic fields.

This is one of the most often used diagrams by means of which the evolution of pulsars can be described.

Pulsars in binaries are marked in figure 2 with open circles and single millisecond pulsars with squares. The figure shows that the majority of pulsars are situated in an "island" with spin periods between 0.2 seconds and several seconds and magnetic fields $\sim 10^{11.5}$ to $10^{13.5}$ G.

It is immediately clear from this diagram that the binary and millisecond pulsars behave differently from the majority of the pulsars: most of the binaries and all of the millisecond pulsars fall outside of the "island" of ordinary pulsars: the magnetic field strengths of these pulsars are much lower (mostly $\sim 10^8$-10^{11} G) and their pulse periods much shorter (mostly between 0.001 and 0.2 seconds) than those of ordinary pulsars. The fact that some 50 per cent of the millisecond pulsars are found in binaries while less than 3 per cent of all pulsars in the galactic disk are in binaries suggest strongly that the binary and millisecond pulsars form one group with a common origin.

As we will explain in section 5, the peculiar combination of rapid spin and weak magnetic fields of the binary pulsars is a result of their evolutionary history as a member of a close binary sytem: the mass transfer that took place in these systems has strongly affected the strengths of their magnetic fields and rates of their spins.

The single millisecond pulsars are thought to have followed the same evolutionary path. They somehow have lost their companions, in a number of cases by disrupting them, tidally or otherwise.

The latter fact is of great importance for our understanding of the formation of planets around pulsars, since the only pulsar known so far to have planets is a "single" millisecond pulsar: PSR 1257+12 (Wolszczan and Frail, 1992).

In view of this we will largely concentrate our attention in this paper on the formation and evolution of binary and millisecond radio pulsars.

To this end we briefly describe, in section 2, the general evolution of pulsars through the B_s vs. P diagram of figure 2, and the ways in which neutron stars may have formed.

In section 3 we review our knowledge of binary and millisecond radio pulsars. Section 4 is devoted to the accreting neutron stars in binaries (X-ray binaries) and the ways in which these systems may have formed. We especially consider the mechanisms driving the mass transfer in them, and the spin evolution of their neutron stars.

In section 5 we describe how the various types of X-ray binaries are expected to terminate their evolution and may form binary and millisecond pulsars.

Special attention is given to the ways in which single millisecond pulsars may be formed from binary sytems, in the galactic disk as well as in globular clusters. For a more detailed reference on the subject we refer to the reviews by Taylor and Stinebring (1986), Srinivasan (1991) and Bhattacharya and van den Heuvel (1991), and to the recent symposium proceedings on the subject (Ögelman and van den Heuvel, 1989; Kundt 1990; Ventura and Pines, 1991; van den Heuvel and Rappaport, 1992).

2. BIRTH AND EVOLUTION.

2.1. Formation mechanisms of neutron stars.

Stellar evolution calculations together with the occurence of white dwarfs in galactic clusters (Romanishin and Angel, 1980; Reimers and Koester, 1988; Weidemann, 1990) suggest that single stars with masses up to about 6 to $8M_\odot$ lose their hydrogen-rich envelopes when they are on the Asymptotic Giant Branch, and leave white dwarfs with masses $\lesssim 1.2 - 1.3$ $M_\odot$ as remnants.

Single stars with masses in the range $8M_\odot$ to at least 40 $M_\odot$ (van den Heuvel and Habets, 1984) are expected to leave neutron stars as remnants, after either an electron-capture collapse or an iron-photodisintegration collapse of their degenerate heavy-element cores (cf. Hillebrandt, 1987; Woosley and Weaver, 1986; Woosley et al., 1992). The event associated with this collapse is expected to be observable as a type II supernova, or if the star lost its hydrogen-rich envelope before core collapse, as a supernova of type Ib or Ic (cf. Nomoto, 1991, 1992; Muller et al. 1992).

For binary components there are still other ways in which neutron stars may be formed. Primary stars in binaries in the mass range ~ 4 $M_\odot$ to 8 $M_\odot$ after one or more phases of mass transfer will leave a C-O white dwarf; in the mass range 8 to $\sim 11M_\odot$ they may leave an O-Ne-Mg white dwarf. Such white dwarfs may, during a reverse phase of mass transfer, be triggered to collapse to a neutron star. This type of evolution requires a great deal of fine-tuning of many parameters, such as the white dwarf mass and chemical composition, the accretion rate and the composition of the accreted matter (see e.g. Nomoto, 1992; Canal et al., 1990; Woosley et al., 1992). It will therefore occur only rarely; probably fewer than one out of 10^4 to 10^5 neutron stars was formed in this way (cf. Webbink, 1992). Nevertheless, this mechanism may be one of the main formation channels for the Low Mass X-ray binaries, which are so rare and long lived that they require an extremely rare formation mechanism (Webbink, 1992).

2.2. The evolution of a neutron star through the B_s *vs.* P diagram of radio pulsars and the number of "dead pulsars" in the Galaxy.

The tracks in Figure 2 depict schematically the evolution of a newborn pulsar the cases that the magnetic field of the neutron star does not decay (full lines), and that it decays on a relatively short time scale, of order 5.10^6 yrs (dashed lines).

In the first case (B_s=const) the pulsar moves to the right along a horizontal track, as indicated. In this case the pulse period varies with time as:

$$P = 3.1 \times 10^{-20} B_s \sqrt{2t}, \tag{4}$$

as follows from eq. (3), and the characteristic spindown-age of the pulsar is simply given by:

$$t = (3.2 \times 10^{19})^2 P^2 / 2B_s^2 = P/2\dot{P} \tag{5}$$

In the case that the field decays exponentially on a timescale of ~ 5.10^6 yrs the track begins to bend downward after a few million years, as indicated by the dashed curves. As the braking torque decays rapidly with time in this case, the rotation period finally becomes constant, i.e. the track becomes vertical. To the right of the diagonal line labelled "deathline" in the $B_s - P$ diagram of Figure 2 no pulsars are found. Apparently, after crossing this line, a pulsar stops pulsing. This is understandable because for a certain combination of B and P the electric field generated by the rotating magnetic dipole, becomes too weak to draw charges (electrons) out of the crust of the neutron star.

As the acceleration of these charges along the magnetic field lines is thought to be the source of the beamed radio emission from the pulsar, one expects the pulsar to stop emitting pulsed radiation after it has crossed the deathline. As lines of constant electric field are lines of $B/P^2 = const$, the equation for this line:

$$log B_s = 2 log P + const. \tag{6}$$

The region of the $B_s - P$ diagram to the right of the deathline is called the "graveyard".

A typical ordinary ("garden variety") pulsar such as the Crab-pulsar, with $B_s \sim 3.10^{12}$ G joins the "island" of pulsars within 10^5 years after its birth and reaches the deathline within ~ 10^7 years after its birth.

Most of the "garden variety" pulsars in the island of normal single pulsars in Figure 2 therefore were born between 10^5 and 10^7 years ago as is also indicated by their characteristic spindown ages given by eq.(5).

As the type II supernova rate in a late type spiral galaxy such as our own is about one in 30-50 yrs, one expects the total number of "living" garden variety pulsars in our galaxy to be about 2-3 × 10^5.

This number agrees well with estimates derived from the observed pulsar population within a few kiloparsecs from the sun (cf. Taylor and Stinebring, 1986) if a "beaming factor" of order 3 to 5 is assumed. (The latter means that, due to the limited opening angle of the radiation beams of the pulsar, we can observe on Earth only about 20 to 30 per cent of all pulsars; for the others the radiation beams miss the Earth).

It will be clear from the above that the vast majority of the neutron stars that were formed since the birth of our Galaxy, some 2.10^{10}yrs ago, have long since disappeared into the graveyard and thus have become unobservable. Assuming a SN rate of one per 30 to 50 yr throughout the lifetime of the galaxy the total number of defunct pulsars -in the graveyard- will be $\sim$ (4-7) $\times 10^8$. This number should, however, be corrected upwards to take into account the rapid heavy-element enrichment which the Galaxy underwent in the first 2-5 billion years of its life (cf. Sandage, 1986). In that time the supernova rate must have been very much larger than than at present. Judging from the heavy-element enrichment in globular clusters in that period a correction factor of 2 or more is needed which brings the total number of "dead pulsars" (old neutron stars) in the graveyard to at least $\sim 10^9$.

For comparison, the number of white dwarfs in the galaxy is probably of order 10^{10}, while the total mass in the form of "normal" stars is $\sim (1-2) \times 10^{11}$ $M_\odot$. Hence, neutron stars make up of order one per cent of the mass of the Galaxy -and (very) old neutron stars outnumber active radio pulsars by a factor of order $10^{3.5}$.

2.3. Neutron Star magnetic fields and their evolution.

Magnetic fields of order $10^{11.5}$ to $10^{13.5}$ G must have been generated in all neutron stars within the first few hundred years following their birth. This follows from the fact that if newborn neutron stars also were to have fields weaker than about $10^{11.5}$ G, one would expect to find many ordinary pulsars in the region below $10^{11.5}$ G in the B_s *vs* P diagram of figure 2, because pulsars with such fields move much slower towards the right through the diagram than pulsars with $B_s \sim 10^{12} - 10^{13.5}$ G (see equation 4). Hence, they remain observable tens to hundreds of times longer than the strong-field pulsars. Consequently, even if only a few per cent of pulsars were born with such weak fields, they would still in the B_s *vs.* P diagram outnumber the much shorter lived pulsars with $B_s \simeq 10^{12} - 10^{13.5}$ G.

Since this is clearly not the case the vast majority (99 per cent or more) of the neutron stars must be born with $B_s \geq 10^{11.5}$ G.

With "born" we mean here that fields of this order are present in them within less than 10^3 yrs after their birth -since all three pulsars known in supernova remnants of that age (the Crab-pulsar, PSR 1509-58 and PSR 0540-693, see Figure 2) have fields in the range $10^{12.5}$ to $10^{13.5}$.

How these fields were generated is not known with certainty although some plausible models have been suggested, such as magnetic flux conservation during core collapse or magneto-hydrodynamic field amplification by dynamo action in the differentially rotating hot liquid neutron star during the first few seconds to minutes after its birth, before its crust solidified (Flowers and Ruderman, 1977).

After the crust solidified the electric currents associated with the fields will have anchored the fields in the crust; also a quantized very strong field may exist in the super-conducting core (cf. Pines, 1991). Ohmic dissipation of the crustal currents may cause a gradual decay of the observable surface field but although the timescale for this decay was originally thought to be short ($\sim 5.10^6$ yrs, cf. Gunn & Ostriker, 1970), modern work shows that the time for decay below 10^{12} G is probably very long, $\geq 10^8$- 10^9 years (Chanmugam, 1992). A very different picture for field generation and evolution has been proposed by Applegate et al. (1984) and Blandford et al. (1983). They suggest that pulsar magnetic fields are generated during the first few centuries of life by a thermal heat flux, and do not penetrate the superconducting core of the star. They suggest that the fields can decay to remnant fields of $\sim 10^8$ G in a few million years.

The recent finding of fields as strong as 10^{10} to 10^{12} G in neutron stars as old as 10^9 yrs (PSR 0655+64, Kulkarni, 1986; Hercules X-1, cf. Verbunt et al., 1990) seems to rule out spontaneous magnetic field decay of isolated neutron stars, on timescales much shorter than 10^9 yrs. Also recent statistical analyses of the space distribution and magnetic field distribution of neutron stars -using an improved model for determining pulsar distances- no longer require field decay on timescales shorter than 10^8 years and, in fact, do well without any field decay (Bhattacharya et al., 1992; Wakatsuki et al., 1992).

On the other hand, neutron stars in binaries clearly often show evidence for considerably weakened fields. This, however, can be interpreted in terms of external causes such as accretion of matter from a companion star or a special rotation history (cf. Taam & van den Heuvel, 1986; Shibazaki et al., 1989; Romani, 1990; Srinivasan et al., 1990). See Bhattacharya and van den Heuvel (1991) for a recent review.

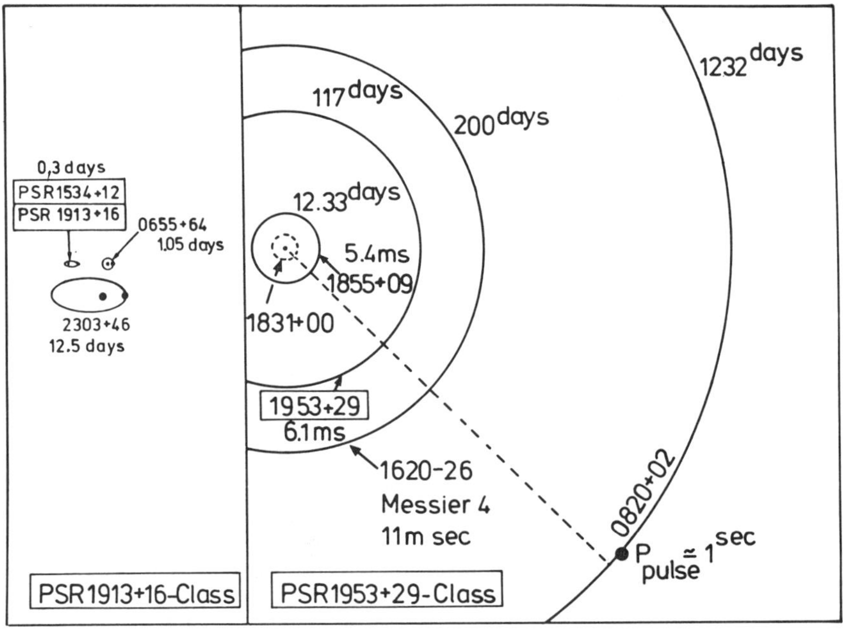

Figure 3. The two classes of binary radio pulsars. Left: The PSR 1913+16-class of systems tend to have narrow and very eccentric orbits; the companion of the pulsar is itself a neutron star or a massive white dwarf. Right: The PSR 1953+29-class systems tend to have wide and circular orbits; here the companion stars always have a low mass, in the range 0.2-0.4 $M_\odot$, and most probably are helium white dwarfs (see text).

3. THE BINARY AND MILLISECOND RADIO PULSARS.

Table I lists the binary and millisecond pulsars known to date. Globular star cluster membership is indicated. The -on average- much faster spin and much weaker magnetic fields of the binary pulsars as compared to the "garden variety" single pulsars were already mentioned in section 1. The number of binary pulsars known to date in the galactic disk is about a dozen. The binary pulsars appear to fall into two groups that differ in a number of characteristics (van den Heuvel and Taam, 1984; van den Heuvel, 1984) as depicted in Figure 3 :

a. those with short orbital periods, of hours to days; relatively large mass functions, and in most cases very eccentric orbits (the "PSR 1913+16-class"), and

b. those in wide, nearly circular orbits with periods of weeks to months, and small mass functions (the "PSR 1953+29-class").

The mass functions of the systems in the first category indicate companion masses in the range 1.0 to 1.5 $M_\odot$, in the second category: in the range 0.2-0.4 $M_\odot$.

For two systems in the first category, PSR 1913+16 and 1534+12, the masses of both components have been very accurately determined from special and general relativistic effects, and indicate that the companions themselves are also most likely neutron stars (see e.g. Wolszczan, 1992). The same is expected to be the case in PSR 2127+11C and 2303+46. On the other hand, in PSR 0655+64 the companion has been optically identified as an about 1 $M_\odot$ white dwarf, with a cooling age > 10^9 yr (Kulkarni, 1986).

In the systems of the second category the companions appear to be low-mass white dwarfs, as follows from the optical identification of PSR 0820+02 (Kulkarni, 1986) as well as evolutionary considerations (see below). Most of the binary pulsars in the second category are millisecond pulsars and their magnetic fields are abnormally weak: $B_s < 10^9$ G.

Figure 2 shows that the positions of these pulsars in the B_s *vs.* P diagram are very similar to those of the isolated millisecond pulsars PSR 1937+21 (which has the extraordinary short period of 1.55 ms) and PSR 1821-24, in the globular cluster M28, which has a period of 3 msec. The extreme stability of the pulse period of PSR 1937+21, which has a spin-down timescale of $\sim 5.10^8$ yrs, has allowed a variety of tests of general and special relativity in the solar system that were previously impossible (Backer and Hellings, 1986).

The table shows that about about half of all known millisecond pulsars (75% of the ones in the field and 45% of those in globular clusters) are in binaries. This is an extraordinarily large percentage since only 2 per cent of the pulsars in the galactic disk are found in binaries (we exclude here the pulsars in the globular clusters since these are thought to all be very old, and underwent a history different from that of the ordinary pulsars (cf. Bhattacharya and van den Heuvel, 1991)).

It thus seems that the millisecond pulsars and the binary pulsars are closely related to one another and have a similar origin.

Table I. Binary Pulsars, Millisecond Pulsars and Pulsars in Globular Clusters (as known on January 31, 1991). (Detailed source references to these pulsars are given in Bhattacharya and van den Heuvel 1991 and Lyne 1992).

PSR	P ms	$\log \dot{P}$	$\log B_s$ (G)	P_{orb} days	e	$f(M)$ $M_\odot$	M_c $M_\odot$	location
0021-72C	5.76			-	-	-	-	GC 47Tuc
0021-72D	5.36			-	-	-	-	GC 47Tuc
0021-72E	3.536			~ 2				GC 47Tuc
0021-72F	2.624			single				GC 47Tuc
0021-72G	4.040			single				GC 47Tuc
0021-72H	3.210			~ (?)				GC 47Tuc
0021-72I	3.485			~ (?)				GC 47Tuc
0021-72J	2.101 or 4.201			0.12				GC 47Tuc
0021-72K	1.786			Binary				GC 47Tuc
0021-72L	4.346			?				GC 47Tuc
0655+64	195.67	-18.17	10.06	1.03	7.5×10^{-6}	0.0712	0.7-0.8	disk
0820+02	864.87	-15.98	11.48	1232.5	0.012	0.003	0.2-0.4	disk
1257 +12	6.22	-	-	-	-	-	-	disk
1310+18	33.16			255				GC M53
1516+02A	5.55			-	-	-	-	GC M5
1516+02B	7.95			6.9				GC M5
1534+12	37.90	-17.61	9.98	0.42	0.2737	0.3146	1.36	disk
1620-26	11.08	-18.09	9.48	191.4	0.025	0.008	0.3-0.6	GC M4
1639+36A	10.38			-	-	-	-	GC M13
1639+36B	30			1.25				GC M13
1744-24A•	11.56	< -19.4	< 8.83	0.075	$< 10^{-3}$	3.2×10^{-4}	> 0.1	GC Ter5
1746-20	288.62			-	-	-	-	GC NGC6440
1802-07	23.10			2.6	0.22	0.0097	> 0.25	GC NGC6539
1820-11	279.83	-14.86	11.79	357.8	0.795	0.068	0.8-1.0	disk
1820-30A	5.44			-	-	-	-	GC NGC6624
1820-30B	378.61			-	-	-	-	GC NGC6624
1821-24	3.05	-17.81	9.34	-	-	-	-	GC M28
1831-00	520.95	-16.84	10.94	1.81	< 0.004	1.2×10^{-4}	0.06-0.13	disk
1855+09	5.36	-19.77	8.48	12.33	2.1×10^{-5}	0.0052	0.2-0.4	disk
1908+00	3.6							GC NGC6760
1913+16	59.03	-17.06	10.35	0.32	0.617	0.1322	1.4	disk
1937+21	1.56	-18.97	8.61	-	-	-	-	disk
1953+29	6.13	-19.53	8.63	117.35	3.3×10^{-4}	0.0027	0.2-0.4	disk
1957+20•	1.61	-19.79	8.21	0.38	$< 2 \times 10^{-5}$	5.2×10^{-6}	~ 0.01	disk
2127+11A	110.66	?[a]		-	-	-	-	GC M15
2127+11B	56.13	-17.05	10.35	-	-	-	-	GC M15
2127+11C	30.53			0.34	0.68	0.15	> 0.9	GC M15
2127+11D	4.65			-	-	-	-	GC M15
2127+11E	4.8			-	-	-	-	GC M15
2303+46	1066.37	-15.24	11.89	12.34	0.658	0.2463	~ 1.5	disk

Notes:
PSR0021-72A and PSR0021-72B are not thought to be located in 47Tuc because of their discordant dispersion measures.

- • Companion in the process of evaporation.

a Observed $\dot{P}$ is negative, $-2 \times 10^{-17} ss^{-1}$. This apparent $\dot{P}$ is almost certainly due to a gravitational acceleration of the pulsar.

4. ACCRETING NEUTRON STARS IN BINARIES.

4.1. The two types of binary X-ray sources.

A second category of neutron stars in binaries is that of the binary X-ray sources. These sources, with a typical X-ray luminosity of $L_x \sim 10^{35} - 10^{38} erg/s$ appear to fall into two categories, the so-called High Mass X-ray Binaries (HMXBs) and Low-Mass X-ray Binaries (LMXBs). Figure 4 depicts this schematically. In an HMXB the companion to the neutron star is a massive early-type star; mostly of spectral type O or early B and with a mass between 8 and 40 $M_\odot$. Most of these sources are regular X-ray pulsars, indicating that the neutron star is rotating and has a strong magnetic field ($B_s \geq 10^{11}$ G).

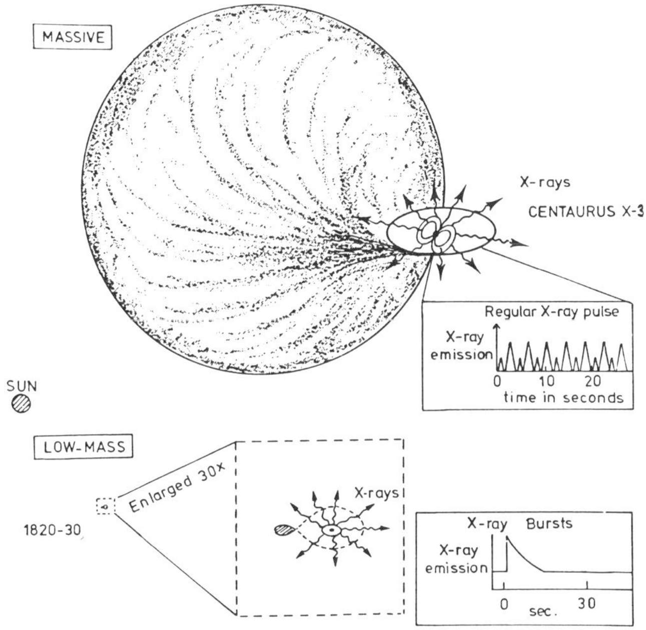

Figure 4. The two basic types of accreting binary neutron stars in the Galaxy. The High Mass X-ray Binaries (top) consist of a massive (8 to 40 $M_\odot$) star together with a neutron star. In almost all sytems the neutron star is a regular X-ray pulsar, indicating that it has a relatively strong surface dipole magnetic field (B $\gtrsim 10^{11}$ G) which channels the inflowing matter towards the magnetic pole. In the Low Mass X-ray Binaries (lower panel) the neutron star has a companion of mass $\lesssim 1.2 M_\odot$. These sources almost never show regular X-ray pulsations, but many sytems show thermonuclear X-ray bursts. Characteristic examples of each type of system are Centaurus X-3 and XB 0820-30, respectively, depicted here to scale with the Sun.

Clearly, since their OB companions cannot live longer than about 10^7 years , the HMXBs are young objects belonging to extreme stellar population I, as is evidenced also by their location in or close to the galactic plane.

On the other hand, in the LMXBs the companion of the neutron star is a low-mass star, $M \lesssim 1.2 M_\odot$; almost none of these X-ray sources shows regular X-ray pulsations, but many of them show thermonuclear X-ray bursts, which implies that they cannot be strongly magnetized, as a surface magnetic field of a neutron star stronger than $\sim 10^{11}$ G is expected to suppress thermonuclear bursts (cf. Lewin and Joss, 1983).

The concentration of LMXBs in the galactic bulge and their rather wide spread around the galactic plane, indicate that they belong to a rather old stellar population with a typical age $\sim 10^{10}$ years. This is confirmed by the presence of a dozen LMXBs in globular star clusters, the oldest objects in our galaxy.

4.2 The abnormally high incidence of X-ray binaries in globular clusters.

The incidence of strong binary X-ray sources (per unit mass) in globular clusters is about 10^3 times higher than in the Galaxy at large (Gursky, 1973). With a total mass of $\sim 2.10^7$ $M_\odot$ the ~ 120 globular clusters in the Galaxy contain about 10 per cent of all strong low-mass binary X-ray sources of a Galaxy that has a total mass of $\sim 2.10^{11}$ $M_\odot$. The same is true in M31 (cf. Long and van Speybroeck, 1983).

This abnormally high incidence of LMXBs in globular clusters is due to a special formation process that can take place only in a globular cluster, but not in the galactic disk: binary formation by close encounters -either by tidal capture or by 3-body interactions (see, e.g. Verbunt, 1990; Bhattacharya and van den Heuvel,1991; Hut, 1992). The reason why globular clusters are favourable places for this process is their combination of very high star densities (up to 10^5 pc^{-3}) with low relative velocities ($\sim$ a few tens of km/s). The discovery of many LMXBs in globular clusters was the first evidence of the presence of a population of very old neutron stars in the Galaxy. These neutron stars must date from some 2.10^{10} yrs ago, from the first few tens of millions of years of the life of the globular clusters, when massive stars were still present in them.

4.3 Observed spin-evolution of neutron stars with accretion disks in X-ray binaries; equilibrium spin.

In view of the orbital motion of the binary, matter that flows from the normal star to the neutron star will have angular momentum. As a result of this the inflowing matter forms an accretion disk around the neutron star and spirals inwards through this disk. The accretion of matter with angular momentum, causes the rotation period of the accreting neutron star to gradually shorten which is indeed clearly observed in all X-ray binaries with an accretion disk.

Figure 5 (after Nagase, 1989) shows the observed "spin-up" graphs of a number of characteristic pulsating binary X-ray sources. One observes that the spin-up timescales in HMXBs such as SMC X-1 and Cen X-3 are very short, of order of only a few thousand years. For the low-mass systems X1626-67 and Her X-1 the timescales range from $\sim 10^4$ to 3.10^5 years. The shortest spin period which a magnetized neutron star can reach for a given surface dipole field strength B_9 (in units of 10^9 Gauss) and accretion rate $\dot{M}$ is the so called equilibrium spin period

$$P_{eq} = (2.4ms)B_9^{6/7}R_6^{16/7}M^{-5/7}(\dot{M}/\dot{M}_{Edd})^{-3/7} \tag{7}$$

(cf. van den Heuvel, 1977, 1984). R_6 is the radius in 10^6 cm, M the mass in solar units, $\dot{M}_{Edd}$ the Eddington limit accretion rate of 2.10^{-8} $M_\odot$/yr. P_{eq} is the period at which the matter at the Alfvén radius (the radius where the inflowing matter couples to the magnetic field lines) co-rotates with the neutron star with just the Kepplerian velocity. Were the neutron star to rotate faster, the matter would be swung out of the magnetosphere by centrifugal forces and accretion would be impossible. So, for a given value of $\dot{M}$, P_{eq} is the shortest possible rotation period for an accreting neutron star.

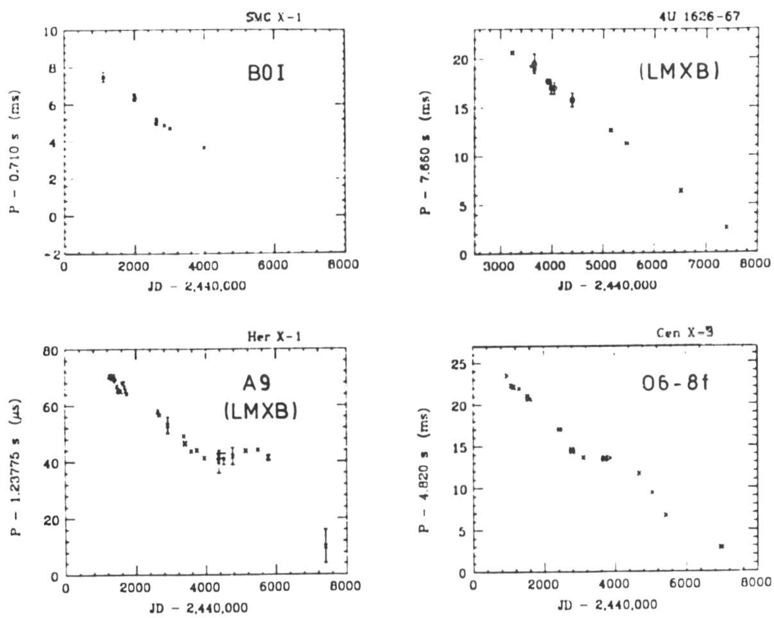

Figure 5. Pulse-period histories of four disk-fed accreting binary X-ray pulsars, after Nagase (1989). The pulsars have relatively short pulse periods (~ 1 s) and show a more or less monotonic decease in pulse period on relatively short timescales, ranging from 10^3 to 10^5 years, so-called "spin-up".

With $M = M_\odot, \dot{M} = \dot{M}_{Edd}, R_6 = 1$ one has:

$$P_{eq,max} = (2.4ms)B_9^{6/7} \tag{8}$$

which is depicted as the "spin-up" line in Figure 6. This line represents the absolutely shortest possible rotation period that an accreting magnetized neutron star can reach for a given field strength B_9, since $\dot{M}_{Edd}$ is the maximum possible accretion rate (at still higher accretion rates X-ray luminosity becomes so high that the excess inflowing matter is blown away by the radiation pressure).

5. SPIN EVOLUTION OF A NEUTRON STAR IN A BINARY; RECYCLING OF PULSARS.

When a single neutron star has crossed the deathline and entered the graveyard, its life as an active pulsar has ended forever.

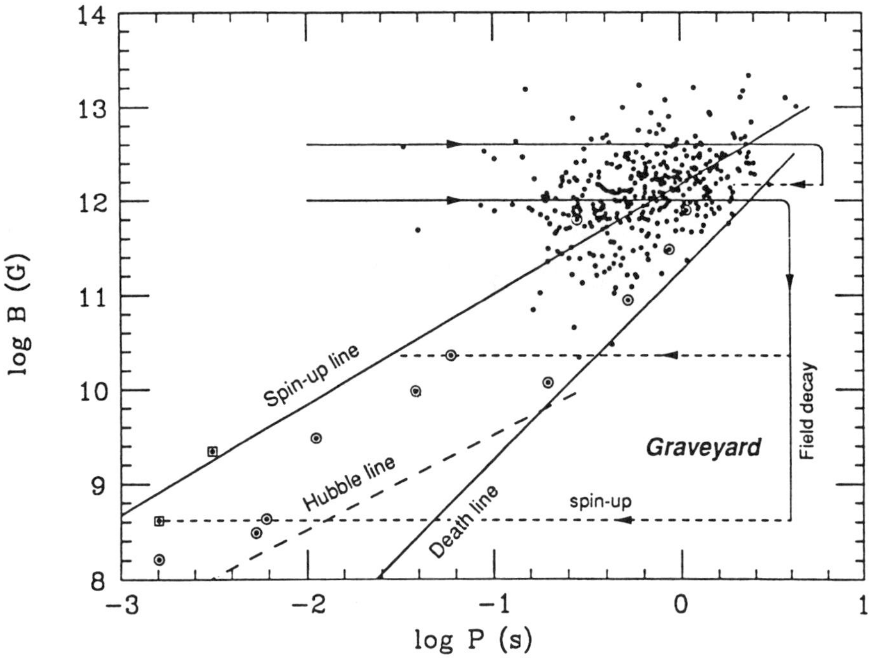

Figure 6. Same diagram as Figure 1 with possible evolutionary tracks of pulsars in binaries indicated. Pulsars are born in the upper left part of the diagram and -if no field decay occurs- move towards the right along horizontal tracks (fully drawn). In the graveyard the field of a single pulsar probably does not decay (see text), but that of a binary neutron star does decay, presumably due to external circumstances (accretion, spin down). Furthermore, the accretion of matter with angular momentum causes the neutron star to be "spun up" towards the left in the diagram along the dashed "spin-up tracks". The shortest period to which a neutron star can be spun up is indicated by the Spin-up line. After the companion has itself become a compact star (neutron star or white dwarf) or has disappeared, the spun-up neutron star becomes observable as a radio pulsar. It will then slowly spin-down, i.e. move towards the right again in the diagram.

Not so for a neutron star that is born as a companion to an ordinary star in a binary system. Also this neutron star will at first spin down and disappear into the graveyard, as depicted in figure 6. However, in this case, once the companion star begins to overflow its Roche lobe and starts tranferring matter, the system will turn into an X-ray binary. As soon as this happens, the accretion of matter with angular momentum (through an accretion disk, which forms in the case of

Roche-overflow) will cause the spin of the neutron star to be accelerated (see Figure 5), causing the neutron star to move back to shorter pulse periods in the B_s *vs.* P diagram, as depicted schematically in Figure 6. It is assumed in figure 6 that during its stay in the graveyard the neutron star's magnetic field has decayed. This decay is probably not spontaneous, but rather due to the interaction of the neutron star with the companion, in the form of accretion or otherwise (see section 2).

Two spin-up tracks are indicated in the figure: one for almost no field decay and one for field decay by a factor of order 10^3.

The decay effects are expected to be smallest if the companion is a massive star, i.e. in the case of an HMXB, since here the neutron star will stay in the graveyard for less than a few times 10^7 yrs (in some cases it may even not yet have reached the graveyard when the spin-up begins). As here the X-ray phase in most cases does not last longer than 10^4 to 10^5 years (cf. Savonije, 1983) the neutron star cannot have accreted here more than $\sim 2.10^{-3}$ $M_\odot$.

On the other hand, field decay may be much larger in the case of LMXBs, where the neutron stars in most cases have remained in the graveyard for billions of years before the accretion started. Since here the duration of the accretion phase is some 10^8 yrs or longer (Savonije, 1983a) the amount of matter accreted in an LMXB may be as much as 0.7 or 0.8 $M_\odot$, which may have caused the surface magnetic field to have decayed by a larger factor, of order 10^3 or so (Shibazaki et al., 1989; Romani, 1990).

Thus the upper spin-up track in figure 6 may be representative for a neutron star in a HMXB and the lower one for an LMXB.

That indeed neutron stars in LMXBs may have rotation periods in the millisecond range and field strengths in the order of 10^9 G is strongly suggested by the occurrence of intensity-dependent Quasi Periodic oscillations in the X-ray flux from these systems (van der Klis, 1987; Alpar and Shaham, 1985).

The spin of the neutron star in an accreting binary cannot be accelerated to periods that are to the left of the spin-up line in the B_s *vs.* P diagram (Alpar et al., 1982; Srinivasan and van den Heuvel, 1982).

The spun-up neutron star in a binary can in general become observable as a pulsar only after its companion has terminated its evolution and has itself become a compact star, i.e. a neutron star or a white dwarf. This is because as long as the companion is an ordinary star, its circumstellar matter (corona, stellar wind, etc.) is expected (in most cases) to disperse the pulsar signal beyond recognition. So, only when the system has been emptied of gas, i.e. when the companion has become a compact star, will it be possible to observe the spun-up neutron star as a radio pulsar.

Such a spun-up pulsar that has experienced an accretion history in a binary sytem is called "recycled pulsar" (cf. Radhakrishnan, 1986).

5.1. Predictions about recycled pulsars.

From the above one predicts that pulsars that have been recycled in binaries will have the following characteristics:

- in the B_s *vs.* P diagram they must be located between the spin-up line and the deathline;

- if they have retained a binary companion, this companion must itself be a compact star; i.e.: a white dwarf or a neutron star.

As figures 3 and 6 show both these predictions have been confirmed by the observations: all binary pulsars are located in the predicted region of the B_s *vs.* P diagram and in all binary pulsar systems where the companions have been identified they appear to be neutron stars or degenerate dwarfs.

A third prediction of the recycling model is that recycled pulsars will be most abundant where X-ray binaries are most abundant, i.e.: in globular star clusters -as these have, per unit mass, over 10^3 times more LMXBs than the Galaxy as a whole (see section 4). Since the LMXBs in globular clusters are almost all bursters, the magnetic fields of their neutron stars are expected to be weak (see section 2), as also follows from the absence of regular X-ray pulsations in them. Therefore, their neutron stars may in many cases have been spun up to millisecond periods and their remnants may therefore well be millisecond radio pulsars. Motivated by this idea Helfand started a VLA search for radio pulsars in globular clusters which turned out extremely successful. The first radio source which he detected was soon found by Lyne et al.(1987) to be a 3 millisecond radio pulsar, PSR 1821-24, in the globular cluster M28 (see table I). Since that time more than two dozen globular cluster pulsars have been discovered, most of them millisecond pulsars (see table I). The record is held by in globular cluster 47 Tucanae in which 11 millisecond pulsars were discovered (Manchester et al. 1991; Lyne, 1992).

The prediction and subsequent discovery of the globular cluster pulsars is one of the most beautiful confirmations of the recycling model.

5.2. Origin of the two types of binary pulsars.

From the above it is clear that the binary pulsars represent a later evolutionary stage of the X-ray binaries.

The presence of the two categories (a) and (b) of binary pulsars depicted in figure 3 and described in section 3 can be understood as a result of the evolution of binary systems in which the neutron star was orbited by a companion that was consiberably more massive (a) or less massive (b) than itself (van den Heuvel, 1984; van den Heuvel and Taam, 1984).

It is well known that as a result of conservation of mass and orbital angular momentum during mass transfer, the binary orbit will shrink if mass is transfered from the more massive to the less massive component (as is the case in HMXBs) and widen in the opposite case (cf. Paczynski, 1971).

A massive star with a neutron star companion (HMXB) will therefore undergo runaway mass transfer when it evolves, forming a common envelope in which the neutron star rapidly spirals down, as the result of the large frictional drag. In this process the orbit will rapidly circularize and the hydrogen-rich outer layers will be ejected, leaving behind a very close system consisting of the neutron star and the heavy-element core of the massive star; this core will finally undergo core collapse and and a supernova explosion, processes that (if the system is not disrupted) will leave behind a second neutron star in a highly elliptical orbit. This is the way in which PSR 1913+16, 1534+12 and 2303+46 have probably formed (Flannery and van den Heuvel, 1975; Smarr and Blandford, 1976; Srinivasan and van den Heuvel, 1982).

Companions in the mass range 3-8 $M_\odot$ do not explode but leave behind relatively massive (0.7-1.0 $M_\odot$) CO white dwarfs in a close circular orbit, as probably has occured in PSR 0655+64 (van den Heuvel and Taam, 1984).

This evolution of massive systems has been depicted in figure 7. A requirement for this type of evolution, in which a close binary pulsar (or two runaway pulsars) remains is probably that the initial orbit must have been fairly wide, $\gtrsim 100^d$ (cf. Taam et al., 1978), since otherwise the neutron star will have spiralled down completely into the center of the massive star and no binary remains (cf. Bhattacharya and van den Heuvel, 1991).

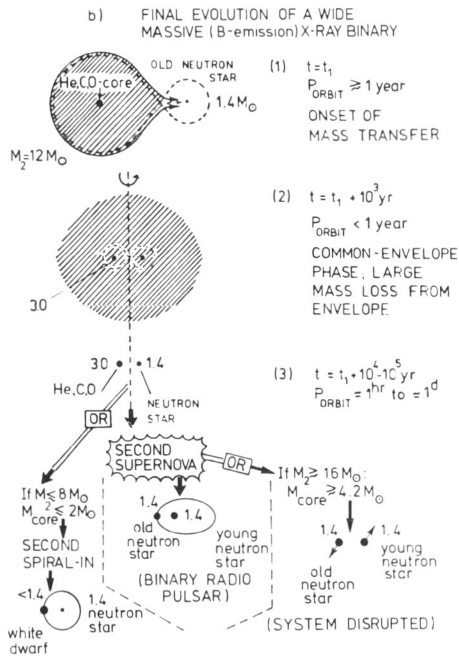

Figure 7. The various possibilities for the final evolution of a High-Mass X-ray Binary. In all cases the onset of Roche-lobe overflow leads to the formation of a common envelope and the occurrence of spiral-in. In systems with orbital periods $\lesssim$ about 1 year there is most probably not enough energy available in the orbit to eject the common envelope, and the neutron star spirals down into the core of its companion. In systems with orbital periods $\gtrsim$ about a year the common enevelope is ejected during spiral-in, and a close binary can be left, consisting of the neutron star and the core (consisting of helium and heavier elements) of the companion. Companions initially more massive than 8-12 $M_\odot$ leave cores that will explode as a supernova, leaving an eccentric-orbit binary pulsar or two runaway pulsars. Systems with companions less massive than ≈ 8-12 $M_\odot$ leave close binaries with a circular orbit and a massive white dwarf companion, similar to PSR 0655+64.

If the secondary is less massive than the neutron star, it will transfer mass to the neutron star in a self-stabilizing manner, as the mass transfer makes the orbit widen. In this case, if the orbital period is $\gtrsim 12^h$, the mass transfer will, like in the case of a massive companion, be driven by the envelope expansion resulting from the interior nuclear evolution of the star. (Losses of orbital angular momentum by gravitational radiation and other mechanisms are negligible in this case, contrary to the case of closer systems, see below).

Well known LMXBs with such nuclearly driven mass transfer are Sco X-1 ($P= 19.2^h$), Cyg X-2 (P=9.8 days) and 4U0921-63 (P=9.0 days). The low-mass companions in these cases are (sub)giants which have already terminated hydrogen-burning in the center and have a small degenerate helium core. Energy is generated in these stars by hydrogen burning in a shell around this core, which causes the core mass to gradually increase while at the same time the outer radius of the star gradually expands. This drives the mass loss, which causes the orbit to expand gradually.

Mass transfer will continue until the entire hydrogen-rich envelope of the companion has been depleted, leaving a low-mass (0.2 to 0.4 $M_\odot$) helium white dwarf in a wide and circular orbit around the neutron star, as observed for the PSR 1953+29-class of binary pulsars (see figure 3). This scenario for PSR 1953+29 was suggested immediately after its discovery (Paczynski, 1983; Savonije, 1983b; Joss & Rappaport, 1983). Figure 8 depicts the above-described evolution of a wide LMXB-system that started out with $M_1 = M_2 = M_\odot$ and $P_o = 12.5^d$ (Joss & Rappaport, 1983). Its final orbital period of 117^d is reached after about 8.10^7 yrs when the companion has become an 0.31 $M_\odot$ helium white dwarf. The resulting system shows a striking resemblence to PSR 1953+29.

In the much wider system that produced PSR 0820+02 the mass transfer took only 4.10^6 years.

As the accretion rate is limited by the Eddington limit to 2.10^{-8} $M_\odot$/yr, PSR 0820+02 cannot have accreted more than 0.08 $M_\odot$ (and probably much less, cf. van den Heuvel, 1984), while PSR 1953+29 may have accreted as much as 0.7 $M_\odot$.

This difference may perhaps explain the disparity of the rotation periods and the magnetic fields in these neutron stars. In fact these systems are very good candidates for neutron star formation by the accretion-induced collapse (AIC) of a white dwarf (van den Heuvel, 1984), as the accretion rates were in the suitable range 10^{-6} to 10^{-8} $M_\odot$/yr (cf. Nomoto, 1991, 1992), and also their magnetic field strengths and spins are consistent with such an explanation (van der Heuvel, 1984).

5.3. Origins of isolated millisecond pulsars.

Although one can never completely discard the possibility that isolated millisecond pulsars were just born with weak magnetic fields and ultra-rapid spin, by the usual process of stellar collapse, their weak fields and rapid spin would still make them problematic in that case. Because, then immediately the questions arise of:

I. Why is there a huge gap in field strength between them and the "island" of ordinary single pulsars with $B_s > 10^{11.5}$ G?

II. Why are single millisecond pulsars so similar to the millisecond pulsars in binaries, which make up about half of all millisecond pulsars? Of the latter ones we know from the above that they have been recycled in binaries.

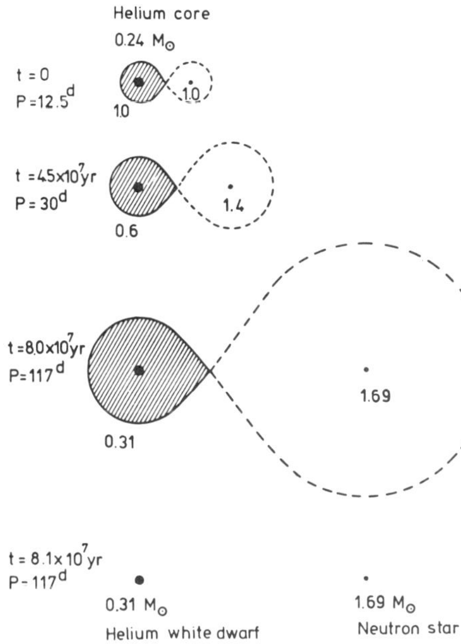

Figure 8. Evolution of a wide low-mass X-ray binary as such as Cygnus X-2 into a wide radio-pulsar binary with a circular orbit and a low mass helium white dwarf companion, like as PSR 1953+29. At the onset of the mass transfer, the low-mass companion is a (sub)giant with a degenerate helium core of 0.24 $M_{\odot}$. Its light is generated by hydrogen fusion in a shell around the core. The mass-transfer from the giant to the neutron star is due to the slow expansion of the giant, driven by the hydrogen-shell burning. During the mass transfer the orbit gradually expands (due to angular momentum conservation) and after 8.10^7 year the system terminates as a wide radio-pulsar binary (after Joss and Rappaport 1983).

It is because of these facts that from the time of the discovery of the first millisecond pulsar, PSR 1937+21, binary scenarios have been put forward to explain their existence (Alpar et al., 1982).

A very strong argument in favour of the binary scenario is that it has a very strong predictive value, as was demonstrated by the case of the discovery of the many millisecond pulsars in globular clusters. The fact that more than half of the millisecond pulsars in globular clusters are single demonstrates that binary recycling can, apparently, also produce single millisecond pulsars. In globular clusters the latter is not so difficult to achieve, as binaries may be disrupted by encounters with another star.

For a variety of reasons origins from an HMXB seems rather unlikely for the millisecond pulsars. For example, the amount of accretion required to spin-up a neutron star to $\lesssim 3ms$ is $\gtrsim 0.1 M_\odot$, which -due to the accretion lifetime $\lesssim 10^6$ yrs- can never be reached in an HMXB.

The two remaining possibilities are :

a. Origin from an intermediate-mass binary which resulted in a close binary (P $\lesssim 15^h$) consisting of a neutron star plus a massive white dwarf, similar to the system of PSR 0655+64,

b. Origin from a close LMXB.

As to the first model, if PSR 0655+64 had been born with P $< 15^H$ it would merge by gravitational radiation losses within a Hubble time. As shown by Bonsema and van den Heuvel (1985) the mass transfer which ensues when the white dwarf overflows its Roche-lobe is unstable if $M_{wd} \gtrsim 0.65 M_\odot$, and the white dwarf will be tidally disrupted and form a (massive) disk around the neutron star. Subsequent accretion of only 0.05 $M_\odot$ from the disk will spin up the rotation of the neutron star to P $<$ 10 ms.

Since there is enough mass and angular momentum present in the disk to form a few Earth-like planets in orbits of $\sim$ 0.5 AU, this model seems attractive for explaining the formation of planets around a millisecond pulsar such as PSR 1257+12 (van den Heuvel, 1992a).

5.4. Origin from a close LMXB.

The most abundant class of LMXBs is that of systems with orbital periods $<$ 12^h, and of such systems no binary remnants are known in the Galaxy. The occurrence of quasi-periodic oscillations in the X-ray fluxes of a number of them (van der Klis, 1987) indicates that they are likely to be rapidly spinning (P $<$ 10 ms), weakly magnetized ($B_s \lesssim 10^9$ G) objects (Alpar and Shaham, 1985).

It seems therefore attractive to identify the majority of the isolated millisecond pulsars as remnants of close LMXBs in which in some way the companion was destroyed. The mass transfer in close LMXBs (P $< 12^h$) is driven by angular momentum losses by the emission of gravitational waves, and "magnetic braking" (see e.g. Verbunt, 1990; van den Heuvel, 1991, 1992b; Bhattacharya and van den Heuvel, 1991). This makes their orbit shrink continously, while the mass of the companion becomes smaller and smaller.

It has been made plausible by Ruderman and Shaham (1983) that once the mass of the companion falls below $\sim$ 0.02 $M_\odot$ it may become tidally disrupted on a short timescale, upon which a ring of asteroid-sized objects or a gas disk will form. Although there has been some discussion as to whether such disruption is likely to occur (Jeffrey, 1986; Hut and Paczynski, 1983; Bonsema and van den Heuvel, 1985) recent considerations, taking into account mass loss through the L_2 point (Kluzniak, 1991, 1992) have revived this possibility and made it quite plausible. Alternatively, the X- and Gamma-ray flux from the accreting neutron star may heat the low-mass companion so much that an evaporative wind is driven from its heated side (Ruderman et al., 1989a, b; Kluzniak et al., 1988), causing the star to lose mass at such a rate that it is driven out of thermal equilibrium, becomes more and more bloated and finally enters runaway mass transfer which causes it to be completely disrupted. It is also possible in the "evaporation" case, that the mass transfer is temporarily interrupted for some reason. In that case the X-ray source will be turned off and the neutron star will now appear as a strong millisecond radio pulsar (cf. van den Heuvel and van Paradijs, 1988). The relativistic pulsar wind will continuously heat and ablate the outer layers of the companion, which at that time has a mass of only 0.3 $M_\odot$ or less, causing it to be "evaporated". The timescale for this kind evaporation is expected to be $\sim 10^6$-10^7 yrs (cf. Shaham, 1992).

The eclipsing binary millisecond radio pulsar PSR 1957+21 (Fruchter et al., 1988; Fruchter and Goss, 1991) has been interpreted as a system with an evaporating companion. The shape of its radio eclipses shows that the companion has comet-like tail of outflowing matter, directed away from the pulsar. As initially it was thought that the companion star is deep inside its Roche lobe, it was thought that evaporation is the only possible explanation for the existence of its tail. However, with its revised distance (Fruchter and Goss, 1991) the companion may well be filling its Roche lobe, and mass ejection through the L_2 point may be an alternative explanation for the presence of the tail. Thus, it is not clear at present how strong the the evidence for evaporation really is.

Be this how it may, it appears clear from the PSR 1957+21 system that once a companion in a close LMXB reaches a mass $\lesssim 0.1 M_\odot$ it is disrupted in one way or another. As the outflow velocities are rather low (Tavani and Brookshaw, 1992ab) it seems likely that the resulting matter will form a disk around the system and finally, around the single millisecond pulsar. Tavani and Brookshaw (1992ab) have suggested that a few Earth-like planets may form from such a disk.

6. SUMMARY AND CONCLUSIONS.

- Pulsars are born in supernova events and at birth have a strong magnetic field ($B_s \simeq 10^{11.5}$ -$10^{13.5}$ G) and rapid spin. Single pulsars spin down rapidly and disappear into the graveyard in $\lesssim 10^7$ yrs. There are some 10^9 "dead" single pulsars in the Galaxy.

- There is no strong evidence that magnetic fields of isolated pulsars do decay spontaneously.

- Binary pulsars tend to have much faster spin and much weaker magnetic fields than ordinary (single) pulsars. Their companions (almost) always are dead stars themselves (neutron stars or white dwarfs), which indicates that the binary pulsars underwent accretion and spin-up during an X-ray binary phase. This is likely to be the cause of their weakened surface magnetic fields.

- The binary pulsars fall into two broad categories:

  1. The PSR 1913+16-class: systems with narrow and , in most cases, very eccentric orbits; these systems consist of two neutron stars and are descendants of HMXBs; in those cases that the narrow orbit is circular (e.g. PSR 0655+64), the companion is a massive white dwarf and descended from an intermediate mass ($\sim$ 3-10$M_\odot$) star.

2. The PSR 1953+29-class: systems with relatively wide and circular orbits; here the companion is a low-mass (0.2-0.4 $M_\odot$) helium white dwarf and the recycled neutron star often is a millisecond pulsar. These systems are descendants of relatively wide ($P \gtrsim 12^h$) low-mass X-ray binaries.

- About 50 per cent of all millisecond pulsars are single. Most of these are probably the remnants of close low-mass X-ray binaries ($P < 12^h$). In such systems the companions remain unevolved and are finally disrupted, either tidally or by a radiation-driven mass-transfer instability ("evaporation"). The remnants of the companion are expected to form a hydrogen-rich disk around the millisecond pulsar with a mass of a few times 10^{-2} $M_\odot$.

  A certain fraction of the single millisecond pulsars is expected to be formed by mergers of close binaries ($P < 15^h$) consisting of a recycled neutron star and a massive white dwarf (like PSR 0655+64). Such mergers produce a millisecond pulsar surrounded by an about 1 $M_\odot$ disk consisting of helium and heavier elements.

  The single millisecond pulsars with disks formed from either of these evolutionary paths seems attractive progenitors for millisecond pulsars with planetary systems, such as PSR 1957+12.

ACKNOWLEDGMENTS

This research was supported in part by the Lunar and Planetary Institute, Houston, Texas.

REFERENCES

Alpar, M.A. and Shaham, J. 1985, *Nature* **316**, 239

Alpar, M.A., Cheng, A.F., Shaham, J. and Ruderman, M. 1982, *Nature* **300**, 728

Applegate, J.H., Blandford, R.D. and Hernquist, L. 1984, *Greenbank Workshop on Millisecond Pulsars* (eds. S.P. Reynolds and D. Stinebring), Greenbank W. Va., p. 205

Backer, D.C. and Hellings, R.W. 1986, *Ann. Rev. Astron. Astrophys.* **24**, 537

Beskin, V.S. Gurevich, A.B. and Istomin, Ya. No 1984, *Astrophys. Sp. Sci.* **102**, 301

Bhattacharya, D. and van den Heuvel, E.P.J. 1991, *Phys. Reports* **203**, 1-124

Bhattacharya, D., Wijers, R.A.M.J., Hartman, J.W. and Verbunt, F. 1992, *Astron. Astrophys.*, **254**, 198-212

Blandford, R.D., Applegate, J.H. and Hernquist, L. 1983, *M.N.R.A.S.* **204**, 1025

Bisnovatyi-Kogan, G.S. and Komberg, B.V. 1975, *Soviet Astron.* **18**, 217

Bonsema, P.F.J. and van den Heuvel, E.P.J. 1985, *Astron. Astrophys.* **146**, L3

Canal,R., Isern, J. and Labay, J. 1990, *Ann. Rev. Astron. Astrophys.* **28**, 183

Chanmugam, G. 1992, *Ann. Rev. Astron. Astrophys.* **30**, (in the press)

Flannery, B.P. and van den Heuvel, E.P.J. 1975, *Astron.Astrophys.* **39**, 61

Flowers, E. and Ruderman, M.A. 1977, *Ap.J.* **215**, 302

Fruchter, A.S, Stinebring, D.R. and Taylor, J.H. 1988, *Nature* **333**, 237

Fruchter, A.S. and Goss, W.M. 1991, *Ap.J.* **384**, L37

Gold, T. 1968, *Nature* **218**, 731

Gold, T. 1969, *Nature* **221**, 25

Goldreich, P. and Julian, W.H. 1969, *Ap. J.* **157**, 869.

Gunn, J.E., and Ostriker, J.P. 1970, *Ap. J.* **160**, 979

Gursky, H. 1973, Lecture presented at NATO Advaced Study Institute on Physics of Compact Objects (Cambridge U.K., 1973), unpublished

Hewish, A., Bell, S.J., Pilkington, J.D.H., Scott, P.F. and Collins, R.A. 1968, *Nature* **217**, 709

Helfand, D.J., Ruderman, M.A., Shaham, J. 1983, *Nature* **304**, 423

Hillebrandt, W. 1987, in: *High Energy Phenomena Around Compact Stars* (ed. F. Pacini), Dordrecht: Reidel, p. 73

Hut, P. and Paczynski, B. 1983, *Ap. J.* **284**, 675

Hut, P. 1992, in: *X-ray Binaries and Recycled Pulsars* (eds. E.P.J. van den Heuvel and S.A. Rappaport) Dordrecht: Kluwer Acad. Publ., p. 317

Jeffrey, L.C. 1986, *Nature* **319** 384

Joss, P.C. and Rappaport, S.A. 1983, *Nature* **304**, 419

Kluzniak, W. 1991, in: *Proceedings 15th Texas Conf. on Relativistic Astrophysics* Annals N.Y. Acad. Sci. **647**, 587

Kluzniak, W. 1992, in: *X-ray Binaries and Recycled Pulsars* (eds. E.P.J. van den Heuvel and S.A. Rappaport), Dordrecht: Kluwer Acad. Publ., p. 413

Kluzniak, W., Ruderman, M., Shaham, J. and Tavani, M. 1988, *Nature* **334**, 225

Kulkarni, S.R. 1986, *Ap. J.* **306**, L85

Kundt, W. (ed.) 1990, *Neutron Stars and Their Birth Events*, Dordrecht: Kluwer Acad. Publishers.

Lewin, W.H.G. and Joss, P.C. 1983, in: *Accretion-driven Stellar X-ray Sources* (eds. W.H.G. Lewin and E.P.J. van den Heuvel), Cambridge: Cambridge Univ. Press, p. 41

Long, K.S. and van Speybroeck, L.P. 1983, ibid., p 117

Lyne, A.G., Brinklow, A. Middleditch, J., Kulkarni, S.R., Backer, D.C. and Clifton, T.R. 1987, *Nature* **328**, 399

Lyne, A.G. 1992, in: *X-ray Binaries and Recycled Pulsars* (eds. E.P.J. van den Heuvel and S.A. Rappaport), Dordrecht: Kluwer Acad. Publ., p. 79

Manchester, R.N., Lyne, A.N., Robinson, C., D'Amico, N., Bailes, M. and Lim, J. 1991, *Nature* **352**, 219

Muller, R.A., Newberg, H.J.M., Pennypacker, C.R. et al. 1992, *Ap. J.* **384**, L5

Nagase, F. 1989, *Publ. Astron. Soc. Japan* **41**, 1

Nomoto, K. 1991, in: *Neutron Stars, Theory and Observation* (eds. J. Ventura and D. Pines), Dordrecht: Kluwer Acad. Publ., p. 143

Nomoto, K. 1992, in: *X-ray Binaries and Recycled Pulsars* (eds. E.P.J. van den Heuvel and S.A. Rappaport), Dordrecht: Kluwer Acad. Publ., p. 143

Ögelman, H. and van den Heuvel, E.P.J. (eds.) 1989: *Timing Neutron Stars*, Dordrecht: Kluwer acad. Publ., 774 pp.

Ostriker, J.P. and Gunn, J.E., 1969, *Ap. J.* **157**, 1395

Pacini, F., *Nature* **216**, 567

Paczynski, B. 1971, *Ann. Rev. Astron. Astrophys.* **9**, 183

Paczynski, B. 1983, *Nature* **304**, 421

Radhakrishnan, V., 1986, *Highlights of Astronomy* (ed. J.P. Swings), **Vol. 7**, Dordrecht: Kluwer Acad. Publ., p. 3

Reimers, D. and Koester, D. 1988, *ESO Messenger* **54**, 47

Romanishin, W. and Angel, J. R.P. 1980, *Ap. J.* **235**, 992

Romani, R. 1990, *Nature* **347**, 741

Ruderman, M.A. and Shaham, J. 1983, *Nature* **304**, 425

Ruderman, M.A., Shaham, J. and Tavani, M., 1989, *Ap.J.* **336**, 507

Ruderman, M.A., Shaham, J., Tavani, M. and Eichler, D. 1989, **Ap.J. 334**, 292

Sandage, A. 1986, *Ann. Rev. Astron. Astrophys.* **24**, 421

Savonije, G.J. 1983a, in: *Accretion Driven Stellar X-ray Sources* (eds. W.H.G. Lewin and E.P.J. van den Heuvel), Cambridge: Cambridge University Press, p. 343

Savonije, G.J. 1983b, *Nature*, **304**, 422

Shaham, J. 1992, in: *X-ray Binaries and Recycled Pulsars* (eds. E.P.J. van den Heuvel and S.A. Rappaport), Dordrecht: Kluwer Acad. Publ. p. 375

Shibazaki, N., Murakami, T., Shaham, J.and Nomoto,K. 1989, *Nature* **342**, 656

Smarr, L.L. and Blandford, R.D. 1976, *Ap.J.* **207**, 574

Srinivasan, G. and van den Heuvel 1982, *Astron. Astrophys.* **108**, 143

Srinivasan, G. 1989, *Astron. Astrophys. Rev.* **1**, 209

Srinivasan, G., Bhattacharya, D., Muslimov, A.G. and Tsygan, A.I. 1990 *Curr. Sci.* **59**, 31

Taam, R.E., Bodenheimer, P., and Ostriker, J.P. 1978, *Astrophys. J.* **222**, 269

Taam, R.E. and van den Heuvel E.P.J. 1986, *Ap. J.* **305**, 235

Taylor, J.H. and Stinebring, D.R. 1986, Ann. Rev. Astron. Astrophys. **24**, 285

Tavani, M. and Brookshaw, L. 1992a, *Nature* **356**, 320

Tavani, M. and Brookshaw, L. 1992b, *Astron. Astrophys. Lett.* (in press)

van den Heuvel, E.P.J. 1977, *Ann. N.Y. Acad. Sci.* **302**, 14

van den Heuvel, E.P.J. 1984, *J. Astrophys. Astron.* **5**, 209

van den Heuvel, E.P.J. 1991, in: *Neutron Stars, Theory and Observation* (eds. D. Pines and J. Ventura), Dordrecht: Kluwer. Acad. Publ., p. 171

van den Heuvel, E.P.J. 1992a, *Nature* **356**, 668

van den Heuvel, E.P.J. 1992b in: *X-ray Binaries and Recycled Pulsars* (eds. E.P.J. van den Heuvel, and S.A. Rappaport), Dordrecht: Kluwer Acad. Publ., p. 233

van den Heuvel, E.P.J. and Taam, R.E. 1984, Nature **309**, 235

van den Heuvel, E.P.J. and Bonsema, P.F.J. 1985, *Astron. Astrophys.* **139**, L16

van den Heuvel, E.P.J. and Habets, G.M.H.J. 1984, *Nature* **309**, 598

van den Heuvel, E.P.J. and van Paradijs, J.A., 1988, Nature, **334**, 227

van den Heuvel, E.P.J. and Rappaport, S.A. (eds.) 1992, *X-ray Binaries and Recycled Pulsars*, Dordrecht: Kluwer Acad. Publishers, 566 pp.

van der Klis, M. 1987, Ann. Rev. Astron. Astrophys. **27**, 517,

Ventura, J. and Pines, D. (eds.) 1991, *Neutron Stars, Theory and Observation*, Dordrecht: Kluwer Acad. Publ., 592 pp.

Verbunt, F. 1990, in: *Neutron Stars and Their Birth Events* (ed. W. Kundt), Dordrecht: Kluwer Acad. Publ. , p. 179

Verbunt, F., Wijers, R.A.M.J. and Burm, H. 1990, *Astron. Astrophys.* **234**, 195

Wakatsuki, S., Hikita, A. Sato, N. and Itoh, N. 1992, *Ap. J.* **392**, 628

Webbink, R.F. 1992, In: *X-ray Binaries and Recycled Pulsars* (eds. E.P.J. van den Heuvel and S.A. Rappaport), Dordrecht: Kluwer Acad. Publ., p.269

Weidemann, V. 1990, *Ann. Rev. Astron. Astrophys.* **28**, 103

Woltjer, L. 1964, Ap.J. **140**, 1309

Wolszczan, A. 1992, in: *X-ray Binaries and Recycled Pulsars* (eds. E.P.J. van den Heuvel and S.A. Rappaport), Dordrecht: Kluwer Acad. Publ., p. 93

Wolszczan, A. and Frail, D.A. 1992, Nature **355**, 145

Woosley, S.E. and Weaver, T.A. 1986, Ann Rev. Astron. Astrophys **24**, 205

Woosley, S.E., Timmes F.X. and Baron, E. 1992, in: *X-ray Binaries and Recycled Pulsars* (eds. E.P.J. van den Heuvel and S.A. Rappaport), Dordrecht: Kluwer Acad. Publishers, p. 167

Planets Around Pulsars
ASP Conference Series, Vol. 36, 1993
J.A. Phillips, J.E. Thorsett, and S.R. Kulkarni

PLANET FORMATION SCENARIOS

PHILIPP PODSIADLOWSKI
Institute of Astronomy, Cambridge CB3 0HA

ABSTRACT We review theoretical models for the formation of pulsars with planet-mass companions. We first summarize observational constraints a successful model has to fulfill and discuss potential tests of the models. We then assess the merits and uncertainties of the various models in the context of the putative planet pulsar PSR 1257+12. At the moment, several of the models appear to be consistent with the available constraints. We conclude the review with a list of "best-bet" models.

1. INTRODUCTION

The possible discovery of a planetary system around the millisecond pulsar PSR 1257+12 (Wolszczan & Frail 1992) has been the latest major discovery in the field of pulsar astronomy and has stimulated the imagination of astronomers and the general public alike. While there may still be some isolated, residual doubts, whether the timing data really imply the presence of planet-mass objects (Gil 1992, this conference; Dolginov & Stepinski 1992, this conference), the evidence presented by Wolszczan at this conference and the independent confirmation of the pulsar timing by Backer 1992 (this conference) appear so convincing, that we conclude that the available data most likely imply that PSR 1257+12 is orbited by at least two planet-mass objects (referred to as "planets" henceforth). In any case, future detection of the predicted secular variations of the planet orbits (Rasio et al. 1992; Malhotra et al. 1992; also Malhotra 1992, this conference) should be able to verify this hypothesis beyond reasonable doubt.

This therefore raises the question of how planets could have survived or formed around such an exotic object. In this review, we summarize the various models that have been advanced over the last year to explain such an unusual system. We will concentrate on discussing the evolutionary scenarios up to the stage where planet formation may occur and assess their merits and problems in the context of PSR 1257+12; the details of the actual planet formation process will be discussed in the accompanying article by Phinney. As we will see, a large number of models seem to be feasible at the moment. However, this does not necessarily mean that planets should be common around pulsars. The relatively large number of possible models is in part due to the limited number of observational constraints presently available and in part due to our poor theoretical understanding of some of the input physics that goes into the various models. In addition, many of the models — even if they were feasible in principle — require such special circumstances that we would be lucky to have discovered even a single planet pulsar.

2. OBSERVATIONAL CONSTRAINTS AND POTENTIAL TESTS

The orbital parameters and the masses of the two planets are some of the few relatively firm constraints that can be inferred from the timing data for PSR 1257+12. A model for PSR 1257+12 has to explain the presence of planets in near-circular orbits at a distance of only $\sim 0.4\,\mathrm{AU}$ from the central pulsar. The orbits imply a total amount of orbital angular momentum for the planets of $J_{1257+12} \simeq 1.3 \times 10^{48}\,\mathrm{erg\,s}\,(\sin i)^{-1}$, where i is the the inclination between the orbital axis and the line of sight (assuming coplanar planetary orbits). This value is only about five times larger than the orbital angular momentum of the Earth and can easily be satisfied by most of the models discussed in this article. On the other hand, the existence of a more distant, third planet (as has even been suggested by Wolszczan & Frail 1992) could seriously constrain available models. A single planet like Jupiter, with $J_{\mathrm{Jup}} \simeq 2 \times 10^{50}$ erg s, would immediately rule out a number of models for PSR 1257+12. In addition, the eccentricity of an outer planet would also be useful for distinguishing among available models, since some of the models predict that outer planets, unlike their inner counterparts, should have eccentric orbits (see §3.2 & 3.3).

A second important constraint for a successful model of PSR 1257+12 is that it has to explain why the central pulsar in this system is an old, millisecond pulsar with a spin period $P_{\mathrm{spin}} \simeq 6.2\,\mathrm{ms}$, which in every respect resembles a typical re-cycled pulsar and not a canonical radio pulsar (for a review of millisecond pulsars see, e.g., Bhattacharya & van den Heuvel 1991). Presumably, this fact provides one of the key clues for the formation of this particular system.

The statistical frequency of planet pulsars provides an additional, although weaker constraint for planet-pulsar models. Most models predict that planet pulsars should be relatively rare (mainly because their evolutionary history requires rather special circumstances), while in some models planets could be quite common. The observational statistics of planet pulsars has been addressed by various speakers at this workshop, but it is too early to draw any conclusions based on this argument.

Most models for PSR 1257+12 also have to explain, how planets were able to form in an exotic environment, either around a massive star or a luminous neutron star; this may be difficult or even impossible (see, e.g., Nakano 1987). The near resonance of the orbits of the two inner planets may also provide an important hint for the planet formation process. Furthermore, the existence of a comet cloud (Chakrabarti & Krishna Swamy 1992) or asteroid belts (e.g., Joss & Rappaport 1984) would not only provide additional constraints for planet-pulsar models, but could also have important implications for other astrophysical phenomena, as, for example, cosmic gamma-ray bursts (see the contributions by Salpeter and Woosley at this workshop).

Finally, we note that, while we will review all of the models in the context of PSR 1257+12, some of the models which can be ruled out for this system may still be applicable to other systems and could, for example, be able to explain planets around ordinary radio pulsars, should they be discovered.

3. FORMATION SCENARIOS

The various planet-pulsar models can be broadly grouped into three classes. In the first class of models, the planets formed around an ordinary star, very much like planets in our solar system, and either survived the subsequent evolution or were captured by a neutron star. In the second class, the planets formed immediately after the pulsar formed in a supernova explosion, and, in the third class, planet formation constitutes the final stage in the evolution of (some) binary millisecond pulsars.

3.1 The Planet Survival Model

In many ways the simplest model to explain planets around around a neutron star is to assume that the planets formed around the progenitor of the observed pulsar, just like planets in our solar system, and survived the subsequent evolution (Bailes, Lyne & Shemar 1991; Field, Roger & Rybicki 1992). This model is truly a survival model, since the planets would have to survive a number of very destructive phases between their formation and the present epoch.

Firstly, they have to avoid being engulfed by the expanding envelope of the progenitor during its red-supergiant phase, since this would lead to the spiral-in and ultimate destruction of the planets (either because of evaporation or dynamical disruption). Since red-supergiant envelopes typically reach maximum extensions larger than $\sim 4\,\mathrm{AU}$, the initial orbital separation of the planets has to be larger than ~ 4 AU. Closer initial separations are, in principle, possible, if the progenitor is a very massive star ($M \gtrsim 40 M_\odot$), because such massive stars do not become red supergiants (see, e.g., Humphreys 1984). On the other hand, it would probably be substantially more difficult to form close planets around such a luminous and short-lived central star.

Secondly, the planets have to survive the supernova explosion which lead to the formation of the observed pulsar. Since during the supernova explosion more than one half the total mass of the system is ejected, all planets would become unbound, if the explosion were spherically symmetric (Blaauw 1961). Thus, planet survival requires that the remnant neutron star received a kick during the supernova with just the right properties: the kick velocity has to be of the order of the orbital velocity of the planet and the kick has to be in a direction similar to the direction of motion of the planet at the time of the supernova. In addition, these conditions have to be fulfilled for both planets simultaneously. While this is possible in principle (supernovae appear to be asymmetric; for references and discussions, see Bhattacharya & van den Heuvel 1991), the overall probability for this is extremely small.

Apart from the overall low probability, this scenario has several other, severe problems. Since the orbits of the observed planets are much smaller than the initial separations for which the planets would have avoided spiral-in and have very low eccentricities (despite a supernova kick), the model requires an additional mechanism for shrinking and circularizing the orbits after the supernova. An asymmetric, evaporative wind driven by the radiation of the young pulsar (similarly to the evaporative wind observed in PSR 1957+20 [Fruchter et al. 1988]) could provide such a mechanism (Field, Roger & Rybicki 1992). However, this wind would have to be rather fine-tuned, since the planetary orbits have to shrink by at least a factor of ten, which implies that the original planets have to lose a large fraction of their initial mass. Thus, planet evaporation must be very efficient. On the other hand, it cannot be too efficient, or the planets would have been evaporated completely. Finally, the most severe problem is that this model cannot explain why the central pulsar should look like a re-cycled pulsar instead of an ordinary radio pulsar.

In summary, we conclude that this model cannot explain the planets around PSR 1257+12 and is unlikely to explain many planetary systems around ordinary radio pulsars (if they existed). In addition, it would be exceedingly improbable to have additional planets in such systems.

3.2 The Planet Capture Model

A second model in which planets formed in a more or less traditional fashion has been proposed by Podsiadlowski, Pringle & Rees 1992. In this model, the observed planets formed around a solar-type system similar to our own; the central star then experienced a direct collision with a neutron star. While the probability for such a direct collision is low, it is similar to the probability of the tidal capture of a neutron star by a main-sequence star (Fabian, Pringle & Rees 1975), the process by which low-mass X-ray binaries are believed to form in globular clusters. The immediate consequence of such a collision is that the central star of the solar system is completely disrupted (Ruffert & Müller 1990). However, most of the material of the disrupted star remains bound to the system and will settle on the neutron star, forming a Thorne-Żytkow object (TŻO; Thorne & Żytkow 1975, 1977), i.e. a supergiant-like star (with a radius $R_{\mathrm{T\dot{Z}O}} \approx 4\,\mathrm{AU}$) with a neutron core. The orbits of the inner planets, which are likely to remain bound to the system, will be highly eccentric after the collision. However, the orbits of planets with periastron distances less than $R_{\mathrm{T\dot{Z}O}}$ will start to decay and will be recircularized due to drag forces inside the short-lived TŻO envelope. For typical planet parameters, the spiral-in time is of the same order as the lifetime of the TŻO phase ($\sim 10^5$ yr). Planets for which the spiral-in time is shorter than the TŻO lifetime will be destroyed. Thus, the observed planets would be the innermost planets that survived the TŻO phase. On the other hand, planets with periastron distances larger than $R_{\mathrm{T\dot{Z}O}}$ after the collision are little affected by the TŻO phase, and hence there orbits should still reflect the parameters they attained immediately after the collision. Thus, this model predicts that outer planets (with separations $a \gtrsim 4\,\mathrm{AU}$) should have highly eccentric orbits, and that the planetary orbits are not coplanar (unless the neutron star's trajectory of approach was in the ecliptic of the original solar-type system).

One of the main problems with this model is that the required collisions between a neutron star and a main-sequence star are very rare in the Galaxy. We estimate that only ~ 100 such collisions would have occurred in the lifetime of the Galaxy. On the other hand, such collisions should be much more common in globular clusters, and one could expect to form planet pulsars at a similar rate as low-mass X-ray binaries in globular clusters (provided that planetary systems are very common and are able to survive dynamical interactions in globular clusters). The second major problem with this model is that it would not naturally produce a millisecond pulsar (there will be some spin-up of the central neutron core during the TŻO phase, but it is unlikely to be sufficient).

In summary, this model is also unlikely to explain PSR 1257+12, although it could potentially explain planets around long-period pulsars in globular cluster, should they be discovered.

3.3 The Massive Binary Model

Wijers et al. 1992 have suggested a model for PSR 1257+12 which shares many similarities with the previous model, even though it starts with a completely different, initial configuration. It was motivated by a similar model for the formation of low-mass X-ray binaries in the galactic disk by Eggleton & Verbunt 1986. In this model, the planets formed around a massive, central binary at a separation of a few AU. The central binary, which initially consisted of two massive stars and had an orbital period of a few days, then evolved into a massive X-ray binary (as described, e.g., by Bhattacharya & van den Heuvel 1991; van den Heuvel 1992, this conference). In the process, it experienced a supernova explosion. However, the system (including the planets) is likely to remain bound after the explosion, since, due to prior mass exchange, less than half the total mass of the system would be ejected during the supernova, and since even a (moderate) supernova kick would not lead to the loss of most inner planets. When, during the massive X-ray binary phase, the secondary fills its Roche lobe, mass transfer is expected to be so dramatic that it leads to the spiral-in of the neutron star and again the formation of a TŻO. The further evolution is similar to the previous model, except that, in this scenario, the TŻO has a more massive envelope and that the spiral-in time scale for any engulfed planet would consequently be much shorter. The final outcome of the TŻO phase is again that all inner planets should be on almost circular planets, while outer planets could still be mildly eccentric.

The major advantage of this model over the previous model is that the basic progenitor systems, massive X-ray binaries, are known to be quite common. Hence, in this model, planet pulsars could be relatively common (provided that it is possible to form planets around massive binaries).

The main problem with this scenario is that it is unlikely to produce a 6 ms pulsar. While some spin-up of the central neutron star can be expected during the TŻO phase, the total mass accreted during this short-lived phase is several orders of magnitudes less than the $\sim 0.02\,M_{\odot}$ required to spin the core up to 6.2 ms (Thorne & Żytkow 1977; Cannon et al. 1992b). In addition, whether significant spin-up occurs, also depends on the specific angular momentum of the accreted material. This may be substantially less than Keplerian, because the large, convective TŻO envelope is likely to act as an efficient brake for the inner part of the envelope. On the other hand, the details of the TŻO phase are not very well understood. It is conceivable that, near the end of the TŻO phase a neutrino runaway will take place, which could be accompanied by substantial accretion onto (and spin-up of) the central neutron core (Cannon et al. 1992b).

In summary, a massive binary model with a planetary system is probably not the appropriate model for PSR 1257+12. However, it might be a good candidate for planets around long-period pulsars in the galactic disk (if they are discovered).

3.4 The Fall-Back Scenario

In the models discussed so far, the planets formed in a relatively conventional way around a normal star (although not necessarily around a solar-type star). Their characteristic feature was that they had to provide a mechanism by which planets could either survive or avoid the supernova that lead to the formation of the observed neutron star. This problem can be completely circumvented, if the planets actually formed after the supernova explosion. Planet formation after the supernova, usually out of a disk that was left over for one reason or another, is the key feature in almost all scenarios discussed in the rest of this review.

The simplest post-supernova formation model is based on the idea that the planets formed out of material that did not escape during the supernova explosion, either because it did not attain escape velocity during the explosion or because it was pushed back during the early hydrodynamical mixing phase (Bailes et al. 1991; Lin, Woosley & Bodenheimer 1991). Lin et al. 1991 suggest a fall-back mass of $M \sim 0.02 M_{\odot}$ as reasonable. They also estimate that this material could have sufficient angular momentum ($J \sim 10^{49}$ erg s) to form a compact disk of size $R \sim 10^{8}$ cm, sufficient to account for close planets. Some observational support for fallback during a supernova may be provided by observations of SN 1987A. The late bolometric lightcurve, as observed by ESO (Suntzeff et al. 1991), shows a transient "bump" at a luminosity $L \sim 10^{38}\,\mathrm{erg\,s^{-1}}$. This luminosity roughly corresponds to the Eddington luminosity of an accreting neutron star. Several theoretical models (Meyer-Hofmeister 1992; Mineshige et al. 1992) have modeled this feature in terms of an accretion-disk instability in a fall-back disk (note, however, that the mass in the disk required to explain the lightcurve bump is many orders of magnitudes less than the mass required to produce the planets in PSR 1257+12).

The main uncertainty in a fall-back scenario is the actual amount and the specific angular momentum of the material that is re-accreted by the central neutron-star remnant after the explosion. A previous study of this problem (Chevalier 1989) concluded that very little mass would fall back ($M \lesssim 10^{-3} M_{\odot}$), if the progenitor was a typical red supergiant (the generally believed precursor for most type II supernovae). On the other hand, Chevalier 1989 also found that the fallback mass could be as large as $\sim 0.1 M_{\odot}$, if the progenitor was a blue supergiant (because of a different density structure near the core), as, for example, the apparent progenitor of SN 1987A. Thus, if these ideas are correct, SN 1987A might provide an ideal test case for this model.

However, a fallback scenario is unlikely to explain the planets around PSR 1257+12, because we would expect the pulsar to be a typical radio pulsar instead of a re-cycled millisecond pulsar. This argument breaks down, if only a special subset of supernovae experienced significant fall-back and if these supernovae preferentially produced millisecond radio pulsars with weak magnetic fields (again SN 1987A could potentially provide some clues).

Finally, if planets were found to be common around ordinary radio pulsars, this would be consistent with a fall-back model, since naively one would expect that fall-back is either unimportant or a generic feature of ordinary type II supernovae. In this case, all planets should be in circular orbits and outer (very distant) planets would be unlikely.

3.5 The White Dwarf (Neutron Star) + White Dwarf Merger Model

The main problem with a simple fall-back scenario is that it is not clear, whether a sufficiently massive disk can form out of the material that is re-accreted. This was the motivation for Podsiadlowski et al. 1991 to suggest that planet pulsars form as the result of the merger of two white dwarfs, because this process naturally leaves a massive disk behind.

The initial configuration in this scenario consists of two relatively massive white dwarfs (with a combined mass larger than the Chandrasekhar mass of 1.4 $M_{\odot}$) with an orbital period less than $\sim$ 10 hours. (In a very similar scenario, the more massive white dwarf is replaced by a neutron star [van den Heuvel 1992; also see Bonsema & van den Heuvel 1985].) Due to gravitational radiation, the binary orbit shrinks, until the lighter white dwarf fills its Roche lobe. If the mass ratio is sufficiently large, the lighter white dwarf is now dynamically disrupted, forming, in the process, a disk around the more massive component (Benz et al. 1990). The central star (a massive white dwarf or a neutron star) subsequently accretes from this disk and is being spun up. In the case of a CO white dwarf, the central star is likely to experience a series of off-center carbon shell flashes (if the accretion rate is large enough; Nomoto & Iben 1985) and is transformed into a O-Ne-Mg white dwarf. Eventually, when its mass reaches the Chandrasekhar mass, it undergoes core collapse and forms a rapidly rotating neutron star (this neutron star will not spin down, if its magnetic field is low). However, even after the explosion, the neutron star is surrounded by a relatively massive disk of material ($M \sim 0.1\, M_{\odot}$), in which planets might form. These planets would all be terrestrial planets, since they would consist completely of white dwarf material (probably mainly carbon and oxygen).

One of the main advantages of this model is that one would naturally expect a millisecond pulsar from the collapse of a rapidly rotating white dwarf or a recycled pulsar in the case of a WD + NS merger. In addition, these mergers could be relatively common. In fact, WD + WD mergers have previously been considered as good candidates for the production of type Ia supernovae (Iben & Tutukov 1984). While it now appears that the progenitor systems of white dwarf mergers are not common enough to produce the observed frequency of type Ia supernovae, there may still be enough systems to produce a few planet pulsars.

The main uncertainty in a WD + WD merger scenario (this uncertainty does not exist in the case of a NS + WD merger) is that the merger of two white dwarfs could also lead to a thermonuclear runaway (i.e., a type Ia supernova; Iben & Tutukov 1984). This will be the case, if carbon ignition in the accreting CO white dwarf occurs under highly degenerate conditions. In this case, a thermonuclear runaway would take place, which would completely disrupt the accreting white dwarf (and, thus, would not leave a pulsar remnant). Detailed theoretical calculations (Nomoto & Iben 1985) have shown that a neutron star forms, if the accretion rate is larger than $\sim 2 \times 10^{-7} M_{\odot}$ yr^{-1}, roughly 1/6 the Eddington rate of an accreting white dwarf.

In summary, a WD + WD or a NS + WD merger could naturally produce systems like PSR 1257+12. In this case, all planets should consist exclusively of heavy elements and be in circular orbits.

3.6 The Be Binary Model

Another possibility for the origin of a relatively massive disk around a neutron star is that it is a remnant disk left over from an earlier binary mass-transfer phase (Fabian & Podsiadlowski 1991). The evolutionary scenario by Fabian & Podsiadlowski 1991 starts with a typical Be neutron-star binary, consisting of a massive spun-up Be star orbiting a neutron star with an orbital period larger than $\sim 5\,\mathrm{yr}$ (such systems are believed to be common; see, e.g., van den Heuvel & Rappaport 1987; Johnston et al. 1992). Because of the long period, the massive star never fills its Roche lobe (so as to avoid spiral-in of the neutron star). However, when the massive star becomes a red supergiant, it develops a very strong stellar wind ($\dot{M} \sim 10^{-5}\,M_\odot\,\mathrm{yr}^{-1}$). For a typical mass ratio of 10 between the massive star and the neutron star, the neutron star accretes $\sim 1\,\%$ of this wind. The resulting accretion rate of $\sim 10^{-7}\,M_\odot\,\mathrm{yr}^{-1}$ exceeds the neutron star's Eddington accretion rate by an order of magnitude. Thus, not all of this material can be accreted by the neutron star; instead, most of it may accumulate around the neutron star and form a massive disk. The mass in the disk depends entirely on the lifetime of this phase and could be substantial (larger than $\sim 10^{-3}\,M_\odot$, if the accretion phase lasts longer than $\sim 10^4\,\mathrm{yr}$). In addition, the specific angular momentum of this material will be very large because it should be of the same order as the specific orbital angular momentum of the binary. Eventually, the massive companion may experience a supernova explosion. This is likely to lead to the disruption of the binary system, but the old spun-up neutron star would still be surrounded by a massive disk, from which planets may form.

One of the advantages of this model is that potential progenitor systems are probably quite common (a few %). In addition, after the explosion the neutron star will be surrounded by a rather large disk ($R \sim 10^{13}\,\mathrm{cm}$), unlike most other disk scenarios in which the initial disk size tends to be very small ($\sim 10^8$ – $10^{11}\,\mathrm{cm}$) and has to expand by several orders of magnitude (see Phinney 1992, this conference).

However, there are several uncertainties in this model. In particular, it is not certain whether a massive disk can form from the accretion of a slow red-supergiant wind and whether the neutron star can actually be spun up to milliseconds (in contrast, wind accretion from a fast wind tends to spin down a neutron star [Illarionov & Sunyaev 1975]). In addition, it is not clear whether the disk around the neutron star can survive the supernova blast wave. This depends on the mass and the geometry of the disk. If the disk is massive and subtends only a small solid angle (as seen from the exploding star), most of the disk is likely to survive.

An alternative possibility, which would avoid the latter problem, is that the massive companion ends its evolution as a white dwarf instead of a neutron star (which is the case, if its initial mass is less than $\sim 8\,M_\odot$). This would have the additional advantage that the distant white dwarf companion should still be present today and could provide an unambiguous test of the model.

In summary, this model could potentially explain PSR 1257+12, although it is probably more likely to produce slower, re-cycled pulsars. It predicts that all planets are circular and low-mass, outer planets are possible (or even a distant white-dwarf companion).

3.7 The TŻO Envelope Deflation Model

A rather exotic way of producing a massive disk around a neutron star is to form the disk by the deflation of the envelope of a massive Thorne-Żytkow object (Podsiadlowski et al. 1991; Rees 1992, this conference; Cannon et al. 1992b). TŻOs, which have already been mentioned several times in this article, probably form mainly by two basic mechanisms (see Cannon et al. 1992a): one is the collision of a neutron star with a main-sequence star (e.g, §3.2); the other is the spiral-in of a neutron star in a massive X-ray binary (e.g, §3.3). It appears likely that most observed, massive X-ray binaries with periods less than a few days evolve into TŻOs (also see van den Heuvel 1992, this conference). In this case, the TŻO consists of a neutron core with a very massive envelope ($M \sim 10\,M_{\odot}$). As was first suspected by Zimmermann 1979 and confirmed by the most recent and most detailed calculations (Cannon et al. 1992b), these objects have to generate the luminosity required to keep their massive envelopes in hydrostatic balance by unusual and non-standard reaction chains (e.g., the r-p process [Biehle 1991; Cannon et al. 1992b]; ordinary hydrogen or helium burning reactions are not efficient enough, since the mass in the region which is sufficiently hot to generate energy by these standard reactions is too small). One of the main consequences of non-standard nuclear burning in TŻOs is that the fuel reservoir is very limited and that a massive TŻO may run out of nuclear fuel, before it has lost its envelope in a powerful stellar wind. If this occurs, the massive, extended envelope (with a radius $\sim 5\,$AU) can no longer be maintained and has to start to contract onto the central neutron core (on the Kelvin-Helmholtz time scale of the envelope). If, however, the envelope has still some angular momentum left (as is likely because of its formation from a binary), this deflation will not be spherically symmetric. Instead, it is most likely that the envelope deflates initially into a hybrid object, consisting of a neutron star surrounded by spherical inner envelope and a disk-like outer structure. Eventually, the system probably evolves into a neutron star surrounded by a massive disk. In fact, the disk is likely to be very massive (several $M_{\odot}$) and it is not at all clear what will happen to this disk. One possibility is that planets form in this disk, but it is also conceivable that one or more low-mass, stellar objects emerge out of it (in fact, this might be a good alternative model for forming low-mass X-ray binaries in the galactic disk, for which no widely accepted model exists at the moment [van den Heuvel 1992, this conference]). It is also not obvious what happens in this phase to the central neutron star. If this phase is accompanied by a neutrino runaway, significant accretion onto the neutron star may take place, and the neutron star may be spun up and become a millisecond pulsar. However, if it accretes too much, it could also collapse to a black hole.

In summary, the deflation of a TŻO envelope could potentially produce systems like PSR 1257+12. However, at the moment this scenario is too uncertain to allow any real conclusions. Indeed, it is also possible that such a scenario could produce completely different (but nevertheless still interesting) systems. It could well explain the origin of low-mass X-ray binaries or soft X-ray transients in the galactic disk, for which no uncontroversial formation scenario exists.

3.8 The Evaporation Model

The pulsar in PSR 1257+12 looks like a typical, old, re-cycled, millisecond pulsar. Therefore, one could argue that perhaps the most natural model for explaining this system would be a model which links its history with the evolution of ordinary millisecond pulsars. Several models of this type have been advanced.

The simplest model, which was first suggested for PSR 1829-10 (Bailes et al. 1991; Krolik 1991; Rasio, Shapiro & Teukolsky 1992), is that the apparent "planet" around the pulsar was not a planet but rather the remnant of a low-mass star that had been (almost completely) evaporated by the pulsar radiation. This model was motivated by observations of PSR 1957+20 (Fruchter et al. 1988) and PSR 1744-24A (Lyne et al. 1990), which show that the companions in these systems are being evaporated by the radiation from their millisecond pulsar companions.

One of the problems in this scenario is that it requires a significant amount of fine-tuning, because most of the companion star has to be evaporated, but not quite all. Krolik 1991 showed, that this is possible in principle, for a proper combination of the evolution of the spin-down luminosity and the orbital evolution of the companion star (provided that the pulsar is a young pulsar).

However, this model cannot apply to PSR 1257+12, because it cannot possibly produce more than one planet-mass object. In addition, we do not expect it to be responsible for many other pulsars with single planets, because the required fine-tuning should make this channel rare.

3.9 The Disrupted Companion Model

A variation of the last model has been proposed by Stevens, Rees & Podsiadlowski 1992. Their model starts with a typical binary millisecond pulsar, where the low-mass companion is being evaporated by the pulsar's radiation (just as PSR 1957+20 and PSR 1744-24A). If the evaporation timescale of the secondary is shorter than its Kelvin-Helmholtz time (as it seems to be for PSR 1957+20), the secondary will respond adiabatically to mass loss. For a low-mass star, this means that it will expand, when it loses mass ($R \propto M^{-1/3}$). Because of this expansion, the secondary will eventually fill its Roche lobe. Stevens et al. 1992 argue that, at this point, the low-mass star is dynamically disrupted, forming a massive disk ($M \sim 0.1\,M_\odot$) around the primary, where planets may be able to form subsequently.

This model is attractive, because it seems to explain all the major features of PSR 1257+12. The main uncertainty in this scenario is, whether the low-mass star is actually disrupted, when it fills its Roche lobe or whether the system will enter into a low-mass X-ray binary phase. The criterion for disruption is that, upon mass loss, the Roche lobe radius of the secondary has to increase less slowly than its stellar radius (in this case, the secondary will overfill its Roche lobe by an increasingly larger amount, which leads to runaway mass loss and dynamical disruption). The detailed criterion depends mainly on how much mass is accreted by the pulsar primary. If a large fraction of the matter lost by the secondary is accreted onto the primary (typically, if this fraction is larger than 1/3), the binary system widens rapidly enough and the Roche lobe radius increases as fast as the stellar radius. In this case, mass transfer is stable and the system becomes a low-mass X-ray binary. On the other hand, if most of the mass is lost, mass transfer is unstable and the secondary is disrupted.

This scenario can also be applied to young pulsars (Stevens et al. 1992). In fact, if the neutron star is a young, powerful pulsar (i.e., a pulsar which just emerged from a supernova explosion), dynamical disruption may be inevitable, since the pulsar's radiation may prevent any accretion onto the pulsar. This requires that the secondary fills its Roche lobe soon after the supernova explosion of the primary. This is possible (provided that the pre-supernova system was a sufficiently close binary), either if the secondary expands because of the heating of its surface by the pulsar's radiation (Podsiadlowski 1991), or if the newborn neutron star receives a kick in the right direction because of an asymmetric supernova explosion (Stevens et al. 1992; Phinney 1992, this conference).

In summary, a disrupted companion model can explain system like PSR 1257+12, but could also explain planets around young radio pulsars, should they exist. In this model, all planets are circular and distant, outer planets are possible.

3.10 The Circumbinary Disk Model

Another variation of the basic evaporation scenario was suggested by Tavani & Brookshaw 1992 and Banit, Ruderman & Shaham 1992. Their model also starts with a typical binary millisecond pulsar like PSR 1957+20 and PSR 1744-24A, in which the companion is being evaporated (alternatively, this process might also take place in a low-mass X-ray binary phase [Tavani & Brookshaw 1992]). However, unlike the simple evaporation model, the authors argue, some of the evaporated material may not escape from the system and form a circumbinary disk instead, in which planet formation may take place. Eventually, after the companion star has been completely evaporated, the system would resemble PSR 1257+12.

The millisecond pulsar binary PSR 1957+20 may provide some indirect observational support for such a scenario. Firstly, mass outflow from the companion appears to be confined to a region near the orbital plane (Fruchter 1992, private communication), a necessary requirement for the formation of an excretion disk. Secondly, the orbital period of PSR 1957+20 is decreasing on a timescale of 3×10^7 yr (Ryba & Taylor 1991), which is significantly shorter than expected. A circumbinary disk could potentially explain the rapid shrinking of the orbit, because it could provide a means by which angular momentum is transferred from the binary to the circumbinary disk.

There are several uncertainties in the model. Since it requires that material can be evaporated from the star, but not from the system, some fine-tuning of the mass-loss process is obviously necessary. Banit et al. 1992 argue that this condition is generally fulfilled for irradiation-driven winds. However, an additional requirement is that the same (efficient) radiative driving process which is responsible for the evaporative mass loss from the secondary is not efficient in exciting a wind from the inner parts of the excretion flow. A second uncertainty is whether it is actually possible to form a circumbinary disk which is sufficiently massive to allow the formation of planets. Since the mass flow rate into the disk is generally very small ($\dot{M} \sim 10^{-10}\, M_\odot\, \mathrm{yr}^{-1}$ for PSR 1957+20) and since mass is also effectively lost from the disk on a viscous timescale ($t_{\mathrm{visc}} \sim 10^6 - 10^7$ yr), the surface density at any given radius is always relatively small (much smaller than in other disk scenarios). Therefore, it may be more difficult to form planets in this scenario. Finally, the effects of a circumbinary disk on the orbital evolution of the binary have not yet been well studied. In one study, Artymowicz et al. 1991 found that the inner binary (in their case an equal mass binary) should shrink rapidly and its eccentricity should increase. These results would be at variance with the suggested scenario for PSR 1957+20, although they may not be directly applicable.

In summary, a circumbinary disk scenario also provides a promising model for PSR 1257+12. It can, in principle, explain the circular orbits of the two planets and a millisecond pulsar.

3.11 Other Scenarios and Ideas

A number of other scenarios or ideas have been suggested for PSR 1257+12. In many cases, these scenarios have not been developed in sufficient detail to warrant a separate, detailed discussion.

The most detailed of these was proposed by Nakamura & Piran 1991. In their scenario, planets form in a disk around a newborn neutron star. In contrast to a simple fall-back scenario (§3.4), they suggest that this material was ablated from a close binary companion (which has the advantage that this material would have relatively large specific angular momentum). While the binary is disrupted in the supernova explosion, some of the ablated material may be captured by the neutron star and form a large disk (with a radius comparable to the initial binary separation). In order to make this disk sufficiently massive, the model requires that the orbital separation of the binary is in a very limited range (with a separation $\sim 3 \times 10^{12}$ cm). If the system were too wide, very little material could be ablated from the companion. If it were too close, the ablated material would have too high a velocity and would escape from the system. A separation of $\sim 3 \times 10^{12}$ cm requires, for the vast majority of conceivable binary models, that the progenitor system had to pass through a mass-transfer phase, most likely a common-envelope phase (this was not considered by Nakamura & Piran 1991). While binary models can probably be constructed to fulfill these constraints, they will be relatively rare. In addition, such a scenario could not produce a millisecond pulsar.

Another idea for the formation of a planet pulsar is that the pulsar passed through a star-forming region and interacted with a protostar, thereby capturing a protostellar disk (Clarke 1991, private communication). While this appears to be possible in principle, it would be a rather rare event. It would probably also predict that planet pulsars should be more common near star-forming regions.

A final idea that has been mentioned is that the planets might have pre-existed around an overmassive white dwarf (i.e., a white dwarf that exceeds the classical Chandrasekhar mass because of rapid rotation; Dawson 1991). When the white dwarf slows down, it will eventually experience core collapse and form a millisecond pulsar. Since only a small amount of mass is lost during core collapse, planets are likely to remain bound. The main problem with this idea is to find an evolutionary scenario that leads to the formation of an overmassive white dwarf and that allows planets to exist within a fraction of 1 AU. All evolutionary scenarios we can think of involve at least one red giant significantly larger than 1 AU. This scenario may therefore not be possible.

4. DISCUSSION

The large number of models for the formation of planet pulsars (see the summary table) is proof of how much the discovery of one such system (initially of PSR 1829-10, then of PSR 1257+12) has stimulated the imagination of astronomers. At the moment, several models appear to be feasible. To conclude this review, we shall now try to order the models that are most likely to explain planet pulsars like PSR 1257+12. This ordering is necessarily subjective and should therefore not be taken too seriously. It will also undoubtedly evolve in time, as our observational and theoretical understanding of such systems matures. The two models that, at the moment, appear to face the fewest theoretical and observational objections are the WD + WD (or NS + WD) merger model (§3.5) and the disrupted companion model (§3.9). Both scenarios can explain the main features of PSR 1257+12 and are likely to occur at least occasionally. The circumbinary disk model (§3.10) follows closely in the list of most-favoured models. It may also be able to explain systems like PSR 1257+12, but it involves more uncertainties than the other two. The two scenarios involving massive binaries (§3.3 and §3.6) should perhaps also be included in the list of more distant favorites. While they may have difficulties in explaining a 6 ms pulsar, they have the advantage that they could be relatively common.

Some of the other models may also be able to produce planet pulsars, although they are unlikely to explain systems like PSR 1257+12. The fall-back scenario (§3.4) could account for a large fraction of planetary systems around ordinary radio pulsars, should they be common. The collision of a neutron star with a solar-type system (§3.2) is also likely to produce a planet pulsar. However, such events should be very rare in the galactic disk, and the model could account for only very few such systems (this situation could, however, be different in globular clusters). Finally, some of the models (for example, the TŻO envelope deflation model; §3.7) are too speculative at the moment to allow any real firm conclusions.

In summary, while initially it seemed surprising to find planets around a neutron star, it is by no means impossible to explain such systems. In fact, there is probably more than one way of producing planet pulsars. Even if it should turn out that the interpretation of the timing data for PSR 1257+12 does not involve the presence of two or more planets, we conclude that such systems should exist and should then be discovered.

	Description	Frequency	ms Pulsar?	Tests
Planet Survival Model	pre-existing planets which survived evolution and supernova	very low	no	third planet very unlikely
Planet Capture Model	neutron star collides with solar system and replaces central star	low	no?	outer planets in eccentric, oblique orbits
Massive Binary Model	circumbinary planets dragged in during Thorne-Żytkow phase	common?	no?	outer planets in eccentric, coplanar orbits
Fall-Back Model	planets formed out of fall-back material after supernova	common?	no	outer planets unlikely
WD+WD or NS+WD Merger	(lighter) WD disrupted to form planetary disk	relatively common?	yes	outer planets in circular orbits, planets out of WD material
Be Binary Model	massive disk accreted from massive companion	relatively common?	?	all planets in circular orbits, WD companion possible
TŻO Deflation Model	TŻO envelope deflates after fuel exhaustion to form massive disk	moderately common?	?	planets (low-mass stars) in circular orbits

Evaporation Model	"planet" is remnant of (almost completely) evaporated low-mass star	rare	yes	only one planet possible
Disrupted Companion Model	low-mass companion is dynamically disrupted and forms massive disk	moderately common?	yes	all planets in circular orbits
Circumbinary Disk Model	evaporated wind forms circumbinary, planet-forming disk	moderately common?	yes	all planets in circular orbits
Ablation Scenario	planetary disk formed out of material ablated from close binary companion during supernova	rare	no	all planets in circular orbits
Disk-Capture Scenario	ms pulsar captures protostellar disk	low?	no	all planets in circular orbits
Overmassive WD Scenario	planets survive around overmassive WD	impossible?	no	

REFERENCES

Artymowicz, P., Clarke, C. J., Lubow, S. H. & Pringle, J. E. 1991, ApJ, 370, L35

Bailes, M., Lyne, A. G., & Shemar, S. L. 1991, Nature, 352, 311

Banit, M., Ruderman, M., & Shaham, J. 1992, preprint

Benz, W., Bowers, R. L, Cameron, A. G., & Press, W. H. 1990, ApJ, 348, 647

Bhattacharya, D., & van den Heuvel, E. P. J. 1991, Phys. Rep, 203, 1

Biehle, G. T. 1991, ApJ, 380, 167

Blaauw 1961, Bull. Astron. Inst. Netherlands, 15, 265

Bonsema, P. J. F., & van den Heuvel, E. P. J. 1985, A&A, 146, L3

Cannon, R. C., Eggleton, P. P., Żytkow, A. N., & Podsiadlowski, Ph. 1992a, ApJ, 386, 206

Cannon, R. C. et al. 1992b, in preparation

Chakrabarti, S. K., & Krishna Swamy, K. S. 1992, preprint

Chevalier, R. A. 1989, ApJ, 346, 847

Dawson, P. C. 1991, Nature, 352, 763

Eggleton, P. P., & Verbunt, F. 1986, MNRAS, 220, 13p

Fabian, A. C., & Podsiadlowski, Ph. 1991, Nature, 353, 801

Fabian, A. C., Pringle, J. E., & Rees, M. J. 1975, MNRAS, 172, 15p

Field, R. G., Roger, R. D., & Rybicki, G. B. 1992, preprint

Fruchter, A. S., Stinebring, D. R., & Taylor, J. H. 1988, Nature, 333, 237

Humphreys, R. 1984, in Observational Test of the Stellar Evolution Theory, ed. A. Maeder, & A. Renzini (Dordrecht, Reidel), p279

Iben, I., Jr., & Tutukov, A. V. 1984 in Stellar Nucleosynthesis, ed. C. Chiosi, & A. Renzini (Dordrecht, Reidel), p. 181

Illarionov, A. F., & Sunyaev, R. A. 1975, A&A, 39, 185

Joss, P. C., & Rappaport, S. 1984, in High Energy Transients in Astrophysics, ed. S. E. Woosley (AIP, New York), p. 555

Johnston, S. et al. 1992, ApJ, 387, L37

Krolik, J. H. 1991, Nature, 353, 829

Lin, D. N. C., Woosley, S. E., & Bodenheimer, P. H. 1991, Nature, 353, 827

Lyne, A. G. et al. 1990, Nature, 347, 650

Malhotra, R., Black, D., Eck, A., & Jackson, A. 1992, Nature, 356, 583

Meyer-Hofmeister, E. 1992, A&A, 253. 459

Mineshige S., Nomoto, K., & Shigeyama, T. 1992, preprint

Nakamura, T., & Piran, T. 1991, ApJ, 382, L81

Nakano, F. 1987, MNRAS, 224, 107

Nomoto, K. & Iben, I., Jr. 1985, ApJ, 297, 531

Podsiadlowski, Ph. 1991, Nature, 350, 136

Podsiadlowski, Ph., Pringle, J. E., & Rees, M. J. 1991, Nature, 352, 783

Rasio, F. A., Nicholson, P. D., Shapiro, S. L., & Teukolsky, S. A. 1992, Nature, 355, 325

Rasio, F. A., Shapiro, S. L., & Teukolsky, S. A. 1992, A&A, 256, L35

Ruffert, M., & Müller, E. 1990, A&A, 238, 116

Ryba, M. F., & Taylor, J. H. 1991, ApJ, 380, 557

Stevens, I. R., Rees, M. J., & Podsiadlowski, Ph. 1992, MNRAS, 252, 19p

Suntzeff, N. B., Phillips, M. M., Depoy, D. L, Elias, J. H., & Walker, A. R. 1991, AJ, 102, 1118

Tavani M., & Brookshaw, L. 1992, Nature, 356, 320

Thorne, K., & Żytkow, A. N. 1975, ApJ, 199, L19

Thorne, K., & Żytkow, A. N. 1977, ApJ, 212, 832

van den Heuvel, E. P. J. 1992, Nature, 356, 668

van den Heuvel, E. P. J., & Rappaport, S. A., in Physics of Be stars, ed. A. Slettebak, & T. P. Snow (Cambridge University Press), P 291

Wijers, R. A. M. J., van den Heuvel, E. P. J., van Kerkwijk, M. H., & Bhattacharya, D. 1992, Nature, 355, 593

Wolszczan, A., & Frail D. A. 1992, Nature, 355, 145

Zimmermann, M. E. 1979, Ph.D. thesis, California Institute of Technology

Planets Around Pulsars
ASP Conference Series, Vol. 36, 1993
J.A. Phillips, J.E. Thorsett, and S.R. Kulkarni

PLANET FORMATION AROUND MILLISECOND PULSARS

MENASHE BANIT, MALVIN RUDERMAN and JACOB SHAHAM
Physics Department and Columbia Astrophysics Laboratory.
Columbia University, New York, NY 10027.

ABSTRACT We present a model (Banit & Shaham 1992; Banit, Ruderman & Shaham 1992, henceforth BRS) for the formation of planets in circular orbits around millisecond pulsars. We propose that the planets originate from a circumbinary excretion disk around a Binary-Millisecond-Pulsar (BMP), and show how physical conditions in such a disk lead to the eventual formation of planets.

INTRODUCTION

Timing measurments of the millisecond pulsar PRS1257+12 shows pulse arrival time delays that can be analysed in terms of two modes of stable oscillations with a combined amplitude of about half the pulsation period (Wolszczan & Frail 1992). Generally, oscillatory delays in pulse arrival time may be a result of two kinds of effects: the pulsar's motion along the line of sight due to gravitational interaction with objects in orbit around it, or an angular oscillation of the emission region in the rotating frame. The latter may result from either free precession (Pines & Shaham 1972) or Tkachenko oscillation (Tkachenko 1966; Ruderman 1970). For PSR1257+12, however, it is hard to understand how these type of modes remain stable with the large pulse arrival time delay amplitude (BRS). Therfore, it seems natural to attribute the pulse arrival time delays to the former effect.

The delays can be fit by assuming that two planets are in orbit around the pulsar (Wolszczan & Frail 1992). The fit uses masses that corresponds to $m \sin i$ of $2.8M_E$ and $3.4M_E$ and periods of 98.2 and 66.6 days respectively. Here, m is a planet mass and i is the inclination angle of the orbital axis to the line of sight. Fitting the orbital parameters gives an eccentricity lower than 0.03 for both planets. Future pulse arrival times measurments, which should exhibit the expected orbital parameters variation due to inter planetary gravitational interaction, can confirm the presence of the planets (Rasio *et al.* 1992).

The planets' low orbital eccentricities challenge the usual NS formation theory. NS's are born during the core collapse of massive stars $8 < \frac{M}{M_\odot} < 20$ in a supernova (SN) explosion. The progenitor star is usually a Red-Giant (RG) with an extended envelope of radius larger than $2AU$ (Woosley & Weaver

1986; Meader & Meynet 1988). Whether planets that are at a smaller orbital radius could survive this phase without spiraling into the RG's center is not yet clear. More importantly, since more than half the system's mass is ejected in the SN explosion, the planets will unbind if the explosion is symmetric. The gravitational binding energy increase of the collapsing core alone is expected to produce an eccentricity of about 0.1. On the other hand pulsar proper motions at velocities of about 10^2Km s^{-1} are attributed, in part, to an asymmetry in the SN explosions (Narayan & Ostriker 1990). A well designed NS kick velocity (Hills 1983) could keep a given planet in a bound circular orbit. However, if the planetary orbits are almost circular before the SN explosion, the probability that *both* observed planets will remain bound with such low eccentricities is negligible. For this reason, a model in which the planets preexist has problems. This suggests that either the planetary orbits were more eccentric after the pulsar's birth and circularized later, or that the planets were formed after the pulsar's birth.

According to one proposal (Field, Roger & Rybicki 1992) the planetary orbits after the SN explosion were, in fact, eccentric, but were circularized by a planetary wind fueled by the pulsar radiation (Ruderman, Shaham & Tavani 1989). Such a process, we find, is unlikely to be efficient enough (BRS).

In another wind-related model for planet-pulsar formation (Krolik 1991), the circular orbit is guaranteed: in this model, the planet is the residual of an ablated (single) companion in a BMP. This model does not explain the presence of two planets.

In yet another model (Wijers *et al.* 1992) circularization occurs via drag in the envelope of a Thorne-Zytkow (TZ) star (Thorne & Zytkow 1977). Here, circularization processes may be sufficiently efficient; however, it is hard to understand how could PSR1257+12 spin up to its present period or faster during a TZ object's lifetime (BRS).

These considerations seem to favor models in which the planets are formed after the pulsar's birth. Since canonical planetary formation puts their origin in a disk, most of the models for the origin of the PSR1257+12 planetary system include a phase with a disk around the pulsar. In such models the planets are likely to have low eccentricities, similar to those observed in PSR1257+12.

In one of these models (Lin, Woosley & Bodenheimer 1991), the disk forms from a fallback of a fraction of the SN explosion debris. The disk's mass is $\simeq 10^{-3} M_\odot$ and its typical initial radius is about $10^8 cm$. In another model (Podsiadlowski, Pringle & Rees 1991) the disk (and the pulsar) form by the coalescence of two white dwarfs (WDs) in a binary. The lighter WD overflows its Roche-Lobe (RL) and forms a disk around the heavier component (Benz, Bowers & Cameron 1990) and the latter forms a NS by accretion induced collapse (Saio & Nomoto 1985). Here the disk's mass is $\simeq 0.1 M_\odot$ and its initial radius is typicaly $\simeq 10^9 cm$. Such models suffer from the difficulty of a relatively low initial disk's specific angular momentum. According to standard disk evolution scenarios (Pringle 1981) it is hard to see how the disk's surface density at $\simeq 0.5 AU$ can build up to the value required for planetary formation, assuming that it starts at such low radii (Banit *et al.* 1992).

Moreover, most models seem to ignore the standard theory for millisecond pulsar formation. PSR1257+12 belongs to a class of pulsars that are believed to have been spun-up to short periods by accretion in LMXBs (Alpar *et al.*

1982; Bhattacharya & van den HEUVEL 1991). In later evolution accretion stops, possibly because of the strong pulsar radiation flux (Iliarionov & Sunyaev 1975; Ruderman, Shaham & Tavani 1989). That, in turn, may ablate a very low mass companion (Ruderman *et al.* 1989) and leave behind an isolated millisecond pulsar.

We suggest, that the evaporation of the companion star in such BMPs, may supply matter to a circumbinary exretion disk (Banit & Shaham 1992; BRS). Such a disk may have a mass $0.1 - 0.01 M_{\odot}$ and a typical initial radius of a few times $10^{11} cm$. Here, therfore, the disk has a relatively large initial specific angular momentum and its evolution may provide the necessary surface density for the planets to form. Furthermore, in this model, the disk formation is an integral part of what is believed to be the most probable scenario for the formation of isolated millisecond pulsars.

FORMATION OF A CIRCUMBINARY DISK

Observations of the BMP systems 1957+20 (Fruchter, Stinebring & Taylor 1988) and 1744-24A (Lyne *et al.* 1990) provide evidence for the possibility of ablation of companions, at least in the BMP evolutionary phase. The orbital period of the BMP1957+20 system appears to be decreasing on a short timescale: $\frac{P_{orb}}{|\dot{P}_{orb}|} \simeq 3 \cdot 10^7$yrs (Ryba & Taylor 1991). A possible reason for both the large eclipse size and the rapid orbital shrinking of this system is that the wind is relatively cool and slow (Banit & Shaham 1992) ($\frac{V_{wind}}{V_{escape}} \lesssim 1$, V_{wind} is a typical wind velocity and V_{escape} is the escape velocity from the companion and/or the binary system). The wind gets away from the binary by the combined effects of pulsar radiation heating and radiation pressure; gravitational interaction of the particles in such a wind with the binary system produces a torque that reduces the system's orbital angular momentum. Particles are delayed in the vicinity of the second Lagrangian point, behind the companion, and their specific angular momentum can increase by a factor $Q \simeq 1.3$ for BMP1957+20. Computer simulations of 3-dimensional ballistic trajectories produced *both* the eclipse shape and the orbital shrinking time-scale (Banit & Shaham 1992). This model fits the observed parameters if the companion is bloated to its RL dimension, as has been independently suggested by recent observations (Fruchter & Goss 1992).

In winds like that, where the particle specific energy can be, on the average, below the escape energy, one therfore expects matter to move out via an excretion disk (Webbink 1976; Pringle 1991), which for 1957+20 would begin at an orbital radius $\simeq 3 \cdot 10^{11}$cm, about twice the systems orbital seperation. The expected surface density of evaporated particles that are fed to the disk, as a function of orbital radius, will be determined by their specific angular momentum distribution (Banit *et al.* 1992). Particles that start to move on eccentric trajectories are expected to rapidly dissipate part of their energy, and merge into circular orbits while keeping their angular momentum. To determine the surface density profile of the disk it seems usefull to use companion wind parameters that agree with observations of BMPs (Banit *et al.* 1992).

EVOLUTION OF THE DISK

In *accretion* disks matter that moves inward accretes onto the compact object. Here, by contrast, matter cannot find itself inside the binary because radiation pressure drives it out. The inner boundary of the resulting *excretion* disk lies just outside of the binary at a specific relative angular momentum that is below Q, and shrinks with the binary. ¿From this inner boundary, angular momentum is transported outwards by the radially expanding disk. At any given moment, the angular momentum lost by the binary up to then is stored in that expanding excretion disk (unless part of it is blown off by the radiation pressure without passing through the disk); the angular momentum balance determines the surface density profile of the disk. Mass loss from the disk is largely due to the pulsar's radiation pressure on its inner surface boundary. From this boundary particles might be blown away from the system with relatively low specific angular momentum.

Most of this circumbinary disk is expected to be locally Keplerian (except, possibly, its inner boundary). The disk's local surface density will vary due to both its viscous evolution and (at its inner region) also due to the rate of evaporated material from the companion. Angular momentum will be transported on the viscous timescale $\tau_\nu \simeq \frac{r^2}{\nu} \simeq 10^6 - 10^7$yrs (Pringle 1981) where r is the orbital radius and ν is the kinematic viscosity. This timescale is shorter than the companion ablation timescale $\tau_m \simeq \frac{m}{|\dot{m}|} \simeq 10^7 - 10^8$ yrs (Banit & Shaham 1992) provided that the system evolve in a way similar to BMP1957+20. The total amount of evaporated material that is fed into the disk is more than two orders of magnitude higher than the observed planetary mass and is relatively heavy in chemical composition, as it originates in the core of an evolved WD. The fraction of mass that leave the companion and sould end up in planets is not very different from the ratio of planetary mass in the solar system to the believed protoplanetary solar nebula mass (Safronov 1969).

The evaporating gas temperature in the companion's vicinity is of the order $10^5 - 10^{6o}K$ (Banit & Shaham 1992), however it is expected to rapidly cool to the local disk temperature by radiative processes (Banit *et al.* 1992). The local disk temperature at $r \gtrsim 10^{12} cm$, where the planets should form, is expected to be of the order of $100^o K$. The effects of illumination of these outer disk regions by the pulsar are likely to be negligible (Banit *et al.* 1992). At such low temperature dust grains will condense fairly rapidly (Takenori 1987; Weidenschilling 1984). Then they sink gradually to the disk's midplane (where they are better hidden from the pulsar illumination) with fast sedimentation (Weidenschilling 1987). The effective viscosity of the so formed thin layer of dust grains may be very low; however, the gaseous disk will continue to evolve on its relatively short viscous timescale.

During the companion's ablation, dust grains and their surface density will continue to grow. Once the companion has completely evaporated, relatively fast fading of the gaseous disk is expected to follow. At this stage the dust grains are no longer hidden from the pulsar's illumination, but, they are likely to be sufficiently large that the Pointing-Robertson drag and gas drag can be neglected (Banit *et al.* 1992). While still in the phase in which the companion is in orbit around the pulsar gravitational instabilities may be suppresed due to the periodic variation of the gravitational field. Whether or not they really are suppresed in that phase, it is clear that instabilities can grow once the companion is sufficiently small and lead to the formation of planetesimals (Goldrich & Ward 1973). Further coalescence (Safronov 1969) will lead to the formation of planets similar to the observed ones, on a relatively short timescale (Wetherill 1980; wetherill 1989; Lin, Woosley & Bodenhcimer 1991).

ACKNOWLEDGEMENTS

We gratefully acknowledge support from NASA (grants NAGW-1618 and NCC 5-37) and the NSF (grant AST89-01681).

REFERENCES

Alpar, M. A., Cheng, A. F., Ruderman, M. A., and Shaham, J. 1982, *Nature*, **300**, 728.
Banit, M., Ruderman, M. A., and Shaham, J. 1992 in *Proceeding of the workshop on Physics of Isolated Pulsars* (Taos New-Mexico); BRS.
Banit, M., Ruderman, M. A., Shaham, J., and Applegate, J. 1992 *In pr eparation.*
Banit, M., and Shaham, J. 1992, *Ap. J. (Letters)*, **388**, L19.
Benz, W., Bowers, R. L., Cameron, A. G. W., and Press, W. H. 1990, *Ap. J.*, f348 647.
Bhattacharya, D. and van den Heuvel, E. P. J. 1991, *Physics Reports* **203**, 1.
Field, R. G., Roger, R. D., and Rybicki, G. B. 1992 *Preprint.*
Fruchter, A. S., and Goss, W. M. 1992, *Ap. J. (Letters)*, **384**, L47.
Fruchter, A. S., Stinebring, D. R., and Taylor, J. H. 1988 *Nature* **333**, 237.
Goldreich, P., and Ward, W. R. 1973, **183**, 1051.
Hills, J. G. 1983, *Ap. J.*, **267**, 322.
Iliarionov, A. F., and Sunyaev, R. A. 1975, *Astr. Ap.*, **39**, 185.
Krolik, J. H. 1991, *Nature* **353**, 829.
Lin, D. N. C., Woosley, S. E., and Bodenheimer, P. H. 1991, *Nature* f353, 827.
Lyne, A. G., 1990, *et al. Nature* **347**, 650.
Meader, A., and Meynet, G. 1988, *Astr. Ap. Suppl.*, **76**, 411.
Narayan, R., and Ostriker, J. P. 1990, *Ap. J.*, **352**, 222.
Nelson, R. W., Finn, L. S., and Wasserman, I., 1990, *Ap. J.*, **348**, 226.
Pines, D., and Shaham, J., 1972a, in Physics of the Earth and Planetary I nterior **6**, 103
——. 1972b *Nature phys. sci.* **253**, 43.
——. 1972c *Comm. Ap.* **6**, 37.
Podsiadlowski, Ph., Pringle, J. E., and Rees, M. J. 1991, *Nature* **352**, 783.
Pringle, J. E. 1991, *M.N.R.A.S.*, **248**, 754.
Pringle, J. E. 1981, *Ann. Rev. Astron. Astrophys.* **19**, 137.
Rasio, F. A., Nicholson, P. D., Shapiro, S. L., and Teukolsky, S. A. 1992, *Nature* **355**, 325.
Ruderman, M. A. 1970, *Nature*, **225**, 619.
Ruderman, M. A., Shaham, J., and Tavani, M. 1989, *Ap. J.*, **336**, 507.

Ruderman, M. A., Shaham, J., Tavani, M., and Eichler, D., 1989, *Ap. J.*, **343**, 292.
Ryba, M. F., and Taylor, J. H. 1991, *Ap. J.*, **380**, 557.
Safronov, V. S. 1969, in *Evolution of the Protoplanetary cloud and the Formation of the Earth and Planets*, (Nauka, Moscow); (English translation 1972) NASA TTF-667.
Saio, H., and Nomoto, K. 1985 *Ap. J. (Letters)*, **150**, L21.
Takenori, N., 1987, *M.N.R.A.S.*, **224**, 107.
Thorne, K. S., and Zytkow, A. N. 1977, *Ap. J.*, **212**, 832.
Tkachenko, V. K. 1966, *Sov. Phys. JEPT* **23**, 1049.
Webbink, R. F. 1976, *Ap. J.*, **209**, 829.
Weidenschilling, S. J. 1984, *Icarus* **44**, 172.
Weidenschilling, S. J. 1987, *Gerlands Beitr. Geophys.* **96**, 21.
Wetherill, G. W. 1980, *Ann. Rev. Astron. Astrophys.* **18**, 77.
Wetherill, G. W., and Stewart, G. R. 1989, *Icarus* **77**, 330.
Wijers, R. A. M. J., van den Heuvel, E. P. J., van Kerkwijk, M. H., and Bhat tacharya, D. 1992, *Nature* **355**, 593.
Wolszczan, A., and Frail, D. A. 1992, *Nature* **355**, 145.
Woosley, S., and Weaver, T. A. 1986, *Ann. Rev. Astron. Astrophys.* f24, 205.

Planets Around Pulsars
ASP Conference Series, Vol. 36, 1993
J.A. Phillips, J.E. Thorsett, and S.R. Kulkarni

PLANETS IN GLOBULAR CLUSTERS?

STEINN SIGURDSSON
Lick Observatory, University of California, Santa Cruz, CA 95064

ABSTRACT

The discovery of planets around PSR 1257+12 suggests that planetary systems may be detected around the recycled pulsars found in globular clusters. Detection of planets around cluster pulsars can constrain planet formation mechanisms, in particular the effects of low metallicity on planet formation in disks. Planetary systems in dense clusters have lifetimes to disruption due to perturbations by passing stars comparable to or shorter than the pulsar lifetime, and observations of planets in the cores of clusters may reveal planetary systems formally dynamically unstable on time scales short compared to the characteristic age, τ_c, of the system. Planets formed around cluster pulsars will most likely be restricted to semi–major axis of $\sim 0.1 - 1.0\,AU$, while "scavenged" planets may be observed in wider orbits, with no stable systems expected in the densest clusters.

PSR 1257+12 is a classic "recycled" millisecond pulsar (MSP), with a short (6.2 ms) period, low inferred magnetic field ($\sim 10^9$ G) and long characteristic age ($\tau_c \sim$ few $\times 10^8$ years) (Wolszczan 1991, Wolszczan and Frail 1992). It is similar to the recycled pulsars observed in galactic globular clusters (GCs) (Phinney and Kulkarni 1992, van den Heuvel 1991), and mechanisms comparable to those that generated the putative planets around PSR 1257+12 may produce planets around around cluster pulsars. Various scenarios have been proposed to account for the presence of planets around pulsars (see summaries of formation scenarios by Phinney and Podsiadlowski, this volume). In clusters formation scenarios involving SNIIs (Lin et al. 1992) or HMXBs (Wijers et al. 1992, Podsiadlowski et al. 1992) can not be at work, only reaccretion from a disrupted companion or an excretion disk, or, possibly, scavenging of a companion's planetary system during a LMXB phase, can apply.

CONSTRAINTS ON FORMATION

Current theories of planet formation require planetesimals to aggregate out of a thin disk of dust sedimented out of the thicker accretion disk. In the case of recycled pulsars the disk may be an "excretion" disk from an ablated companion (Krolik 1991, Nakamura and Piran 1991, Tavani and Brookshaw 1992), or the remnant of a wholly disrupted star that merged with the neutron star (or heavy white dwarf) (Podsiadlowski et al. 1991, Fabian and Podsiadlowski 1991). Such disks may have considerably higher surface densities in early stages of their evolution than the protoplanetary disks normally considered for planet formation. Upon reaching a critical size, the planetesimals undergo runaway growth, reaching masses of order $10M_\oplus$, before gas accretion commences, if the ambient temperature is low enough (Cameron 1988). The timescale for planetesimal growth is a function of metallicity, and with cluster metallicities, Z, ranging from 10^{-1}–10^{-3} solar, the presence of sufficient metals to grow planetesimals is a concern. At $Z = 10^{-3}$, the total mass of ice (rocky material) available for aggregation is only $6\,(1.5)\,M_\oplus$ per $M_\odot$ of disk matter, insufficient to form even a single large planet if the total mass available in the disk is much less than $M_\odot$.

The time scales for planet formation have been studied mostly in the context of the pre–solar nebula. Following Nakano (Nakano 1987) we estimate a conservative timescale, t_{pf}, for planet formation, assuming a surface density profile, $\Sigma(r) = fr^k$, $r \leq r_0$, where f is a normalising constant, canonically $k = -3/2$. For total disk mass m_d, $f = (k+2)m_d/2\pi r_0^{k+2}$, $k \neq -2$. Assuming the planet formation timescale is dominated by the planetesimal aggregation time scale, $t_{pf} \approx 10^3 \frac{r}{\Sigma_s(r)}$, where Σ_s is the surface density of solids. $\Sigma_s(r) \approx 1.8 \times 10^{-2}(4.2 \times 10^{-3})\, Z\, \Sigma(r)$ for ice and C–Si–Fe grains respectively, for r such that the local temperature is low enough for the respective solids to condense out. Hence, for a total disk mass $m_d = M_{-2}10^{-2}M_\odot$,

$$t_{pf} \approx 4(16) \times 10^6 \frac{1}{Z} \frac{1}{(k+2)M_{-2}} r_{AU}^{1-k} \text{ years}, \qquad (1)$$

for nucleation from ice or C–Si–Fe grains respectively, and r_{AU} is the radius in AU at which planetesimal growth is taking place. Grain formation can not take place if the ambient temperature is greater than the vapourisation temperature of the grains, approximately 2000K for C–Si–Fe grains, about 200K for ice grains, providing a lower limit of $r_{min} = 0.02\,(2.0)\,(L_{PSR}/L_\odot)^{1/2}$ AU for C–Si–Fe and ice grains respectively. It is clear that, except possibly for the lowest metallicity clusters, $t_{pf} \lesssim \tau_c$ for $r \gtrsim r_{min}$, for C-Si–Fe nucleated planets, assuming efficient absorption of the pulsar flux, and pulsar luminousities characteristic of recycled pulsars. The effects of metallicity on the surface density profile of solids in disks are not known, it is possible that the low opacity of metal poor accretion disks increases the timescale for disk evolution by turbulent viscosity, leading to higer Σ_s and enhanced planet formation efficiency. Formation scenarios involving AIC or white dwarf collisions (with another white dwarf or a neutron star), provide cases in GCs in which the accreting matter has a very high metallicity and consequently a short timescale for planetesimal aggregation. Simulations suggest that in neutron star–binary interactions up to half the tidal interactions may be with white dwarfs, assuming a reasonable initial mass function and an evolved binary population, with the heavier C–O white dwarfs having a proportionately larger fractional cross–section for tidal encounters (Sigurdsson and Phinney 1992). For $L_{PSR} \sim L_{Edd} \approx 10^5 L_\odot$ expected during pulsar spin–up, $r_{min} = 6\,AU$ for C–Si–Fe nucleated planetesimals, so planetary formation only commences after spin–up. It is possible that any scavenged planets present with semi–major axis $a_p \lesssim 1\,AU$ will be ablated completely during spin–up. Most of the accretion disk may be ejected during spin–up, but if the metals from order

$10^{-2}M_\odot$ or greater are retained, in an annulus from r_{min} to $r_0 \lesssim 2\,AU$ enough matter is present for planet formation to be possible. It is not necessary to retain the bulk of the gas, as long as the metals have opportunity to condense out into a thin disk. If PSR 1257+12 scavenged its planets from a companion main–sequence star with a normal planetary system, circularising the orbits during the accretion phase, a similar mechanism may also operate in clusters, with pulsars retaining primordial planets after disrupting the companion. With the time scales assumed here, primordial planetary formation around main–sequence stars is possible in GCs, and although the encounter between a neutron star and a main–sequence primary may disrupt the planetary system, friction in the envelope of the merged remnant would circularise and shrink the orbits of any surviving planets. This scenario permits the presence of planets in eccentric orbits with $a_p \gg 1\,AU$ around cluster pulsars in low density clusters, whereas the timescales for reaccretion would predict maximum $a_p \lesssim 1\,AU$ and small eccentricities.

If the cluster formation rate is dominated by neutron star–binary star interactions (Phinney and Kulkarni 1992, Sigurdsson 1991), and planets do not form in main–sequence binaries, then scavenging is less likely in GCs, leaving reaccretion from a remnant as the dominant channel for planetary formation in clusters. Scavenging is still possible in the case where the neutron star is in a binary, and encounters a single main–sequence star with a planetary system. If the single star has spent a large fraction of the cluster history outside the cluster core, it may possess an extended planetary system, even in the denser clusters. If pulsar recycling in GCs is dominated by binary encounters, a companion of mass m_c may remain in an eccentric orbit, semi–major axis a_c, about the accreting system (Sigurdsson 1991). The effects on planet formation of a stellar companion in a highly eccentric orbit about a protoplanetary disk are unknown; perturbations from the companion may disrupt planet formation, or may enhance planetary formation by providing density enhancements in the disk and perturbing planetesimal orbits. In either case, the companion will clean out part of the disk, in a zone of width $\sim 3 \times a_c(m_c/M_{PSR})^{(1/3)}$ about periastron and apastron. Recycled MSPs with stellar mass companions of periods $\lesssim 1$ year would not be expected to form planets.

STABILITY OF PLANETARY SYSTEMS IN GLOBULAR CLUSTERS

Most cluster pulsars are found in the cores of clusters, where the stellar density, $n_4 = n/10^4$ pc^{-3}, may be as high as 10^2 for post–core–collapse (PCC) clusters. At such densities close stellar encounters are frequent, and it becomes necessary to consider the lifetime of planetary systems to disruption. The disruption mechanism may be usefully considered in two parts, direct ionisation of the planet by encounters with field stars with pericenters $p \lesssim a_p$, and the perturbation of the outer planets of a system by more distant encounters, leading to chaotic disruption of the planetary system through dynamical evolution of the orbits. The cross-section for a field star of mass $M_* \sim 0.7M_\odot$, to approach to within p_{AU} from a neutron star is $\sigma \approx \pi p_{AU}^2(1 + 36/(p_{AU}v_{10}^2))\;AU^2$, for a relative velocity at infinity of $v_{10} = v_\infty/10$ km s^{-1}. The mean time between encounters, T, is given by

$$T = \frac{3 \times 10^9}{< n_4 \sigma_{AU} v_{10} >} \text{ years}, \tag{2}$$

where $\sigma_{AU} = \sigma/\sigma(p = 1\,AU)$. Thus a system like PSR 1257+12 would be expected to be disrupted on time scales short compared to τ_c in PCC clusters like M15, but would not experience ionising encounters in lower density clusters such as M13 or M53. Interestingly, the time scales for direct disruption of a planet in a $\sim 1\,AU$ orbit in a medium density cluster like 47Tuc are comparable to τ_c in the core. For encounters with pericenter $> a_p$, the planet's orbit is perturbed by the encounter, with the eccentricity, e, and inclination being sensitive to encounters with pericenters of order $3\,a_p$, for wider encounters the perturbation is exponentially small and can be neglected.

The effective cross–section for ionisation and strong perturbations was evaluated by direct integration of 25000 encounters between a $1.4M_\odot$ (M_{PSR}) primary with a $10M_\oplus$ (m_p) secondary in a $a_p = 1\,AU$, $e = 0$ orbit and a $0.7M_\odot$ (m_c) field star, with $v_\infty = 5 - 10$ km s^{-1}. The encounters were uniformly distributed in phase space. Defining a critical velocity v_c, such that for $v_\infty = v_c$ the total energy of the system is zero in the center–of–mass frame,

$$v_c = \sqrt{G \frac{M_{PSR} m_p}{m_c} \frac{(M_{PSR} + m_p + m_c)}{(M_{PSR} + m_p)} \frac{1}{a_p}}, \tag{3}$$

we find for parameters relevant for cluster encounters that $v_\infty/v_c \sim 16 - 32$. The induced eccentricity and inclinations are not sensitive to the exact mass–ratio or relative velocity. The cross-section for direct ionisation was dominated by encounters with pericenter $p < a_p$, with approximately one quarter of those encounters ionising the system. An additional quarter of such close encounters led to an increase in the planetary semi–major axis by a factor of two or greater, leaving a system likely to experience a second close encounter on a timescale short compared to the timescale for the initial encounter.

Perturbating encounters can be usefully parameterised by λ, the ratio of the mean motion of the binary to the mean motion of the perturber. Analytic approximations are possible for $\lambda \gg 1$ and $\lambda \ll 1$ respectively, with $\delta e, \delta i \propto \int p(\lambda) e^{-\lambda}$ for $\lambda \gg 1$ (Murray personal communication), where $p(\lambda)$ is a polynomial in λ, and the integral is over phase angles. For the case of interest, $\lambda \gtrsim 1$, and the simulations were designed to probe the region $\lambda \sim 1$, where the analytic approximations are expected to be inadequate. The induced eccentricity for co–planar encounters was compared to analytic calculations and the shape of the power law was verified for moderate λ, as was the relative magnitude of perturbations due to prograde and retrograde encounters. The simulations were not sufficiently precise to accurately track perturbations of order 10^{-5} or less, but correcting for known numerical errors did show the expected exponential decrease in perturbation amplitude for large λ. Asymptotic estimates for λ for $v_p \gg v_\infty$ and $v_p = v_\infty$, for relative pericenter speed v_p, show $\lambda \propto (p/a_p)^{3/2}$ and $\lambda \propto (p/a_p)/(v_\infty/v_c)$ respectively (see Figure 1). If there are several planets in the system, perturbations in e or i of the outer planets greater than order 10^{-2} may induce instability in the planetary system, with time scales of order $10^3 P$ or greater, where P is the period of the outermost planet (Quinlan 1992). As

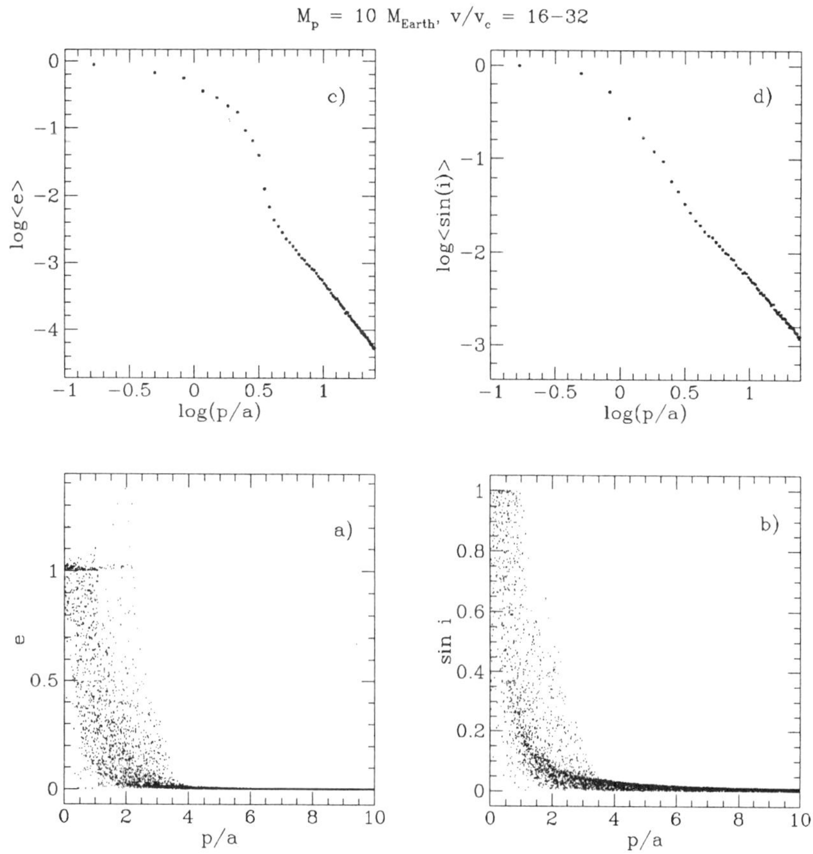

FIGURE I (a) Distribution of induced eccentricity, for 25000 encounters as function of ratio of pericenter, p, to planet semi–major axis, a. (b) Distribution of induced inclination, i, relative to the original binary axis, for same set of encounters as in Figure 1a. If there are several planets in the system, the outer planets may be strongly perturbed in inclination while the inner planets are slightly perturbed, the resulting system may be dynamically unstable. (c) Mean induced eccentricity as a function of pericenter, p. (d) Mean induced inclination as a function of pericenter. The inclination is more sensitive to encounters of order $3a$ than the eccentricity, and this may dominate the rate at which the system becomes unstable.

Figure 1 shows, encounters with $p/a_p \lesssim 3$ are effective in inducing perturbations of that order, with the inclination more sensitive to distant perturbations. More effort is needed to understand the effect of perturbations of inclination of outer planets on planetary system stability. The transition to chaotic internal dynamics may then lead to ejections of planets or collisions between planets on time scales short compared to τ_c. Colliding planets may remain in a (high e) orbit, or may disintegrate. In case of disintegration, the debris may reaccrete into a planet in a near circular orbit on time scales short compared to τ_c.

The probability of finding planets around recycled pulsars in PCC clusters such as M15 is small. In low density clusters such as M13, planets may be found with semi-major axis as large as $10\,AU$, with e increasing with a_p. A metal rich, medium density cluster, such as 47Tuc or Ter 5, offers the most interesting prospect for planet detection, with $a_p \lesssim 1\,AU$ possible, and e increasing with a_p.

Observations of planets about pulsars in 47Tuc type clusters could constrain planet formation mechanisms and determine the presence of primordial planets in GCs. It is possible that formally dynamically unstable planetary systems may be observed, allowing direct comparison with numerical models. With 10 MSPs now reported in 47Tuc (Manchester et al. 1991), detection of planets is highly probable, *if* planet formation or scavenging mechanisms operate in GCs. Failure to detect planets around any MSP in low or medium density GCs would strongly suggest that the formation mechanisms possible in GCs are not at work, and that low metallicity inhibits planetary formation in disks around pulsars and pre main-sequence stars.

ACKNOWLEDGMENTS

I would like to thank E.S. Phinney, D. Lin, P. Bodenheimer, P. Artymowicz and N. Murray for helpful discussion. This research was supported in part by NASA grant NAGW-2422.

REFERENCES

Cameron, A.G.W. 1988, *ARA&A*, **26**, 441

Fabian, A.C. and Podsiadlowski, Ph. 1991, *Nature*, **353**, 801

Grindlay, J.E. and Bailyn, C.D. 1988, *Nature*, **336**, 48

Krolik, J.H. 1991, *Nature*, **353**, 829

Lin, D.N.C., Woosley, S.E. and Bodenheimer, P.H. 1991, *Nature*, **353**, 827

Manchester, R.N., Lyne, A.G., D'Amico, N., Bailes, M. and Lim, J. 1991, *Nature*, **352**, 219

Michel, F.C. 1987, *Nature*, **329**, 310

Nakano, T. 1987, *MNRAS*, **224**, 107

Nakamura, T. and Piran, T. 1991, *ApJ*, **382**, L81

Phinney, E.S. and Kulkarni, S.R. 1992, *Nature*, (submitted)

Podsiadlowski, Ph., Pringle, J.E. and Rees, M.J. 1991, *Nature*, **352**, 783

Quinlan, G.D. 1992 in *Chaos, Resonance and Collective Dynamical Phenomena in the Solar System*, Ferraz-Mello, S

Sigurdsson, S. 1991, *Ph. D. Thesis*, Caltech

Sigurdsson, S. and Phinney, E.S. 1992, (in preparation)

Stevens, I.R., Rees, M.J. and Podsiadlowski, Ph. 1992, *MNRAS*, **254**, 19p

Tavani, M. and Brookshaw, L. 1992, *Nature*, **356**, 320

Thorsett, S.E. and Nice, D.J. 1991, *Nature*, **353**, 731

van den Heuvel, E.P.J. 1991, in *Neutron Stars: Theory and Observation*, Ventura, J. and Pines, D., Dordrecht: Reidel, 99

Wijers, R.A.M.J., van den Heuvel, E.P.J., van Kerkwuk, M.H. and Bhattacharya, D. 1992, *Nature*, **355**, 593

Wolszczan, A. and Frail, D.A. 1992, *Nature*, **355**, 145

Wolszczan, A. 1991, *Nature*, **350**, 688

ORDINARY PLANETARY SYSTEMS: ARCHITECTURE AND FORMATION

E. H. LEVY
Lunar and Planetary Laboratory, Department of Planetary Sciences
The University of Arizona, Tucson, Arizona 85721 USA

ABSTRACT Today we believe ordinary planetary systems to be an unremarkable consequence of star formation. The Solar System, so far the only confidently known example in the universe of a planetary system, displays a set of striking structural regularities. These structural regularities provide fossil clues about the conditions and mechanisms that gave rise to the planets. The formation of our planetary system, as well as its general characteristics, resulted from the physical environment in the disk-shaped nebula that accompanied the birth of the Sun. Observations of contemporary star formation indicate that the very conditions and mechanisms thought to have produced our own planetary system are widely associated with the birth of stars elsewhere. Consequently, it is reasonable to believe that planetary systems occur commonly, at least in association with single, Sun-like stars. Moreover, it is reasonable to believe that many planetary systems have gross characteristics resembling those of our own Solar System.

Keywords: cosmogony, origin of planetary systems, planets, planetary systems, planetary system formation, star formation.

INTRODUCTION

This article discusses the formation of "ordinary" planetary systems in broad conceptual overview. By an ordinary planetary system I mean what has loosely been encompassed under that rubric in the past: a system of several planets in orbit around a main-sequence star, a system that one might conceivably imagine to include one or more planets capable of supporting life. This definition is not rigorous, nor is it unique or complete, but it captures much of the intellectual and emotional motivation for focussing so much attention on planets—which, afterall, occupy an enormous amount of human concentration, far out of proportion to the relatively small fraction of the cosmic mass for which they account—and it has the advantage of distinguishing these traditional planets from the so-called pulsar planets that are the main subject of the present volume.

Human beings have speculated about the prevalence and nature of planets for thousands of years. The idea that planets are numerous and varied dates at least to the atomistic philosophers of Greece. Epicurus, circa 300 B.C., in *Letter to Herodotus*, wrote

> "There are infinite worlds both like and unlike this world of ours. For the atoms, being infinite in number ... are borne out far into space ... [and] have not been used up, either on one world or a limited number of worlds. So that there nowhere exists an obstacle to the infinite ..." (Dick 1982)

Today, more than two thousand years later, our own Solar System is still the only planetary system confidently known to exist. Nonetheless we are in a far better position than was Epicurus to speculate on both the prevalence and the character of planetary systems throughout the universe. In this article, I review the gross physical characteristics of the Solar System, especially as these reveal important clues about the conditions and mechanisms of planetary genesis. Then I describe the principal processes that are believed to have led to the formation of our planetary system. My emphasis throughout is on the most fundamental and basic physical aspects of planetary-system formation. My purpose is to provide a basis for speculating about how widespread and general the formation of similar planetary systems is likely to be. For this reason, I will avoid as much as possible delving into the details, and focus instead on the basic physical principles believed to be associated with the formation of planetary systems, and the ways in which these basic principles are believed to produce the gross architecture of planetary systems.

Pulsar Planets and the Pertinence to Real Planets

At the time of this writing it remains uncertain whether observed periodic pulsar-timing residuals actually signal the existence of "planets", or whether some other explanation will prevail. Moreover, if planets really do occur in orbit around pulsars it remains unknown how they formed. However, the putative existence of two planets in relatively short-period, circular orbits (Wolszczan and Frail 1992) suggests that these bodies formed in orbit after the supernova explosion that produced the pulsar, either out of the supernova debris or from some other material source. In this respect, the existence of pulsar planets would have potentially significant implications for our ideas about the formation of ordinary planets.

Perhaps the major goal of cosmical physics is to understand the nature and causes of cosmic structure. One of the most remarkable features of our universe is its variety of elaborate structures, including, for example: clusters of galaxies, galaxies, stars, planetary systems, and on to perhaps the most remarkable structure in the universe, life. For reasons that are not at all obvious, the universe seems to have been born with an exuberant propensity to generate complex structure. Among astrophysical structures, planets hold a special fascination. In particular, planets are probably uniquely the cosmic locales of life. And, as living things, we humans have a special interest in understanding the physical processes that give rise to planetary systems, and in learning whether such processes and such systems occur commonly in the universe, or whether they are somehow rare.

At present, we are in the peculiar situation that only one "true" planetary system is known to exist, while at the same time we speculate that there exist numerous similar systems throughout the universe. As we try to understand the generality of the processes that make planetary systems, and as we try to ascertain the prevalence and character of other planetary systems, it is of great interest and pertinence to learn whether similar processes operate to produce planetary-type systems even under the different conditions that may occur in so exotic a place as the orbital neighborhood of pulsar.

STRUCTURAL REGULARITIES IN THE SOLAR SYSTEM

The architecture of our own planetary system displays a remarkable set of structural regularities. These regularities provide crucial clues about the nature of conditions and physical mechanisms responsible for the formation of the Solar System. The motions of solar-system objects immediately suggest that the entire system descended from a single, simple object. The planets all revolve about the Sun in the same direction, in nearly concentric, nearly circular orbits; except for Mercury and Pluto—which have orbital eccentricities somewhat larger than 0.2—all of the planet have eccentricities $\lesssim 0.1$. The planetary orbits are closely confined to a single plane, again except for Mercury and Pluto—which have inclinations to the ecliptic of 7° and 17° respectively—none of the orbit inclinations exceed a few degrees. The average plane of the planetary orbits closely coincides with the spin equator of the Sun, which itself has the same sense of rotation as the planetary orbits. Planetary spin vectors show a tendency to be aligned with the angular momentum vectors of the orbits— though significant departures from true alignment suggest the influence of some possibly large random process at work. Altogether, kinematically, the Solar System has the appearance of giant clockwork of wheels turning within wheels, with nearly a single sense of rotation.

Planetary compositions and masses fall into two broad classes. One class, the terrestrial planets (including Mercury, Venus, Earth and Mars), occupies the inner Solar System, within about 1.5 astronomical units (1 AU = 1.5×10^{13} cm) of the Sun; the other class, the jovian planets (including Jupiter, Saturn, Uranus and Neptune), occupy the outer part of the Solar System, outward of about 5 AU from the Sun. The terrestrial planets are composed almost entirely of rock and metal and have relatively low masses, comparable to (in the case of Venus) or somewhat smaller (about an order of magnitude smaller in the cases of Mercury and Mars) than the mass of Earth. In the jovian planets, rock and metal account for a relatively small fraction of the mass, while various ices (H_2O, CH_4, NH_3, etc.) and gas (mostly hydrogen and helium) make up the major part of the total mass.

The known planetary system occupies a region of space extending to about 40 AU. Pluto—the most distant of the known planets—is a small and enigmatic double system composed of rock and ice; it is probably a remnant of the still somewhat mysterious accretion processes that built the planets of our Solar System. Orbiting in the gap between Mars and Jupiter—which marks the transition between terrestrial and jovian planets—are myriad small asteroids. Asteroids range from the largest, Ceres, nearly a thousand kilometers in diameter, to pebbles and specks of dust. The comets—perhaps 10^{12} in number—occupy a large volume of space surrounding the Sun and extending to tens of thousands of astronomical units.

The four jovian planets in the outer Solar System vary in mass from a high (Jupiter) of about 317 times the mass of Earth to a low (Uranus) of about fourteen times the mass of Earth. Despite the large mass differences among these four planets, they share a remarkable similarity in the amount

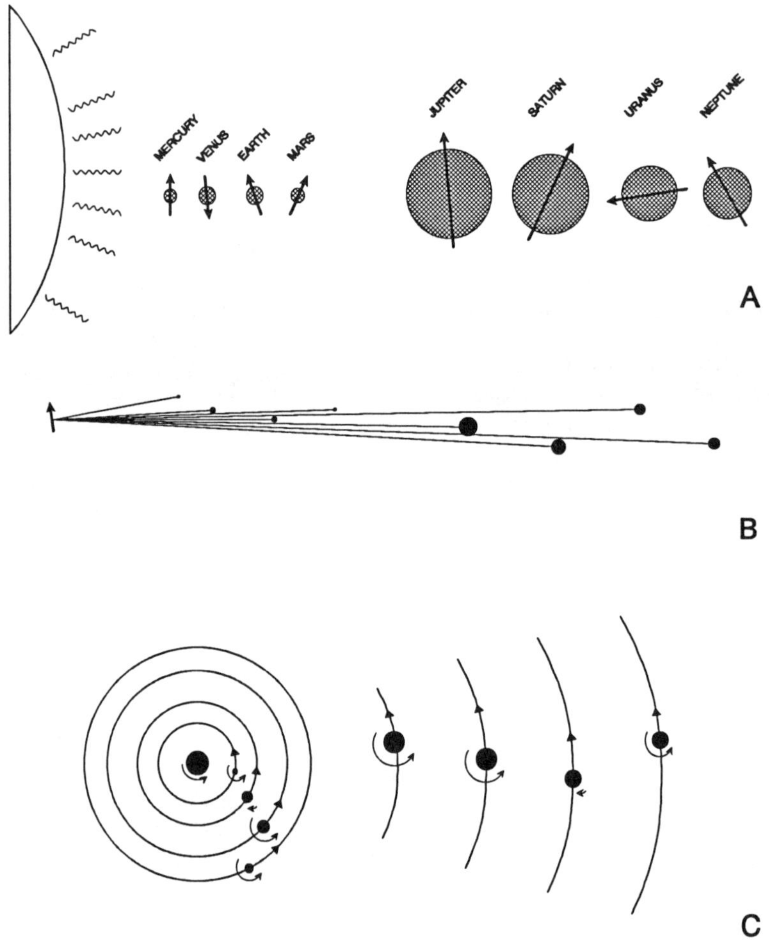

FIGURE 1. Cartoon representation of the major kinematical structural regularities of the Solar System. Planets orbit the Sun predominantly in the same plane (A,B), with a strong tendency for all angular momentum vectors to be aligned and oriented in the same direction (A,C). From the kinematics alone, the Solar System gives a strong signal of having formed from a single, coherently-structured precursor object. (None of the cartoons in this chapter attempt to accurately portray correct sizes or spatial scales.)

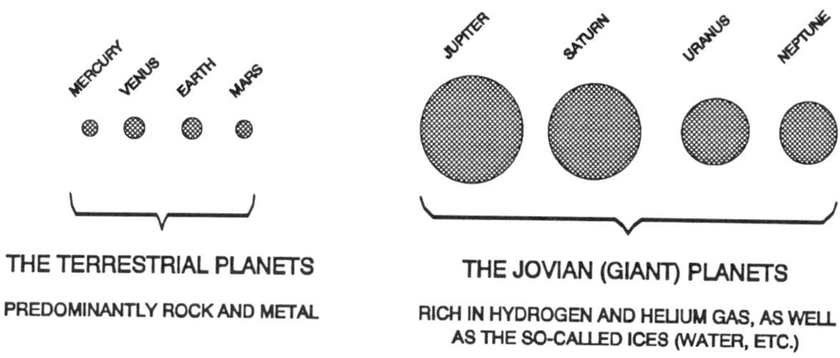

FIGURE 2. The major planets themselves fall into two classes: The terrestrial—or Earth-like—planets which are relatively small, occupy the inner part of the Solar System and consist almost entirely of rock and metal. The jovian—or Jupiter-like—planets, which are much larger, occupy the outer Solar System, and are rich in ice and gas.

of condensible matter in their interiors. Each of the these planets contains condensible material—comprised largely of rock and ice—in the amount of ten to twenty Earth masses, probably largely confined to a central core. The major differences in their sizes and masses are accounted for by massive gas envelopes composed predominantly of hydrogen and helium. This fact—the close similarity of the condensible, presumably accreted, cores of the giant planets—provides an important strategic basis for planet-system searches in the foreseeable future.

All of the planets have their heavy-element fraction highly enriched with respect to normal cosmic abundances ($z \sim 0.02$). For the terrestrial planets $z \sim 1$, while the giant planets have heavy-atom enrichments ranging from about $z \sim 0.15$ for Jupiter to perhaps $z \sim 0.9$ for Uranus and Neptune. This enrichment of condensible material already suggests that the planets formed through processes other than compositionally-nonselective hydrodynamic accumulation.

Altogether, the structural regularities of our Solar System strongly suggest that the system formed from a single, structured, precursor object and as a result of orderly processes.

CLOUD COLLAPSE, ROTATION & PROTOSTELLAR DISKS

As can be seen from the dynamical state of our own planetary system, which depends on the balance between centrifugal force and gravity, rotation is an essential feature. The formation of a planetary system around a star depends on the rotation of the precursory interstellar cloud. Interstellar clouds rotate for two reasons: 1) they partake of the general rotation of the galaxy, at an angular speed of about 10^{-15} radians s^{-1}; and 2) turbulent cloud motions, involving

FIGURE 3. The interstellar clouds, from which stars and planetary systems form, rotate by virtue of the overall rotation of the Galaxy and by virtue of the peculiar rotations associated with turbulent motions in the interstellar medium. Rotation plays a seminal role in the formation of planetary systems. Excess angular momentum prevents the gas from collapsing immediately to a spherically symmetric, pressure-supported star. Instead, the excess angular results in the formation of a disk, and ultimately a centrifugally-supported planetary system.

cloud-cloud collisions can cause individual clouds to have peculiar rotations as high in some instances as 10^{-13} radians s^{-1}.

As an initially extended and diffuse cloud of gas collapses, angular momentum conservation will cause the spin rate to increase. At the same time, processes that remove angular momentum—specifically Lorentz forces associated with the magnetic fields that pervade interstellar matter—will slow the rate of rotation. If angular momentum is lost rapidly in comparison with the rate of collapse, then the collapsing gas will tend to rotate rigidly with the surrounding matter; if on the other hand, the angular-momentum-loss timescale is long in comparison with the collapse time, then angular momentum will tend to be conserved and the collapsing cloud will spin up. To estimate the rate of angular momentum loss, consider a cloud of mass M and linear dimension R, surrounded by tenuous gas of density ρ_o, the whole system threaded by a magnetic field having intensity B in the external gas. All of the gas is highly electrically conducting. Crudely, the timescale for angular momentum transfer between the collapsing cloud and the surrounding medium is (see, for example, Levy *et al.* 1991)

$$\mathcal{T}_{\rm spin} \approx \frac{M}{\sqrt{4\pi\rho_o}BR^2}\,. \qquad (1)$$

$\mathcal{T}_{\text{spin}}$ is the time over which the cloud loses a significant fraction of its angular momentum in the form of torsional Alfvén waves transmitted into the surrounding medium. This timescale is to be compared with the timescale for gravitational collapse of the cloud:

$$\mathcal{T}_{\text{grav}} \approx \frac{1}{\sqrt{G\rho_c}}, \tag{2}$$

where ρ_c is the density of the collapsing cloud gas. When $\mathcal{T}_{\text{spin}} > \mathcal{T}_{\text{grav}}$ the cloud tends to preserve its angular momentum during collapse. Roughly speaking, this occurs when

$$R \lesssim \left[\frac{GM^3}{4\pi\rho B^2}\right]^{\frac{1}{7}}. \tag{3}$$

Applying eq. (3) to one solar mass ($\sim 2 \times 10^{33}$ gm) of collapsing gas, a magnetic field intensity of about 10^{-4} Gauss, and a surrounding density of $\sim 10^{-20}$ gm cm^{-3}, the collapsing protostellar gas will more or less preserve its angular momentum from the time that $R \lesssim 10^{17}$ cm. Although highly simplified (see, eg., Cassen *et al.* 1985) this indicates that interstellar cloud cores are likely to preserve much of their angular momentum while collapsing to form protostars. Such cores, which typically have radii of the order of 10^{17} cm, masses of the order of a solar mass, and angular speeds of the order of 10^{-13} to 10^{-14} s^{-1}, contain some 10^{53} gm cm^2 s^{-1} or more of angular momentum.

The angular momentum of the present Sun is a few $\times\ 10^{48}$ gm cm^2 s^{-1}, a small fraction of the angular momentum that the previous paragraph suggests might have been associated with the Solar System's precursor cloud core. By comparison, the angular momentum of the planetary system is a few $\times\ 10^{50}$ gm cm^2 s^{-1}. Restoring the planetary compositions to match that of cosmic matter—thus accounting for material and angular momentum likely to have been lost from the Solar System—brings the initial Solar System angular momentum into the range from a few $\times\ 10^{51}$ to a few $\times\ 10^{52}$ gm cm^2 s^{-1} (Weidenschilling 1977).

Angular momentum in the amount of 10^{52} to 10^{53} gm cm^2 s^{-1} per solar mass inevitably has significant effects on a compact object such as as a star. This can be seen easily by comparing it with the amount of angular momentum that the Sun could contain without "breaking up". Setting $R\Omega_b^2 \sim GM_\odot/R_\odot^2$, we find the breakup-up angular velocity $\Omega_b \sim (GM_\odot/R^3)^{1/2} \sim 6.3\times10^{-4}$ s^{-1}. Assuming a rigidly rotating (and, for simplicity, uniform density) Sun, this angular velocity implies an angular momentum $\frac{2}{5}MR^2\Omega \sim 2.5 \times 10^{51}$ gm cm^2 s^{-1}. In fact, this crude calculation overestimates the angular momentum of a Sun spinning at its "break-up" angular velocity; because the Sun's density increases rapidly inward, the actual "break-up" angular momentum of a rigidly rotating Sun would be smaller than estimated here. In any case, these illustrative calculations suffice to indicate that protostars are likely to be born with significantly more angular momentum than can eventually end up in a single, approximately-spherical, star with the dynamical equilibrium determined primarily by a balance between pressure and gravity. Indeed, with such a large angular momentum, centrifugal force provides the dominant expansive stress, and one expects that the final state of such a stellar system is likely to have major components that are centrifugally supported against gravity. At least in our own Solar System, the major remnant of this centrifugally-supported component is the planetary system.

The large amount of angular momentum in collapsing protostellar cloud fragments results in the formation of protostars surrounded by accretion disks. To a first approximation, such disks are supported against gravity by the centrifugal force of the approximately Keplerian rotation. Observations suggest that disks with a variety of characteristics actually occur in protostars, obviously depending on the details of the initial conditions and the environment in which the star-forming cloud complexes occur. Although there remain numerous observational, as well as theoretical, uncertainties, it appears that a one-solar-mass star is likely to form surrounded by a disk nebula having a mass in the range of a percent to a few tens of percent of a solar mass and angular momentum of the order of 10^{52} or more gm cm^2 s^{-1}. It is apparent that the formation of our own planetary system involved the loss of significant amounts of angular momentum during the transition from a diffuse interstellar cloud to a compact, gravitationally-bound assemblage of a star and several planets. This angular-momentum-dominated evolution seems to play a critical role in the formation and character of a planetary system.

Astronomical observations of star-forming regions lend strong support to these ideas. Extensive observations of nearby molecular clouds, such as those in Ophiucus and Taurus, show the formation of stars—including stars with masses comparable to the Sun—to be a widespread and continuing process. Specifically, energy reradiated in the infrared from objects believed to be protostars, together with the fact that many of these stars can also be observed directly in optical radiation, indicates that dusty, gaseous circumstellar disks occur commonly, and possibly typically, around stars during their formation (Beckwith *et al.* 1984; Rucinski 1985; Strom *et al.* 1985; Sargent and Beckwith 1987; Bertout *et al.* 1988; Sargent 1989) These disks are thought to be the birthplaces of planetary systems.

PROTOPLANETARY DISKS & THE FORMATION OF PLANETS

Dissipative processes in a collapsed protostellar nebula rapidly deplete the infalling motion of the gas and cause it to settle down into a relatively thin disk, consistent with the approximate conservation of the entrained angular momentum. For simplicity, consider a thin, relatively low-mass disk of gas and dust orbiting an approximately young, one-solar-mass, single star. Because the gravitational potential is dominated by the mass of the central star, the motion of the gas in the disk is approximately Keplerian, so that the angular velocity $\Omega \sim \sqrt{GM_\odot/r^3}$. Deviations of the angular motion from this simple relation, which occur because of the finite gas pressure, are small and can be neglected for now. By the same token, the disk is relatively thin, its finite thickness again due to the gas pressure. The vertical distribution of density in such a disk is approximately $\rho(z) = \rho_c \exp(-z^2/\Lambda^2)$, where the scale height is given by $\Lambda^2 \equiv 2kTr^3/GM_\odot m$; for the purpose of illustration, the temperature, T, is assumed to not vary in the vertical direction. The midplane density is ρ_c, r is radial distance from the center of the nebula, m the mean molecular mass, and k the Boltzmann constant. At a few astronomical units, and with $T \sim 250$K, $\Lambda \approx 4 \times 10^{12}$ cm. Because of the disk's Keplerian motion, dissipative processes can lower the overall energy of the system by transporting angular momentum

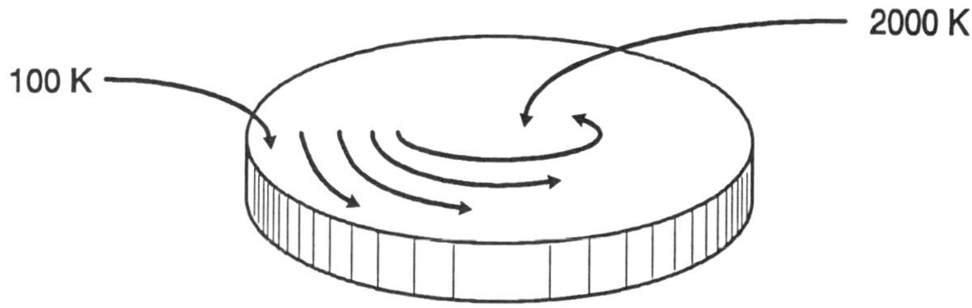

FIGURE 4. Radial temperature variation in protoplanetary disks is thought to play a dominating role in planet formation. At the high temperatures associated with the inner parts of disks, only the most refractory compounds—rock and metal—can persist in the solid form. In the low-temperature outer nebula, even the most volatile ices can persist as condensed solids. Because planet formation is thought to me controlled—at least initially—by the accumulation of these solid materials, the composition and abundance of condensed solids is critically important in determining the nature of the resulting planets.

outward and mass inward (Lynden-Bell and Pringle 1974). The net result is that such a disk will tend to accrete matter onto the central object and expand at the peripheries, carrying both matter and angular momentum outward. This same process will generate heat in the disk. The ultimate source of this heat is the gravitational energy of the disk and the disk's differential rotation.

In Keplerian motion, the rotational rates and gradients are largest in the inner part of the disk and diminish rapidly outward. Thus the rate of heat deposition is sharply peaked toward the center of the disk. If the predominant loss of energy from the system is radiation from the faces of the thin disk, then one expects a radial temperature distribution that reflects the heat deposition profile, with the highest temperatures occurring toward the disk's center. One of the principal uncertainties in understanding the evolution of such disks, and the generation of heat within them, comes from our ignorance about the mechanisms that transport angular momentum and dissipate energy. Although many physical mechanisms have been suggested and explored—including turbulent viscosity, spiral density waves, and magnetic fields—the transport and dissipative processes occurring in real protostellar disks have not been identified. Nonetheless, much progress can be made without detailed knowledge of the dissipation mechanisms: If the dissipation and energy loss are approximately local in a thin disk, then the heat-deposition and temperature profiles are essentially independent of the details of the dissipation, and depend only on the rate of mass transport. In the simplest case of a quasi-steady-state disk, the heat deposition profile depends on only a single parameter, the mass accretion rate, which for typical protostars seems to be of the order of 10^{-8} to 10^{-7} $M_\odot$ per year. Without going into the details here—which include a variety of important physical considerations, including the radiative opacity and convective heat-transport efficiency—such mass accretion rates yield disk temperatures which reach 2000K within the inner few tenths AU of the disk, and which can fall as low as a 100K in the far outer reaches of the disk, beyond ten AU, depending on the details of the disk (Cameron 1985). An important transition in the radial temperature profile occurs at the temperature above which water can persist only as vapor in the nebula and below which water can persist in solid form as ice. This transition—occurring at a temperature of about 170K and radial distance of four or five astronomical units in typical model nebulae—is especially important because water is the most abundant condensible substance in the cosmic mix of matter, approximately four times more abundant than the sum of of the higher-temperature condensates.

Solid matter and the accumulation of planets

At any given temperature, a mixture of matter will have some of its components in vapor form and some in condensed form, either liquid or solid. At the low pressures prevailing in protostellar disks—of the order of 10^{-5} atmospheres—liquid phases are not generally in equilibrium, so that the matter in such a disk nebulae exists as a mix of gaseous and solid matter. Because the partitioning of molecular species between the gas and solid phases is a sensitive function of temperature, this mix varies markedly as a function of radial distance from the center of a protostellar disk. In the innermost parts of the disk, near the star where temperatures exceed 2000K, the matter is entirely vapor. At larger radii, and successively lower temperatures, increasingly larger fractions of the potentially condensible components occur in the condensed solid phase. In the outermost reaches of the disk, where the temperatures fall to the vicinity of 100K, virtually all condensible matter is in the solid phase, leaving only helium and residual hydrogen gas as the major uncondensed matter, with traces of other atoms. The solid matter exists in the form of dust particles, probably ranging initially in size from a tenth to ten microns.

We now believe that the formation of planets differs from the formation of stars in an essential way. Star formation proceeds through hydrodynamic process and is compositionally nonselective. As noted above, planet formation could not have occurred as a result of hydrodynamic processes in undifferentiated cosmic matter. Planetary compositions differ greatly from that of cosmic material, the planets being depleted in the most abundant of cosmic species—gaseous hydrogen and helium—while they are highly overabundant in heavy, condensible matter. It is apparent that planet formation was controlled by the accretion of solid matter, beginning with dust, into successively more massive bodies, eventually culminating in the small number of relatively large planets that today comprise our Solar System (eg., Safronov 1969). A detailed discussion of this process is beyond the scope of this paper. Although significant difficulties remain, especially in understanding the earliest phases of dust accretion (Weidenschilling, Donn and Meakin 1989), a strong case can be made that the formation of all of the planets in our Solar System either occurred through, or was controlled by, solid accretion.

Investigations of this 'planetary accretion' phenomenon suggest that it can account for the formation of the terrestrial planets. Recent numerical simulations (Wetherill 1991) indicate that the formation of a small number—typically about four—of rocky, metallic objects in the inner Solar System is a common outcome of late-stage accretion. In this picture, the rock-and-metal composition of the

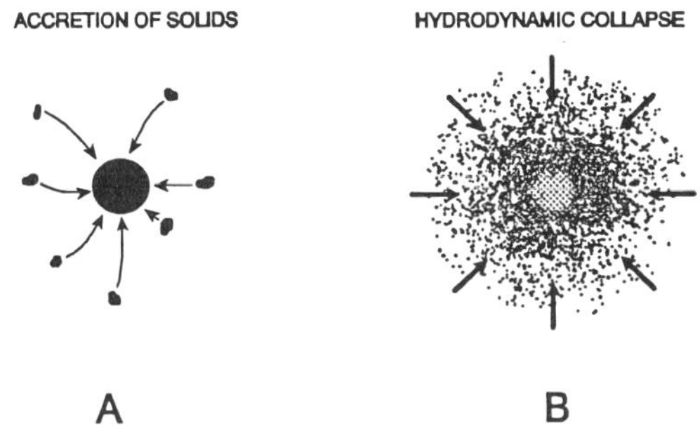

FIGURE 5. A) In the inner Solar System, the terrestrial planets apparently formed entirely as a result of solid body accretion. B) In the outer Solar System planetary cores are thought to have grown to sufficiently large mass that the associated gravity was strong enough to cause thc hydrodynamic capture of surrounding gas, resulting in massive gaseous envelopes surrounding outer-planet cores.

terrestrial planets results because, at the temperatures prevailing in the inner nebula, only rock and metal persisted in solid form; the remaining gas-phase material was not incorporated in these planets. In this picture, the presence of water and other volatiles on terrestrial planets is thought likely to result from a combination of hydration water carried in the accreting rock, gas released from the high-temperature planetary interiors, and impacts of water-rich, comet-like objects. These water-rich planetesimals, formed in the outer Solar System where water was solid and could therefore accumulate in small, ice-rich bodies, are thought to have been scattered into the inner Solar System by gravitational disturbances of the young giant planets.

In the outer Solar System, planet accumulation concluded somewhat differently. A major difference between the inner and outer Solar System was that the outer-solar-system temperature was low enough for water to condense to the solid, ice form. Water being the most abundant condensible substance in cosmic matter, a much larger amount of solid matter was available to drive accretion. In addition, the outer Solar System encompasses a much larger spatial volume, providing a larger material reservoir to form planets. As a result, planetary masses in the outer Solar System grew much larger than in the inner Solar System. Indeed, planetary masses apparently grew large enough that the gravity surrounding them attained sufficient strength to cause hydrodynamic infall and capture of the surrounding nebular gas. This is believed to the be the origin of the massive gaseous envelopes comprising the outer parts of the jovian planets, and especially of Jupiter and Saturn. The major part of the mass variation among the jovian planets is thus thought to have arisen from differences in the speeds of their formation. Other things being equal, the timescale for a planet's accretion should scale as the orbit period at the planet's distance from the central star. One expects that the more distant planets are likely to have accumulated more slowly than those closer to the Sun. Consequently, it is not unreasonable to expect that much of the nebular gas—which was still

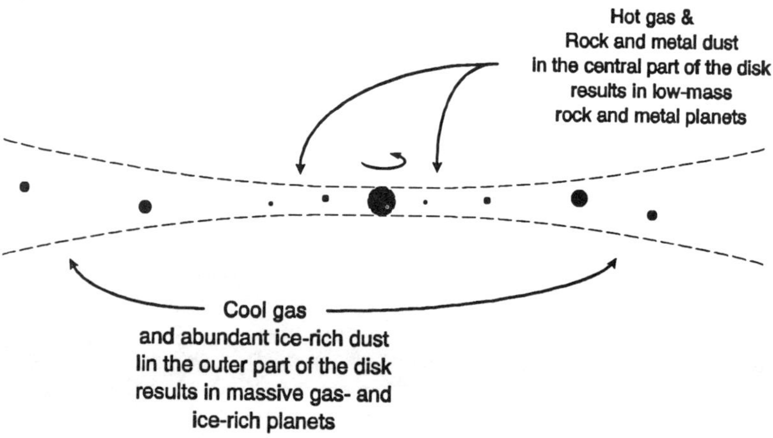

FIGURE 6. The principal regularities of the planetary systems reflect the structure of the protoplanetary nebula. Small, rock and metal planets formed in the inner part of the Solar System, where the temperature was high. More massive ice- and gas-rich planets formed in the cold, outer part of the Solar System. The spatial distribution and motions of the planets reflect the shape and motion of the nebular gas.

apparently present in abundance when Jupiter and Saturn had already formed—had dissipated and been lost from the Solar System by the time Uranus and Neptune had formed. This qualitative picture is consistent with the fact that the gaseous envelopes of Jupiter and Saturn are far more massive than those of Uranus and Neptune.

Origin of the Solar System's Principal Regularities

The picture that we have reviewed here in gross physical outline encapsulates the main ideas underlying our present understanding of the formation of our planetary system. Although many unsolved mysteries still cloud our understanding, and many details remain to be worked out—for example, the apparently fast formation timescales of the giant planets, especially Jupiter, remains to be understood—this picture gives us a relatively simple basis for understanding the principal features of our planetary system. And, combined with astronomical observations, which show protostellar disks to be commonly associated with star formation, this picture gives us a basis for speculations about the prevalence and character of planetary systems more broadly in the universe.

Now consider how the principal characteristics of our planetary system seem to have arisen. The flattened, corevolving distribution of planets, turning in the same direction as the Sun's spin, and coinciding with the Sun's rotational equator, apparently reflects the fact that the planets formed late in the evolution of the protoplanetary nebula. Thus dissipative processes (the equivalent of 'friction') had reduced the 'peculiar' motions of the original infalling material before the planets formed. As a result, the planets and the Sun formed from a single, coherently rotating nebula, with its random motions reduced to a low-energy state constrained by the angular momentum of the system. In this respect, planetary systems seem to form through mechanisms very different from those typical of multiple-star-system formation. Binary stars frequently occur in high eccentricity orbits, the radial motions of which apparently reflect the infalling motions of the cloud material from which the stars formed. For many multiple star systems, it is likely that the stars contracted to compact objects during the infall of the cloud material; thus decoupled from the surrounding gas and isolated from the high dissipation that characterizes gaseous systems, the infalling, eccentric motions are preserved. (It is also possible that some multiple star systems form from massive disks; but it is not clear how frequently such a process actually occurs.)

The primary differences between the terrestrial and jovian planets—one of the striking regular features of our planetary system—seem to have been caused by the fact that the planets formed through processes controlled by the accumulation of solid matter on different sides of the water-ice condensation boundary. Thus jovian-planet accumulation was driven by the presence of abundant water ice, resulting in massive planetary cores able to capture, gravitationally, surrounding nebular gas. On the other hand, the terrestrial planets formed from the relatively sparse rock and metal compounds that could remain solid at the high temperatures prevailing in the inner Solar System, resulting in small, relatively dry, low-gravity planets, without massive gas envelopes.

CONCLUDING REMARKS

What is perhaps most remarkable about our current picture of planetary-system formation is that so much of what would constitute a first-order description of the solar system seems to emerge naturally from the most basic—and seemingly universal—characteristics of the protostellar nebula from which both the Sun and the planets were born. This, together with the fact that similar disk nebulae seem to accompany star formation in many instances where this phenomenon is observed, suggests that planetary systems may be associated with a large fraction of, at the least, single stars similar to the Sun. Indeed, because so much of our planetary system's gross architecture seems to have arisen almost inescapably from basic nebular conditions and physical processes, one can make a persuasive case that many other planetary systems are likely to have gross characteristics similar to the characteristics of this one.

Up to now, actual detection of another planetary system has eluded astronomical observation. However, this situation is likely to change in the foreseeable future. Planetary systems around other nearby stars could be detected through at least three techniques, all of which are at or approaching the state of maturity needed for a definitive search program. While direct imaging of planets is not possible with current instrumentation, new generations of special instruments in space is likely to make direct imaging of other planets a reality in the near future, pending only the commitment to undertake that endeavor. Jupiter-like planets, should they be present, could be imaged around several dozen nearby stars with instruments presently being designed for space.

Orbiting planets perturb the velocity and position of their central stars. A Jupiter-like planet orbiting a Sun-like star induces velocity perturbations with amplitude exceeding 10 m s^{-1} and a period just over a decade. This compares with radial-velocity-measurement accuracies that now range near 5 m s^{-1} for the best measurements. Similarly a Jupiter-like planet orbiting a Sun-like star, at about five parsecs distance from Earth would produce a wobbling disturbance in that star's position approaching a milliarcsecond in amplitude. This is near the range of the best astrometric measurements currently being conducted from ground-based observatories; space-based astrometry is expected to achieve as much as two orders of magnitude improvement in the foreseeable future, again pending only the commitment to undertake that endeavor.

ACKNOWLEDGEMENT

This work was supported in part by grants from the National Aeronautics and Space Administration.

REFERENCES

Beckwith, S., Zuckerman, B., Skrutskie, M.F., and Dyck, H.M., 1984. Discovery of Solar System-size halos around young stars. *Astrophys. J.* **278**, 793.

Bertout, C., Basri, G., and Bouir, J. 1988. Accretion disks around T Tauri stars. *Astrophys. J.* **330**, 350.

Cameron, A.G.W. 1985. Formation and Evolution of the Primitive Solar Nebula, *Protostars & Planets II,* (D.C Black and M.S. Matthews, eds.; University of Arizona Press, Tucson)

Cassen, P., Shu, F.H., and Tereby, S. 1985. Protostellar Disks and Star Formation. In *Protostars & Planets II,* (D.C Black and M.S. Matthews, eds.; University of Arizona Press, Tucson), pp. 448–83.

Levy, E.H., A.A. Ruzmaikin, and T.V. Ruzmaikina 1991. Magnetic History of the Sun. In *The Sun in Time,* (C.P. Sonett, M. Giampapa and M.S. Matthews, eds.; University of Arizona Press, Tucson), pp. 589–632.

Lynden-Bell, D., and Pringle, J.E., 1974. The evolution of viscous disks and the origin of nebular variables. *Mon. Not. Roy. Astron. Soc.* **168**, 603.

Rucinski, S.M. 1985. *Astron. J.* **90**, 3221.

Safronov, V.S. 1969. *Evolution of the Protoplanetary Cloud and the Formation of the Earth and Planets,* (Moscow: Nauka Press), NASA TTF-677 (1972).

Sargent, A.I. 1989. Molecular disks and their link to planetary systems. In *Formation and evolution of planetary systems,* H.A. Weaver, F. Paresce, and L. Danly, eds. (Cambridge University Press).

Sargent, A.I., and Beckwith, S. 1987. Kinematics of the circumstellar gas of HL Tau and R Monocerotis. *Astrophys. J.* **323**, 294.

Strom, S.E., Strom, K.M., Grasdalen, G.L., Capps, R.W., and Thompson, D.A. 1985. High spatial resolution studies of young stellar objects. II. A thick disk surrounding Lynds 1551, IRS 5. *Astrophys. J.* **90**, 2575.

Weidenschilling, S.J., Donn, B. and Meakin, P. 1989. The physics of planetesimal formation. In *The Formation and Evolution of Planetary Systems,* H.A. Weaver and L. Danly, eds. (Cambridge University Press: Cambridge).

Wetherill, G.W. 1991. Occurrence of Earth-like bodies in planetary systems. *Science*, **253**, 535-38.
Wolszczan, A. and Frail, D.A. 1992. A planetary system around the millisecond pulsar PSR 1257+12. *Nature*, **355** 145-47.

For more complete discussions of the topics covered in this article, the reader is referred to several review volumes, which also provide extensive lists of pertinent references.

Protostars and Planets, T. Gehrels, ed. (University of Arizona Press: Tucson), 1978.
Protostars and Planets II, D. Black and M.S. Matthews eds. (University of Arizona Press: Tucson), 1985.
Protostars and Planets III, E.H. Levy, J.I. Lunine, and M.S. Matthews eds. (University of Arizona Press: Tucson), 1992.
The Formation and Evolution of Planetary Systems, H.A. Weaver and L. Danly, eds. (Cambridge University Press: Cambridge), 1989.

Planets Around Pulsars
ASP Conference Series, Vol. 36, 1993
J.A. Phillips, J.E. Thorsett, and S.R. Kulkarni

THE EVOLUTION OF PROTOPLANETARY DISKS

STEVEN P. RUDEN
Physics Department, University of California, Irvine, CA 92717

ABSTRACT Accretion disks evolve by transferring angular momentum outwards, which allows mass to flow inwards onto the central object. The radial structure and time-dependent behavior of low-mass, Keplerian, viscous disks is analyzed. Potential mechanisms for the angular momentum transport in disks are reviewed. An evolutionary model of a low mass disk around a pulsar is presented.

INTRODUCTION

The possible detection of two roughly earth-mass planets in near circular orbits at distances of $\sim 1/2$ AU from the millisecond pulsar PSR1257+12 (Wolszczan and Frail 1992) has led to great interest in how planets form around pulsars. Many of the proposed theoretical scenarios are similar not only to each other but also to models for planet formation in our own solar system. Their common feature is the presence of a disk of matter surrounding the central star that provides the environment in which condensed solids accumulate into planets. The scenarios differ greatly in their suggestions for the origin, composition, mass, and size of the disk (for a review of competing models see the chapters by Podsiadlowski and by Phinney). In this chapter we will review the basic processes that control the evolution of gaseous disks around stars. We will discuss mechanisms that govern the transport of angular momentum in accretion disks and analyze a model for the structure and evolution of a disk around a pulsar. For details on how planetesimals form, sediment to the nebular midplane, and coagulate into planetary bodies, see the chapter by Lissauer.

STRUCTURE AND EVOLUTION OF ACCRETION DISKS

Consider a geometrically thin disk of gas (centered around the $z = 0$ plane) revolving in steady circular orbits in an axisymmetric gravitational potential $\phi(r, z)$. Centrifugal force balance requires that the local angular velocity $\Omega(r)$ satisfy

$$\Omega^2 r = \frac{\partial \phi}{\partial r} + \frac{1}{\rho}\frac{\partial P}{\partial r}, \tag{1}$$

where P and ρ are the pressure and density of the gas. Hydrostatic equilibrium in the z direction requires that

$$\frac{1}{\rho}\frac{\partial P}{\partial z} = -\frac{\partial \phi}{\partial z}. \tag{2}$$

For the case of a disk of mass $M_{\rm d}$ around a star of mass M_* we can ignore the self-gravity of the disk if $M_{\rm d} \ll M_*$ and take the potential to be Keplerian: $\phi = -GM_*/\sqrt{r^2+z^2}$. In the limit that the disk is thin, $|z| \ll r$, equation 2 shows that the vertical scale height H of the disk is approximately

$$H = \frac{c_{\rm s}}{\Omega_K}, \tag{3}$$

where the isothermal sound speed is defined to be $c_{\rm s} = \sqrt{P/\rho}$ and $\Omega_K \equiv \sqrt{GM_*/r^3}$ is the local Keplerian rotation rate. In the same thin-disk limit, equation 1 then shows that the rotation rate of the gas is Keplerian, $\Omega \approx \Omega_K$, to order $(H/r)^2 \ll 1$. For our purposes the neglect of the small radial gas pressure support in equation 1 is acceptable, and we will ignore the small difference between Ω and Ω_K. However, the small difference in rotation rates between the gas and a solid particle on a circular orbit leads to gas drag effects on planetesimals that may be significant to the solid accumulation process (Weidenschilling 1977a; also see the chapter by Lissauer). Equation 3 allows us to rephrase the thin disk condition, $H/r \ll 1$, as the requirement that the gas disk be *cool*, namely, that the sound speed be much smaller than the orbital speed in the disk, $c_{\rm s} \ll \Omega r$.

In the absence of any dissipative mechanism operating between neighboring circular streamlines, the gas would continue to revolve around the star indefinitely. In the presence of dissipation however, Lynden-Bell and Pringle (1974) elegantly showed that if two nearby fluid parcels exchange mass and angular momentum such that mass flows inwards and angular momentum flows outwards (in the usual case where $d\Omega/dr < 0$) then the total energy of the system is lowered while the total angular momentum remains constant. If we now consider a continuous disk of matter in which the angular velocity decreases outwards and the specific orbital angular momentum, Ωr^2, increases outwards, then the inner regions revolve faster than the outer regions. Any dissipative process acting on this velocity shear that transfers angular momentum from the inner regions to the outer regions will cause the inner regions to spiral inwards, deeper into the gravitational potential well of the star, and the outer regions to spiral outwards, to conserve total angular momentum. As the evolution proceeds, the radius separating the inward moving and the outward moving matter will increase, and ever more of the disk matter will be accreted by the central star. The ultimate low energy state of the system occurs when the entire disk has been accreted by the star except for an infinitesimal amount of mass at infinity that contains all the angular momentum. The dissipative process liberates and converts the gravitational potential energy of the star into heat, which is ultimately radiated by the surface of the disk and provides a spectral signature of this process.

The most common case discussed for the transfer of angular momentum in disks is where a local viscous stress acts on the shear between adjacent fluid annuli. In a thin Keplerian disk the Navier-Stokes equations can be reduced to a single partial differential equation for the time evolution of the surface mass density $\Sigma(r,t)$ (Lynden-Bell and Pringle 1974; Pringle 1981)

$$\frac{\partial \Sigma}{\partial t} = \frac{3}{r}\frac{\partial}{\partial r}\left(r^{1/2}\frac{\partial}{\partial r}\left(\Sigma \nu r^{1/2}\right)\right), \tag{4}$$

where ν is the kinematic viscosity in the disk. The mass accretion rate at radius r in the disk is

$$\dot{M} = 6\pi r^{1/2} \frac{\partial}{\partial r} \left(\Sigma \nu r^{1/2} \right) \quad (5)$$

defined such that $\dot{M}$ is *positive* when mass flows *inwards.*

Viscosity

The rate at which a gas disk evolves is governed by the rate at which it can transfer angular momentum outwards. The prime problem in accretion disk physics is to understand the origin and magnitude of the viscous torque in the disk, i.e., what is the value of ν in equation 4? It is easily shown that ordinary atomic or molecular viscosities are too small by many orders of magnitude to lead to astrophysically interesting evolutionary timescales. Thus it is widely assumed that the fluid flow in the disk is turbulent and that angular momentum is transported when fluid elements carrying different amounts of specific angular momentum mix together in turbulent eddies. We must point out however that there is no general proof that thin accretion disks are unstable to fluid turbulence; in fact, according to Rayleigh's stability criterion (Drazin and Reid 1981) Keplerian disks are stable to axisymmetric interchange instabilities because the specific angular momentum increases outwards. Nonetheless, we may posit the existence of a turbulent viscosity and later attempt to discover the origin for the turbulent instability in particular disk models (see sections on magnetic, gravitational, and convective instabilities below).

In a turbulent disk, we expect the largest eddies to be responsible for the angular momentum transport because they couple together regions of the disk with the largest differences in specific angular momentum. If the largest eddy size is H_t (in the radial direction) and its turnover speed is v_t, the turbulent viscosity is of order

$$\nu \sim v_t H_t. \quad (6)$$

Since the eddy size is almost certainly less than the vertical scale height of the disk ($H = c_s/\Omega$ from eq. 3) and since the eddy velocity is less than the sound speed for subsonic turbulence, we can rewrite equation 6 as

$$\nu = \alpha c_s H = \alpha \frac{c_s^2}{\Omega}, \quad (7)$$

which is the famous "alpha" prescription of Shakura and Sunyaev (1973) that lumps all the uncertainties about the turbulence into the unknown parameter $\alpha \approx (H_t/H)(v_t/c_s) \lesssim 1$. If a particular fluid dynamical mechanism or instability is proposed for the origin of the turbulent viscosity, the parameter α may be calibrated by comparing equations 6 and 7. In many treatments, α is simply treated as a constant parameter.

Equation 7 relates the viscosity to the sound speed so that constructing nebular evolution models requires understanding the temperature structure in the disk. The following simplified "one-zone" treatment of the thermal physics is often an excellent approximation to detailed radiative transfer calculations (Pringle 1981; Ruden and Lin 1986; Ruden and Pollack 1991). The density in the midplane of the disk is approximately

$$\rho = \frac{\Sigma}{2H}. \tag{8}$$

For a perfect gas, the midplane pressure is

$$P = \rho c_s^2 = \frac{\mathcal{R}}{\mu}\rho T, \tag{9}$$

where T is the midplane disk temperature, $\mathcal{R}$ is the gas constant, and μ is the mean molecular weight. The optical depth between the center and surface of the disk is

$$\tau = \kappa\rho H, \tag{10}$$

where $\kappa(\rho, P)$ is an appropriately averaged mean opacity. If the optical depth is sufficiently large, the diffusion equation for radiative transfer may be used to relate the midplane temperature to the internally generated viscous heat flux $\mathcal{F}$ radiated by each side of the disk

$$\sigma T^4 = \frac{3}{4}\tau\mathcal{F}, \tag{11}$$

where σ is the Stefan-Boltzmann constant and in a Keplerian disk

$$\mathcal{F} = \frac{9}{8}\Sigma\nu\Omega^2. \tag{12}$$

Along with the definitions of viscosity (eq. 7) and vertical scale height (eq. 3), equations 8 – 12 form a consistent set that allows for the solution of ν in terms of Σ and r. This viscosity is then used in equation 4 to calculate time-dependent nebular evolutionary models. The above set of equations can be easily modified if radiation or degeneracy pressure becomes important or if radiative transfer occurs under optically thin conditions.

General Features of Viscous Disk Evolution

Equation 4 governing the evolution of the disk is a diffusion equation. An annular ring of matter placed in orbit around the star will spread with time: the mass flowing inwards loses energy and the mass flowing outwards soaks up angular momentum. In this process viscosity taps the gravitational potential energy of the star and dissipates it as heat, which is radiated by the disk surfaces. The viscous timescale over which a disk of radial scale r evolves can be found from equation 4 and is

$$t_{\rm vis} \sim \frac{r^2}{\nu}, \tag{13}$$

which exhibits the characteristic quadratic dependence on length scale typical of diffusive systems. If the viscosity can be expressed by the alpha formula (eq. 7), the evolution timescale can be written as

$$t_{\rm vis} = \frac{1}{\alpha}\left(\frac{r}{H}\right)^2 \Omega^{-1}, \tag{14}$$

which shows that even if α is of order unity the viscous time for thin disks is much longer than the orbital time.

Let us consider the structure of a disk in a steady state. The mass flux crossing any radius must be constant, thus equation 5 can be integrated to give

$$\Sigma\nu = \frac{\dot{M}}{3\pi}\left(1-\sqrt{\frac{R_*}{r}}\right) \tag{15}$$

where we have taken the viscous stress ($\propto \Sigma\nu$) to vanish at the stellar surface $r = R_*$ (for a discussion of stellar boundary conditions see Lynden-Bell and Pringle 1974). Far away from the star, the product $\Sigma\nu$ is approximately constant; thus, under steady conditions there is always a tradeoff between having a high viscosity and a low surface density or vice-versa – it is only the product of the two (i.e. $\dot{M}$) that counts. The viscous heat flux radiated by the disk is defined in equation 12, and using the above steady solution we find the effective temperature of the disk surface to be

$$\sigma T_{\rm e}^4 \equiv \mathcal{F} = \frac{3}{8\pi}\frac{GM_*\dot{M}}{r^3}\left(1-\sqrt{\frac{R_*}{r}}\right). \tag{16}$$

The effective temperature is independent of the detailed nature of the viscosity (it only depends on $\dot{M}$), and at large radii it becomes the well known law $T_{\rm e} \propto r^{-3/4}$. If we integrate equation 16 across both disk surfaces we find that the emitted disk luminosity is

$$L_{\rm d} = \frac{1}{2}\frac{GM_*\dot{M}}{R_*}, \tag{17}$$

a value 1/2 as large as expected from the gravitational potential well depth. An equal luminosity is emitted by the stellar surface (from a boundary layer at the equator or from magnetic polar caps) when the accreted mass settles on the central star, assumed slowly rotating (Lynden-Bell and Pringle 1974).

Consider next the nonsteady evolution of a disk. The viscosity, $\nu(\Sigma, r)$, typically depends on the surface density in a nonlinear fashion.[1] After a transient phase in which the initial conditions are diffused away and "forgotten", the evolution becomes roughly self-similar, which means that the solution at one time can be appropriately scaled to match the solution at a different time (see Zel'dovich and Raizer 1967, Ch. X, §9 for an argument why diffusive evolution is generally self-similar). The radial disk structure can be well approximated by assuming that the mass flux, $\dot{M}$, is independent of r (it will change with time as the disk is depleted) and the surface density can be determined by solving equation 15 for $\Sigma(r)$ (for an example see Ruden and Lin 1986).

To further analyze the behavior of the evolution, we will assume the viscosity can be written in the following power law form:

$$\nu(\Sigma, r) = \nu_0\,\Sigma^a\, r^b\,, \tag{18}$$

[1]To avoid viscous instability in the disk the product $\Sigma\nu$ must be an increasing function of Σ (see Pringle 1981).

where ν_0, a, and b are constants. Self-similarity of the solution means that the surface density can be decomposed as

$$\Sigma(r,t) = \Sigma_0(t)\, f\left(\frac{r}{R(t)}\right), \tag{19}$$

where $R(t)$ is the outer disk radius. As long as the disk is large compared to the star, $R(t) \gg R_*$, we can ignore the finite size of the central object and set $R_* = 0$. When the viscosity depends on surface density ($a \neq 0$), we have a nonlinear diffusion problem. In such cases it is easily shown that the outer disk edge "sharpens" (as long as $a > 0$) and propagates outward as a steep front so that $R(t)$ is a well-defined quantity (Zel'dovich and Raizer 1967). This behavior differs from that of linear diffusive systems where there is no edge but instead a long exponential tail. One consequence of this nonlinearity is that the disk edge increases outward *less quickly* than in the linear case (see eq. 22).

We are now in a position to determine how the disk properties scale with radius and time. Substituting equations 18 and 19 into equation 15 and taking $\dot{M}$ to be roughly constant with radius, we find the surface density profile to be

$$\Sigma(r) \propto r^{-b/(1+a)}. \tag{20}$$

The disk mass and angular momentum (for a Kepler disk) scale with time as

$$\begin{aligned} M_\mathrm{d}(t) &\sim \Sigma_0 R^2, \\ J_\mathrm{d}(t) &\sim M_\mathrm{d}\sqrt{GM_*R} \sim \Sigma_0 R^{5/2}. \end{aligned} \tag{21}$$

Two classes of time-dependent solutions can be found (Pringle 1981): in the first, disk angular momentum is conserved as the disk mass is accreted by the star, and in the second, disk mass is conserved but disk angular momentum increases due to an applied torque at the origin. The first class of solution will be discussed here; the second class can easily be constructed and applies to a disk around a pulsar with such a strong magnetosphere that mass accretion is entirely halted. Taking J_d as constant, equation 21 indicates that $\Sigma_0 \propto R^{-5/2}$ and $M_\mathrm{d} \propto R^{-1/2}$. The viscous evolution time (eq. 13) for our power law viscosity is $t \sim R^2/\nu \propto R^2/\Sigma_0^a R^b$, which can be solved for $R(t)$ using the constancy of J_d. The time-dependent behavior of the disk is then

$$\begin{aligned} R(t) &\propto t^{1/(2+5a/2-b)}, \\ M_\mathrm{d}(t) &\propto t^{-1/(4+5a-2b)}, \\ \Sigma(t) &\propto t^{-5/(4+5a-2b)}. \end{aligned} \tag{22}$$

Some general features of equations 22 can be noted. First, if the viscosity is constant ($a = b = 0$) then the disk radius scales as $R \propto t^{1/2}$, which is the standard behavior for (linear) diffusive systems. Second, if the viscosity increases with surface density ($a > 0$) then the disk radius increases with time *less quickly* than $t^{1/2}$. Third, for $a > 0$ the disk mass decreases very slowly with time.

The simple power law behavior described by the above equations can be verified by numerical simulations (for an illustrative example see Lin and Pringle

1987). The main requirements for the validity of the scaling laws seem to be that enough time has passed for the influence of the initial conditions to be negligible and that a single power law viscosity apply throughout most of the disk. We will return to these equations when we discuss the evolution of a pulsar disk model below.

ANGULAR MOMENTUM TRANSPORT MECHANISMS

We next turn to the physical mechanisms that may be responsible for the transport of angular momentum in accretion disks. Although the qualitative nature of disk evolution is not particularly sensitive to the origin for the turbulence, the quantitative evolution timescales do depend on the magnitude of the viscosity. In turn we consider magnetic, gravitational, and convective instabilities.

Magnetic Instabilities

Shakura and Sunyaev (1973) pointed out that the Maxwell stresses associated with a tangled magnetic field in a disk will lead to an effective viscosity where

$$\alpha \approx \frac{v_A^2}{c_s^2}. \tag{23}$$

Here, v_A is the Alfvén velocity. Since magnetic flux tubes are buoyant and can be lost from the disk (Parker 1979), the magnetic pressure is most likely limited to be comparable in magnitude to the thermal pressure, which implies that $\alpha \lesssim 1$.

An intriguing recent development is the suggestion by Balbus and Hawley (1991) (see also Hawley and Balbus 1991) that a disk threaded by a weak vertical magnetic field is unstable to axisymmetric shear instabilities. Based on a local analysis of the disk, they argue the instability grows at nearly dynamical rates ($\sim \Omega$), its maximum growth rate is *insensitive* to the strength of the magnetic field (for weak fields whose magnetic pressure is less than the thermal pressure), the instability is unaffected by small radial or toroidal fields, and the only requirement for instability is that the angular velocity decrease outwards ($d\Omega/dr < 0$). Keplerian disks clearly satisfy the latter condition and may be virulently unstable provided a small poloidal seed field is present. The mechanism for the proposed instability is that as a fluid parcel, threaded by a small, frozen-in vertical magnetic field, is perturbed (outwards, say) the field lines stretch, and their tension tries to restore the parcel to its initial position (which is stabilizing). However, the field lines also resist azimuthal shearing and attempt to enforce rigid rotation of the parcel at its new, larger radius. This magnetic torque on the parcel gives it additional angular momentum, which drives the parcel still farther out (which is destabilizing). For sufficiently long wavelengths,

$$\lambda \gtrsim \frac{v_A}{\Omega} \tag{24}$$

the (stabilizing) magnetic restoring forces are smaller than the (destabilizing) azimuthal stresses, and the instability proceeds with neighboring regions of high and low angular momentum interpenetrating and mixing. This mixing will lead to an effective magnetic viscosity of order the radial mixing length (eq. 24) times the Alfvén speed, which gives an estimate for alpha identical to equation 23.

As the instability progresses and field lines are stretched, the magnetic field energy increases. Balbus and Hawley (1991) suggest two mechanisms for saturation. In the first, the magnetic field increases until the critical wavelength (eq. 24) exceeds the disk height, at which time the shear instability should shut off; this corresponds to $\alpha \approx 1$. In the second, magnetic reconnection limits the maximum field strength, and α is probably significantly less than unity. In either case, the proposed magnetic instability provides an interesting origin for the angular momentum transport in Keplerian disks.

The existence of this powerful shear instability has been cogently questioned by Knobloch (1992), however. He considers a global (as opposed to local) treatment of the magnetized disk and argues that the presence of an azimuthal magnetic field stabilizes the flow. Furthermore, if the instability is present at all he finds that it proceeds via overstable oscillations whose growth rate is proportional to magnetic field strength (and thus it is an exceedingly weak instability in the limit of extremely small fields). Until these theoretical conflicts are resolved, the nature of this instability and its origin for magnetic turbulence in accretion disks is an open question.

Gravitational Instabilities

The most useful measure of the importance of self-gravity in a disk is the Toomre Q parameter, which for a gaseous Keplerian disk is (Toomre 1964)

$$Q = \frac{\Omega c_s}{\pi G \Sigma}. \tag{25}$$

In a pioneering investigation, Toomre (1964) showed that disks are unstable to local axisymmetric gravitational disturbances if Q is less than one. In general, disks can be unstable to non-axisymmetric gravitational disturbances as well if Q is less than a factor of a few (Binney and Tremaine 1987). We will adopt $Q \lesssim 1$ as a rough indication of the susceptibility of a disk to gravitational instability.

We can see how the Q parameter arises by comparing the vertical component of the self-gravity of the disk to that of the gravity provided by the central star. From Poisson's equation, the disk self-gravitational acceleration is $g_{\rm self} \sim \pi G \Sigma$, while the central star contributes $g_* = (GM_*/r^2)(z/r) \sim \Omega^2 H \sim \Omega c_s$, using equation 3 to substitute for H. We expect disk self-gravity to be important when $g_{\rm self} \gtrsim g_*$, which is just the relation $Q \lesssim 1$.

This criterion can be used to estimate the disk mass needed for robust gravitational instabilities to be present. Write the sound speed as $c_s = \Omega H$ (from eq. 3) and substitute into equation 25 using the Kepler relation $\Omega^2 = GM_*/r^3$. If we identify the disk mass as $M_d \approx \pi \Sigma r^2$ then the instability criterion becomes[2]

$$\frac{M_d}{M_*} \gtrsim \frac{H}{r}. \tag{26}$$

---

[2]By similar substitutions, the $Q \lesssim 1$ instability criterion can also be shown to be equivalent to a constraint on the disk central density: $\rho \gtrsim M_*/r^3$, which is complementary to the famous Roche tidal disruption condition.

Using a rather more detailed analysis, Shu, et al. (1990) find that $M_d/M_* \gtrsim 0.314$ is required for one-armed spiral instabilities to be unstable. Since in many astrophysical regimes the disk height is typically $H \sim \frac{1}{10}r$, these last two conditions are roughly equivalent, and we anticipate that disks around pulsars will be gravitationally stable unless they contain more than a few tenths of a solar mass or are cool enough that $H/r \ll 1$.

In disks where self-gravity is important, two distinct regimes occur. In the first, the gravitational instabilities are local in nature and act effectively to mix the angular momentum of neighboring fluid annuli. In this case Lin and Pringle (1987) have suggested that an effective gravitational viscosity will arise with $\alpha \sim Q^{-2}$. As long as disk cooling is effective at maintaining $Q \lesssim 1$, the disk will evolve viscously. In the second, the gravitational disturbances are global, that is, they are spiral density waves that propagate energy and angular momentum between distant disk regions. In this regime, the evolution is not simply diffusive in nature but depends on how and where the waves damp and deposit their energy and angular momentum into the gas. Adams, Ruden, and Shu (1989) have suggested that if global one-armed modes become dynamically unstable the disk may gravitationally fragment and form a binary companion. The distinction between these two regimes depends on the ratio M_d/M_* and on what mechanism damps the spiral disturbances (i.e., whether the waves can propagate for large distances in the disk).

In the models for pulsar disks where a binary companion is disrupted (Podsiadlowski, Pringle, and Rees 1991; Stevens, Rees, and Podsiadlowski 1992), the initial disk may indeed be relatively massive, and gravitational instability may dominate the early nebular evolution. In the alternative models where the initial disk mass is small (as in the supernova fall-back scenario of Lin, Woosley, and Bodenheimer 1991), gravitational instability probably plays little role in the evolution unless the disk can efficiently cool to keep $Q \ll 1$.

Thermal Convective Instabilities

In nebular regions where the temperature is less than $\sim$ 1500K, micron-sized dust grains can condense out of the gas. The dust will dominate the opacity, and its presence will be the key ingredient for determining the thermal structure and viscosity of the disk. In 1980 Lin and Papaloizou demonstrated that the vertical temperature gradient in cool disks will be unstable to convection if the temperature dependence of the opacity is sufficiently large, $\beta > \beta_{\rm crit} = (3\gamma-4)/(\gamma-1)$, where $\beta \equiv \partial \ln \kappa / \partial \ln T$ and γ is the ratio of specific heats of the gas. For molecular gas $\beta_{\rm crit} = 1/2$. In dusty, solar composition disks β lies in the range $1/2 - 2$ (Pollack, McKay, and Christofferson 1986) so these disks should be unstable to thermal convection. Lin and Papaloizou (1980) thus proposed that thermal convection is the origin for turbulent viscosity in disks around protostars. Since it is likely that planets around pulsar(s) formed from the accumulation of dust grains in the outer regions of a disk and since the planetary environment in PSR1257+12 (2 earth-mass planets at radii $\sim$ 1 AU from a solar mass star with [spin-down] luminosity $\sim 10\ L_\odot$) is not much different from our solar system, it is reasonable to expect that convective turbulent viscosity will also play an important role in the evolution of planet-forming pulsar disks.

To estimate the magnitude of the convective viscosity Ruden and Lin (1986) constructed detailed vertical structure models of cool nebular disks and using a mixing-length prescription for convection, determined that the convective viscosity could be well represented by a Shakura and Sunyaev alpha-law with $\alpha \approx 10^{-2}$. Using a different model for convective turbulence Cabot, et al. (1987) found that α is in the range $10^{-3} - 10^{-2}$, while Ruden, Papaloizou, and Lin (1988) estimated α to be in this same range from a linear perturbation analysis of axisymmetric convective modes. In this paper, we will adopt $\alpha \approx 10^{-2}$ as the scale for convective viscosity, however, we must note the following concerns. First, Ryu and Goodman (1992) have analyzed the behavior of linear, nonaxisymmetric, convective modes in a sheared disk and found that there are exponentially growing solutions. Unfortunately, the linear modes seem to carry angular momentum the wrong way – inwards. If this holds true for fully developed, nonlinear convective turbulence then its source as the origin for viscosity in disks will be called into question. Second, Korycansky (1992) has shown that the amplitude of an initially unstable convective disturbance in a sheared disk will decay with time as t^{-2} in the presence of radiative conduction. Such effects could limit the efficiency of convective viscosity.

Using the above calibration for alpha and with a detailed opacity table, self-consistent simulations of nebular diffusion in disks can be made (Ruden and Lin 1986; Cabot, et al. 1987; Ruden and Pollack 1991). These calculations find that (1) actively accreting nebular lifetimes are in the range $10^5 - 10^6$ years, (2) mass accretion rates decrease from 10^{-5} to 10^{-8} $M_\odot$/yr as the disk is depleted, (3) much of the final protostellar mass is accreted through the disk, (4) evolutionary models are insensitive to the initial conditions except for the values of the total disk mass and angular momentum, (5) self-gravity is unimportant ($Q \gg 1$) except at the cool, outer edge of the disk, and (6) disk radii can spread by a factor $\lesssim 10$ during the lifetime of the disk. These conclusions are in general accord with the observational lifetimes and luminosities of protostellar disks (for a detailed review see the chapter by Strom).

As a consequence of mass accretion, the surface density and temperature in the disk decrease with time. Because the dust opacity is relatively temperature sensitive (for convection to occur), the mean optical depth in the disk also decreases with time (see eq. 10). Once τ becomes of order unity, convection becomes inefficient due to radiative losses from the eddies, and the convective viscosity decreases. Ruden and Pollack (1991) have shown that this happens in the outermost part of the disk first because the temperature, surface density, and opacity are lower there. As the disk is depleted further, the radius marking the $\tau \sim 1$ front moves inward. It is unclear what source of turbulence (if any) replaces thermal convection in the outer regions. Upon the termination of convective turbulence the nebula becomes relatively quiescent. Small dust grains, which couple strongly to the turbulent eddies, are no longer stirred up and may begin to coagulate and settle to the midplane of the disk. At this stage planetesimal growth begins in earnest. Ruden and Pollack (1991) have speculated that this transition to marginally optically thin conditions may mark the onset of the planet formation epoch. If this scenario is correct, planets may begin to form in the outer disk regions first (see Fig. II).

Thermal convection is a natural aspect of fluid layers heated from below. In convective disks, viscous energy dissipation is concentrated in the dense midplane regions and provides the internal heating. However, if the *surface* of the disk is heated enough by radiation from the central star then the internal vertical temperature gradient can be reduced enough to stabilize the convection. It is not hard to show that this can happen when the luminosity of the accretion disk (due to viscous dissipation) is comparable to the central stellar luminosity. The inner disk regions are most intensely heated by the star and are most susceptible to this effect. Stellar illumination of the disk there may result in reduced efficiency for the turbulent viscosity.

THE MINIMUM MASS SOLAR NEBULA

Before we examine an illustrative model of nebular evolution, let us ask the following question: how much solar composition gas does it take to make the planets? When we assay the solar system we find of order 50 earth masses of rocky and icy material, mostly in the centers of the four giant planets, and about 400 earth masses of hydrogen and helium gas (Podolak and Reynolds 1985). To reconstitute the heavy element rock/ice material to its original solar composition requires multiplying by the gas-to-dust ratio, a number of order 100. Summing up gives the total amount of solar composition gas required to build the present planets. The mass of this *minimum mass solar nebula* (MMSN) is about 1% – 2% of a solar mass. More detailed evaluations (going as far back as Hoyle in 1946) using data from models of planetary interiors give very similar mass estimates (Weidenschilling 1977b; Hayashi, et al. 1985). If the planet building process was inefficient, the original solar nebular mass must exceed this limit. The total angular momentum in the MMSN is about 4×10^{51} g cm^2 s^{-1}, although this estimate could be in serious error if there is much angular momentum in Kuiper belt comets (Weissman 1989).

Since we know the present distances of the planets from the sun, we can calculate a surface density distribution for the MMSN by suitably smearing out the planetary masses *assuming* they have not moved radially since their birth. An example of this exercise is shown in Figure I, which is adapted from the review article by Hayashi, et al. (1985). These authors find that the distribution of mass in the MMSN falls off as $r^{-3/2}$. Beyond the position of the asteroid belt temperatures were cool enough for ices (especially water) to condense out of the gas so that the solid component (available for planets) was larger in the outer solar nebula. While the MMSN provides an interesting fiducial model of the early solar nebula, we must remain aware that analyses such as this cannot make corrections for planets that were ejected long ago or the systematic radial migration of planets away from their formational locations.

Minimum Mass Pulsar Nebula

Although we do not have as much information about the planetary system around the pulsar PSR1257+12 as we do for the solar system, it is nonetheless the *first* extrasolar planetary system about which we have *any* information. The total mass in the system is $6.2/\sin i \ M_{\oplus}$, and the two planets are at 0.36 AU and 0.47 AU from the neutron star (Wolszczan and Frail 1992). The planetary angular momentum is $\approx 10^{48}$ g cm^2 s^{-1}. To calculate the analog of the MMSN, which we call the minimum mass pulsar nebula (MMPN), we must know what original nebular composition to reconstitute to and what the planets themselves

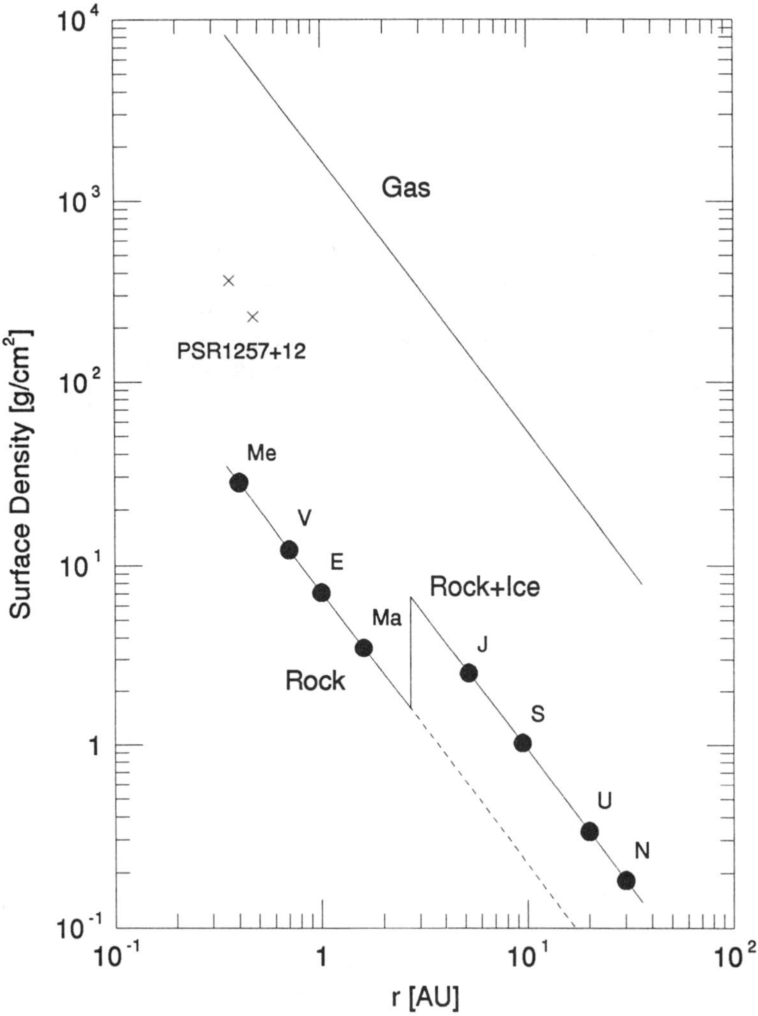

FIGURE I Surface density distribution of rock, ice, and gas in the minimum mass solar nebula (after Hayashi, et al. 1985). Positions of the planets are marked by Me, V, E, Ma, J, S, U, N. The crosses indicate the minimum estimated solid surface density for the two planets observed near PSR1257+12 (no corrections for sin i have been made).

are made of.

If the pulsar planets are similar to the giant planets (with significant gaseous atmospheres and smaller heavy element cores) the inclination angle of the orbital plane to our line-of-sight must be $\lesssim 10°$ for their masses to exceed those of Uranus or Neptune and $\lesssim 1°$ to exceed that of Jupiter. Either possibility (especially the latter) is relatively unlikely. We will therefore assume that the pulsar planets are similar to the rocky terrestrial planets and have a total mass of order 10 $M_\oplus$. In the scenarios where the disk is made from solar composition gas (e.g. Stevens, Rees, Podsiadlowski 1992), the MMPN has a mass of order 0.003 $M_\odot$ and angular momentum of order 10^{50} g cm^2 s^{-1}. In the models where the pulsar disk is rich in metals (e.g. Lin, Woosley, and Bodenheimer 1991) *no* augmentation of the composition must be made, and the MMPN has just the mass and angular momentum observed by Wolszczan and Frail. ¿From the masses and distances of the two planets from PSR1257+12 we can estimate the surface density of solids needed to build them, assuming the planets were formed at their present radii. These estimates lie in the range 230 g cm^{-2} $\lesssim \Sigma_{\rm solid} \lesssim$ 360 g cm^{-2} and are plotted as crosses in Figure I. It is seen that the inferred solid surface densities in the MMPN are an order of magnitude higher than those at the same distance from the sun in the MMSN. Reconstituting these solid surface densities to solar composition results in gas surface densities of order 6×10^4 g cm^{-2} at $\sim 1/2$ AU from the pulsar.

VISCOUS EVOLUTION OF A PULSAR NEBULA

To illustrate some of the features of nebular evolution described above, we have performed a numerical simulation of the evolution of a disk around a pulsar. In the model, the pulsar mass is 1.5 $M_\odot$, and the viscous alpha parameter is $\alpha = 10^{-2}$, a value consistent with the magnitude of convective viscosity in protostellar disks or magnetic turbulence from a weak field ($v_{\rm A} \approx 0.1\, c_{\rm s}$). Because viscous diffusion rapidly causes the detailed form of the initial surface density profile to be forgotten, we will choose values for the total disk mass and angular momentum and then take $\Sigma(r,0)$ to have a step function profile consistent with these global numbers. In most of the theoretical scenarios presented, the initial radial size of the disk is small: of order 10^8 cm in the supernova fallback model (Lin, Woosley, and Bodenheimer 1991) and of order $10^{10} - 10^{11}$ cm for the disruption of a binary stellar companion model (Stevens, Rees, and Podsiadlowski 1992; Podsiadlowski, Pringle, and Rees 1991; however see Tavani and Brookshaw 1992 for a model where the initial disk extends to ~AU distances). The disk mass is ≈ 0.02 $M_\odot$ in the fallback model and ~ 0.1 $M_\odot$ (or larger) for binary disruption.

To be consistent with the opacity table available to us, we have adopted a solar composition for the pulsar disk. We take the initial disk mass and angular momentum to be $M_{\rm d} = 0.1$ $M_\odot$ and $J_{\rm d} = 8.5 \times 10^{50}$ g cm^2 s^{-1}, which corresponds to an initial disk radius of 1.5×10^{11} cm. These initial conditions contain about 30 times the mass and 8 times the angular momentum of the MMPN and correspond roughly to the disruption of a low-mass binary companion near the pulsar. We want to note that in any of the theoretical scenarios proposed the initial disk environment is rather extreme compared to protostellar nebulae. In the model considered here, the surface density is $\Sigma \sim 10^9$ g/cm^2; for $H/r \approx 1/10$ the volume density is $\rho \sim 1$ g/cm^3 and the midplane temperature is $T_c \gtrsim 10^5$K.[3] As the disk expands and cools though, it approaches conditions far more typical of disks around sun-like stars.

In Figure II we plot the surface density distribution in the disk after 10^6 and 10^7 years of evolution. The surface density profiles at these times bear no resemblance to the initial conditions, and the shape of the curves clearly illustrates the self-similarity of the evolution. After 10^7 years the disk has expanded to 30 AU (solar system sized dimensions), and its mass has decreased to $\approx 2 \times 10^{-3}$ $M_\odot$ (comparable to the gas mass in an MMPN). See Table I for further details.

TABLE I Disk parameters as a function of time.

t (yr)	M_d ($M_\odot$)	R_d (AU)	$\dot{M}^a$ ($M_\odot$/yr)
0	10^{-1}	0.01	0
10^3	7.5×10^{-3}	4	1.6×10^{-6}
10^4	4.1×10^{-3}	9	9.4×10^{-8}
10^5	2.7×10^{-3}	14	3.0×10^{-9}
10^6	2.2×10^{-3}	21	1.6×10^{-10}
10^7	1.9×10^{-3}	29	1.4×10^{-11}

[a]Mass accretion rate onto the pulsar

The outer edge of the nebula is a sharp front as is expected from a nonlinear viscosity. Indeed, at late times when the disk is cool enough for water-ice grains to condense, the viscosity becomes $\nu \propto \Sigma^2$ (Lin and Bodenheimer 1982; Ruden and Pollack 1991), i.e., $a = 2$ and $b = 0$ in the notation of equation 18. Using equations 22, we expect (and confirm numerically) that $R \sim t^{1/7}$ and $M_d \sim t^{-1/14}$. Also, equation 20 indicates that the surface density should be independent of r in these regions, which explains the horizontal parts of the $\Sigma(r)$ curves in Figure II.

In Figure II, we also plot the midplane temperature in the disk. Temperatures are cool enough beyond $\sim 10^{-2}$ AU for grain species to condense from the gas phase. In the regions where the planets are observed around PSR1257+12 the midplane temperatures are $\lesssim$ 100K and are cool enough for water-ice to condense at late times. The condensation of ices increases by a factor of ≈ 4 the amount of material that is available in the solid phase for planetesimal formation. Together, the two plots in Figure II show that the outer regions ($r \gtrsim 0.1$ AU) of the pulsar disk form a conducive environment for grain accumulation and planetary growth. Also, the viscous timescale of $\gtrsim 10^6$ years is comparable to that seen (observationally and theoretically) in disks around young stars, hence

[3]In the supernova fallback picture, the initial disk is relativistically degenerate.

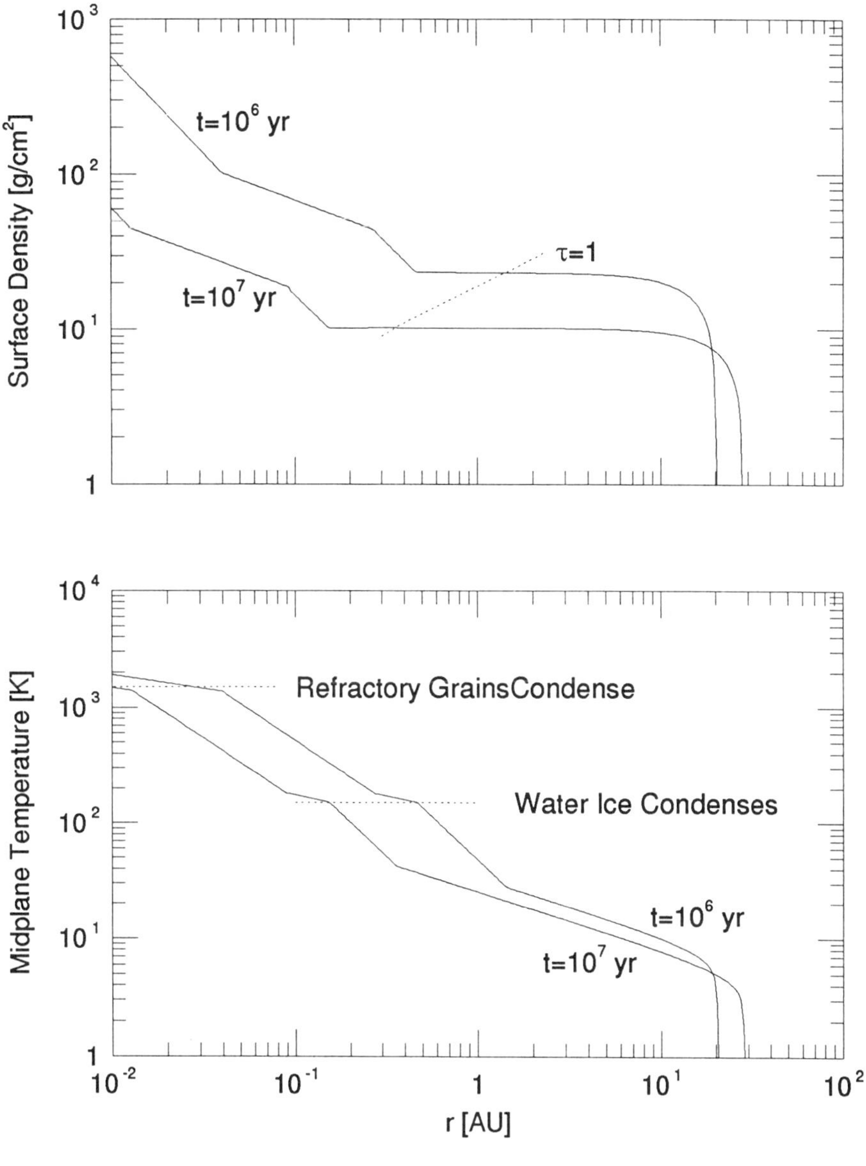

FIGURE II (*Upper*) Surface density distribution in the pulsar nebula after 10^6 and 10^7 years. The dotted curve marks the optically thick/thin transition radius. The disk matter outside this line is only marginally optically thick. (*Lower*) Midplane temperature distribution in the pulsar nebula after 10^6 and 10^7 years. The critical temperatures below which refractory and water ice grains condense are marked by dotted lines.

planet formation in pulsar disks may proceed via substantially the same route as in protostellar disks.

The outer regions of the disk have low surface density, are cool, and are optically thin.[4] In Figure II we have marked the locus where the optical depth $\tau = 1$ by a dotted line. This radius moves inwards with time as the disk is depleted by accretion and cools off. As noted above, the transition to optically thin conditions and the decay of convective turbulence may allow small dust grains to sediment rapidly to the midplane and form planetesimals. It is of interest to note that this transition radius is ~ 1 AU after $\sim 10^6$ years. At these times the outer disk may be actively engaged in building planets, and it is entirely possible in this type of model that more distant planetary companions to PSR1257+12 exist. In the next decade, timing data may be able to tell us whether such planets actually exist.

Figure III shows the mass accretion rate onto the pulsar as a function of time. The value of the disk luminosity corresponding to this accretion rate, $L_d = GM_*\dot{M}/2R_*$ (eq. 17), is also marked. The accretion rate falls off roughly as t^{-1}; equation 22 indicates that at late times when water-ice has formed ($a = 2$, $b = 0$) we should expect $\dot{M} = |dM_d/dt| \propto t^{-15/14}$, consistent with the numerical simulation. Before $\sim 10^4$ years, the accretion is so intense that it exceeds the Eddington limit. Such large accretion rates may be unavoidable consequences of dumping $\sim 0.1\ M_\odot$ of gas near a compact object. At these early times, this large luminosity will strongly affect the surrounding nebular gas (see the review article by Phinney). At the late times when planets form, the disk luminosity still ranges from $10 - 10^3\ L_\odot$. Significant surface heating effects may be present if the inner disk regions do not shadow the outer disk, and thermal convection might be stabilized, which would modify the α–parameter.

SUMMARY

We have reviewed the important physical processes responsible for making thin disks around stars evolve. Low-mass disks around pulsars are relatively stable to gravitational instabilities and most likely evolve due to magnetic or turbulent stresses. Convective viscosity plays an important role in the outer nebular regions where the temperatures are cool enough for dust grains to condense. It is possible for initially small pulsar disks to expand to sizes large enough to make planets at ~AU distances, and the mass in these disks is comparable to the mass needed to make the planets seen around PSR1257+12. The viscous timescale is $\sim 10^6$ years. All in all, the outer regions of pulsar disks seem to be a very conducive environment for the formation of planets.

---

[4]In the optically thin regions of the disk we have reduced the viscosity to account heuristically for the inefficiency of turbulent convection. In our models we take $\alpha = \alpha_0$ for $\tau \geq 1$ and $\alpha = \alpha_0 \tau$ for $\tau < 1$ where $\alpha_0 = 10^{-2}$.

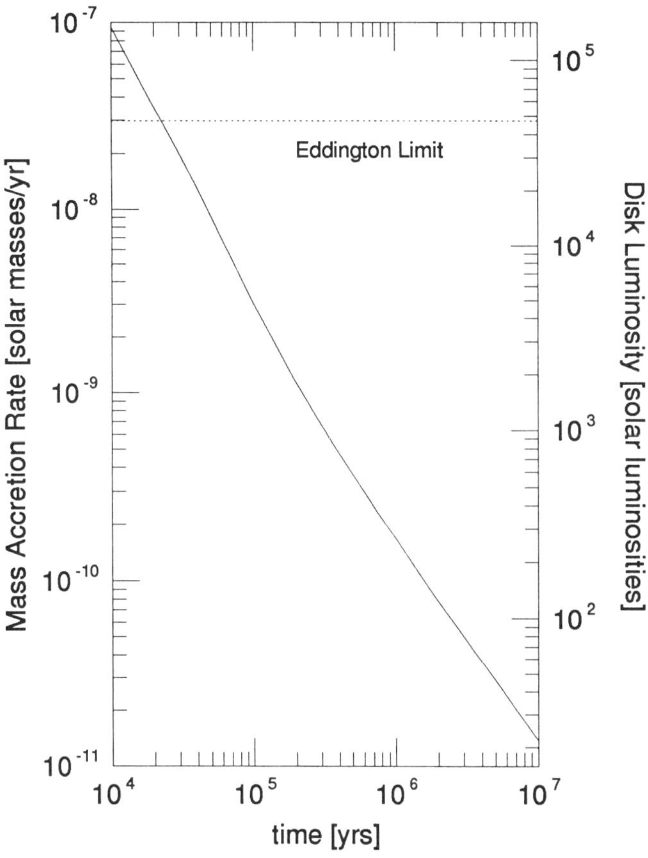

FIGURE III Mass accretion rate onto the pulsar as a function of time. The corresponding integrated disk luminosity (assuming $R_* = 10$ km) is marked. The dotted line indicates the Eddington limit.

ACKNOWLEDGMENTS

This research is partially supported by NSF grant AST-9157420, NASA grant NAGW-2260, and the University of California, Irvine through an allocation of computer resources.

REFERENCES

Adams, F. C., Ruden, S. P., and Shu, F. H. 1989, *ApJ*, **347**, 959

Balbus, S. A., and Hawley, J. F. 1991, *ApJ*, **376**, 214

Binney, J., and Tremaine, S. 1987, *Galactic Dynamics*, Princeton: Princeton University Press

Cabot, W., Canuto, V. M., Hubickyj, O., and Pollack, J. B. 1987, *Icarus*, **69**, 423

Drazin, P. G., and Reid, W. H. 1981, *Hydrodynamic Stability*, Cambridge: Cambridge University Press, 69

Hawley, J. F., and Balbus, S. A. 1991, *ApJ*, **376**, 223

Hayashi, C., Nakazawa, K., and Nakagawa, Y. 1985, in *Protostars and Planets II*, D. C. Black and M. S. Matthews, Tucson: University of Arizona Press, 1100

Hoyle, F. 1946, *MNRAS*, **106**, 406

Knobloch, E. 1992, *MNRAS*, **255**, 25p

Korycansky, D. 1992, *ApJ*, **399**, to appear

Lin, D. N. C., and Bodenheimer, P. 1982, *ApJ*, **262**, 768

Lin, D. N. C., and Papaloizou, J. 1980, *MNRAS*, **191**, 37

Lin, D. N. C., and Pringle, J. E. 1987, *MNRAS*, **225**, 607

Lin, D. N. C., Woosley, S. E., and Bodenheimer, P. H. 1991, *Nature*, **353** 827

Lynden-Bell, D. and Pringle, J. E. 1974, *MNRAS*, **168**, 603

Parker, E. N. 1979, *Cosmical Magnetic Fields: Their Origin and Their Activity*, Oxford: Oxford University Press

Podolak, M. and Reynolds, R. T. 1985, in Protostars and Planets II, D. C. Black and M. S. Matthews, Tucson: University of Arizona Press, 847

Pollack, J. B., McKay, C. P., and Christofferson, B. M. 1986, *Icarus*, **64**, 471

Podsiadlowski, P., Pringle, J. E., and Rees, M. J. 1991, *Nature*, **352**, 783

Pringle, J. E. 1981, *ARA&A*, **19**, 137

Ruden, S. P., and Lin, D. N. C. 1986, *ApJ*, **308**, 883

Ruden, S. P., Papaloizou, J. C. B., and Lin, D. N. C. 1988, *ApJ*, **329**, 739

Ruden, S. P. and Pollack, J. B. 1991, *ApJ*, **375**, 740

Ryu, D., and Goodman, J. 1992, *ApJ*, **388**, 438

Shakura, N. I., and Sunyaev, R. A. 1973, *A&A*, **24**, 337

Shu, F. H., Tremaine, S., Adams, F. C., and Ruden, S. P. 1990, *ApJ*, **358**, 495

Stevens, I. R., Rees, M. J., and Podsiadlowski, P. 1992, *MNRAS*, **254**, 19p

Tavani, M., and Brookshaw, L. 1992, *Nature*, **356**, 320

Toomre, A. 1964, *ApJ*, **139**, 1217

Weidenschilling, S. 1977a, *MNRAS*, **180**, 57

Weidenschilling, S. 1977b, *Ap&SS*, **51**, 153

Weismann, P. 1989, in *Origin and Evolution of Planetary and Satellite Atmospheres*, S. K. Atreya, J. B. Pollack, and M. S. Matthews, Tucson: University of Arizona Press, 230

Wolszczan, A., and Frail, D. A. 1992, *Nature*, **355**, 145

Zel'dovich, Ya. B., and Raizer, Yu. P. 1967, *Physics of Shock Waves and High-Temperature Hydrodynamic Phenomena, Vol. II*, New York: Academic Press, Ch. X

Planets Around Pulsars
ASP Conference Series, Vol. 36, 1993
J.A. Phillips, J.E. Thorsett, and S.R. Kulkarni

PLANETARY ACCRETION IN CIRCUMSTELLAR DISKS

JACK J. LISSAUER
Astronomy Program, Department of Earth and Space Sciences, State University of New York, Stony Brook, NY 11794

GLEN R. STEWART
LASP, Campus Box 392, University of Colorado, Boulder, CO 80309

ABSTRACT The formation of terrestrial planets and the cores of Jovian planets is reviewed in the framework of the planetesimal hypothesis, wherein planets are assumed to grow via the pairwise accumulation of small solid bodies. The rate of (proto)planetary growth is determined by the size and mass of the protoplanet, the surface density of planetesimals, and the distribution of planetesimal velocities relative to the protoplanet. Planetesimal velocities are modified by mutual gravitational interactions and collisions, which convert energy present in the ordered relative motions of orbiting particles (Keplerian shear) into random motions and tend to reduce the velocities of the largest bodies in the swarm relative to those of smaller bodies, as well as by gas drag, which damps eccentricities and inclinations. The evolution of the planetesimal size distribution is determined by the gravitationally enhanced collision cross-section, which favors collisions between planetesimals with smaller velocities. Runaway growth of the largest planetesimal in each accretion zone appears to be a likely outcome. The subsequent accumulation of the resulting protoplanets leads to a large degree of radial mixing in the terrestrial planet region, and giant impacts are probable.

INTRODUCTION

At present, we have confirmed knowledge of only one planetary system, the solar system in which we live. Regular variations in arrival times of pulses from PSR1257+12, which may be interpreted as resulting from perturbations by unseen planetary-sized companions, have recently been observed (Wolszczan and Frail 1992). Such new data offers hope that we may soon have valuable constraints on the properties of planetary systems other than our own. However at present, models of planetary formation are developed using our single example of a planetary system, supplemented by limited astrophysical observations of star forming regions and circumstellar disks.

The nearly circular and coplanar orbits of solar system planets argue for planetary formation in a flattened disk revolving about the Sun (Kant 1745, Laplace 1796). Astrophysical evidence suggests such disks are a natural byproduct of star formation from the collapse of rotating molecular cloud cores (Hoyle 1960, Cameron 1962, Terebey *et al.* 1984). Models for the formation of analogous disks in orbit around pulsars have been presented by Lin *et al.* (1991), Nakamura and Piran (1991), Podsiadlowski *et al.* (1991), Tavani and Brookshaw (1992) and others. Observational evidence for the presence of disks of solar system dimensions around pre-main sequence stars has increased substantially in recent years (Beckwith and Sargent 1992). Planets are likely to grow within such circumstellar disks via binary accretion of solid planetesimals (Safronov 1969, Hayashi *et al.* 1977). Sufficiently massive planetary bodies embedded in a gas-rich disk can gravitationally capture large amounts of gas (Mizuno 1980, Hayashi *et al.* 1985, Bodenheimer and Pollack 1986).

This article reviews the process of planet formation within circumstellar disks around young stars. The process of planetary growth is generally divided into several distinct stages. In the first stage, microscopic grains collide and grow via pairwise collisions during settling towards the midplane of the disk. If the disk is laminar, the solids can form a thin enough layer for gravitational instabilities to occur. If the disk is turbulent, binary agglomeration continues as a result of relative motions. The gaseous component of the protoplanetary disk plays an important role in this stage of planetary growth. Once solid bodies reach kilometer-size (for parameters believed to be appropriate for the terrestrial region of the proto-solar disk), gravitational interactions between pairs of solid "planetesimals" dominate electromagnetic forces, gas drag and collective gravitational effects. Growth via binary collisions proceeds until either a protoplanet becomes massive enough to gravitationally trap substantial quantities of gas, or the solid protoplanets become dynamically isolated from on another. (We use the term protoplanets to refer to exceptionally large planetesimals, not the giant gaseous protoplanets once popularized by Cameron (1962).) The stage of gaseous accretion (if it occurs) is terminated when either the disk is dissipated, or the planet accretes all of the gas within its region of the disk, or gravitational torques from the planet clear a gap in the disk.

We concentrate here on the dynamics of solid body accretion from kilometer size planetesimals to terrestrial type planets; this stage of planetary growth is likely to depend the least on the characteristics of the evolutionary state of the central star. We omit any further discussion of gas accretion by planets, as this phase is likely to be quite dependent on the characteristics of the disk in which the planets are growing; moreover, the masses of planets thus far tentatively observed around pulsars are too small to suggest a substantial gaseous component. See Podolak *et al.* (1992) for a recent review of planetary accretion of gas. More detailed reviews of planetesimal formation and of planetesimal growth can also be found in the book *Protostars and Planets III* (Weidenschilling and Cuzzi 1992, Lissauer and Stewart 1992, Ohtsuki *et al.* 1992).

Strictly speaking, the planet formation process is unlikely to be as purely sequential as we have lain out. Grain growth and even the accretion of large planetesimals may well begin during the epoch when a protoplanetary disk is still accreting and redistributing matter. Given that the theories of each of these epochs are still rather primitive, a sequential study of each stage is probably adequate. However, to the extent that planetary growth depends on, e.g., the initial size distribution of planetesimals, we must recognize that various processes currently being treated as separate events occur simultaneously and affect one another.

PLANETESIMAL FORMATION

As a disk of gaseous matter cools, various compounds condense into microscopic grains. For a disk of solar composition, the first substantial condensates are silicates and iron compounds; in a carbon-rich gas carbides would condense first. At lower temperatures, characteristic of the outer region of our planetary system, large quantities of H_2O and other ices can condense. A detailed discussion of equilibrium condensation of a solar composition gas is given by Barshay and Lewis (1976). The time required to reach equilibrium for the low gas densities expected in protoplanetary disks ($\rho_{tot} \sim 10^{-9}$ g/cm^3) at the low temperatures required for ice condensation may well be longer than characteristic cooling and condensation times, and may even exceed the lifetime of the disk. Thus certain chemical reactions may be kinetically inhibited from reaching equilibrium (Prinn and Fegley 1989). Some preexisting condensates from the ISM and stellar atmospheres may also be present. Growth of solid particles then proceeds primarily by mutual collisions.

The microphysics of the growth of sub-centimeter grains is quite different than the dynamical processes important to later stages of planetary accretion. The mechanical and chemical processes related to grain agglomeration are poorly understood. Data from smokestack studies (Forrest and Witten 1979) and numerical models (Meakin and Donn 1988, Weidenschilling *et al.* 1989) suggest that loosely packed fractal structures may be formed. However, most primitive meteorites contain chondrules, which are small glassy inclusions ~ 1 mm in size (e.g., Grossman *et al.* 1988). The large abundance of chondrules implies that a significant fraction of these fluffy aggregates were rapidly heated and cooled prior to being incorporated into larger bodies. Various models of chondrule formation exist (Grossman 1988, Levy 1988), but researchers in this field have not yet reached a consensus paradigm.

The motions of small grains in a protoplanetary disk are strongly coupled to the gas (Adachi et al. 1976). For the parameter regime believed to have existed in the proto-solar nebula, the coupling between the gas and solid particles smaller than 1 cm is well described by the Epstein's drag law (Weidenschilling 1977). Substantially larger particles experience Stokes drag.

The vertical component of the star's gravity causes dust to sediment out towards the midplane of the disk. The equilibrium settling velocity is determined by a balance between gravity and gas drag. For fixed particle density, the rate of sedimentation is proportional to particle radius (cf. Weidenschilling 1980 for mathematical details).

Sedimentation times for 1μm grains are quite long, ~ 10^7 years, but collisional growth of grains during their decent to the midplane of the disk shortens sedimentation times by several orders of magnitude (Weidenschilling 1980), and differential settling velocities increase the collision rates between particles of differing sizes. For fluffy fractal configurations, the settling rate increases much more gradually with increasing particle size, but the greater collision cross-sections of such low density agglomerates compensates for the slower settling rate (Weidenschilling *et al.* 1989). Recent models suggest that (for the parameters believed appropriate to the terrestrial planet region of the solar nebula) the bulk of the solid material was contained in bodies of radius ~ 10^3 - 10^4 cm, and that most of these bodies were confined to a relatively thin region about the midplane of the disk in which the density of condensed material was comparable to, or exceeded, that of the gas (Weidenschilling and Cuzzi 1992).

The gaseous component of a protoplanetary disk is partially supported against stellar gravity by a pressure gradient in the radial direction, so gas circles the star less rapidly than the Keplerian rate. Large particles moving at (nearly) the Keplerian speed thus encounter a headwind, which removes part of their angular momentum and causes them to spiral inwards towards the star. This effect is greater for smaller particles, which have larger surface area to mass ratios. Substantial inward drift may occur for meter-sized bodies in the inner solar nebula (Weidenschilling 1977). The effect diminishes for very small particles, which are so strongly coupled to the gas that the headwind they encounter is very slow.

Two alternative hypotheses describe the next stage of growth. If the nebula is quiescent, the planetesimals settle into a layer thin enough to be gravitationally unstable to clumping, and planetesimals presumably are formed as a result of this instability (Safronov 1969, Goldreich and Ward 1973). The planetesimals produced by this mechanism have well-defined radii, proportional to the surface density of the particle layer at the time the instability occurs. Planetesimals formed in the inner regions of the solar nebula would have been ~ 1 km in radius, with larger planetesimals forming farther from the Sun.

In a turbulent nebula, growth continues via simple 2-body collisions. Under these circumstances, there is no fine line between planetesimal formation and accretion from planetesimals to planets. Current models suggest that interactions between the gas and solids produced sufficient turbulence to prevent the particulate layer from becoming thin enough to be gravitationally unstable, at least in the terrestrial region of the solar nebula (Weidenschilling and Cuzzi 1992). However, if protoplanetary disks around pulsars contain predominantly condensed matter (Lin *et al.* 1991), they would be more likely to produce planetesimals via gravitational instability.

Whether by turbulent motions or by gravitational instabilities, kilometer-sized planetesimals form, probably on a ~ 10^4 year timescale. How do these planetesimals then grow into planets?

FROM PLANETESIMALS TO PLANETS

The star's gravity is the dominant force upon planetesimals, and other forces are generally treated as perturbations on the planetesimal's Keplerian orbit.

The largest perturbations to a planetesimal's heliocentric trajectory are usually a result of gravitational interactions with other planetesimals and protoplanets. When a single protoplanet is the dominant perturber in a given region of the protoplanetary disk, it is convenient to treat its perturbations separately. The largest non-gravitational forces upon planetesimals are mutual inelastic collisions (which may lead to accretion and/or fragmentation), and gas drag. Whole body magnetic forces are believed to be negligible for the dynamics of bodies of kilometer size and larger. However, if electromagnetic induction heating (Herbert 1989) is sufficient to melt planetesimal interiors, then the outcome of physical collisions could be altered.

Planetesimal Velocities for a Static Mass Distribution

The distribution of planetesimal velocities is one of the key factors which control the rate of planetary growth. The simplest analytic approach for calculating the evolution of planetesimal velocities uses a "particle-in-a-box" approximation in which the evolution of the mean square planetesimal velocities are calculated via the methods of the kinetic theory of gases. A kinetic theory approach appears to be the only viable method for treating the initial stages of planetesimal accumulation because the initial number of planetesimals is extremely large. During the final stages of planetesimal accumulation, the number of planetesimals eventually becomes small enough that a more direct treatment of individual planetesimal orbits is feasible. With a modest number of planetesimals, the most straightforward approach would be a numerical n-body integration of the planetesimal orbits, but the exceedingly long time scales required for the final stages of planetary accumulation (e.g., between 10^7 and 10^8 years in the inner solar system) severely limit the usefulness of this method. An alternative approach that has enjoyed considerable success is to assume that the planetesimals follow slowly precessing elliptic orbits that are occasionally altered by rare close encounters with other planetesimals. A Monte Carlo procedure is then used to choose successive pairs of planetesimals that interact according to the two-body gravitational scattering formula (cf. Wetherill 1990a and references therein).

In the particle-in-a-box approximation, one ignores the details of individual planetesimal orbits and uses a probability density to describe the distribution of orbital elements in the planetesimal population. Specifically, the orbital perihelia and longitudes of the ascending nodes are assumed to be random and the orbital eccentricities, e, and inclinations, i, are assumed to be Rayleigh distributed,

$$f(e, i) = 4\,(\sigma/m)\, e\, i\, (\langle e^2\rangle\langle i^2\rangle)^{-1} \exp\left[-(e^2)/\langle e^2\rangle - (i^2)/\langle i^2\rangle\right] \qquad (1)$$

where m is the mass of the individual planetesimals and σ is surface mass density of planetesimals with a particular semi-major axis; $\langle e^2\rangle$ and $\langle i^2\rangle$ are the mean square eccentricity and inclination.

Another important property of planetesimal orbits is the fact that planetesimals with different semi-major axes orbit the star with different mean velocities. This Keplerian shear is usually introduced into a local velocity distribution by postulating a local mean velocity that reproduces the differential rotation between coplanar circular orbits. Since the orbital phase angles are averaged out in this approximation, the deviations from coplanar circular orbits are reduced to a distribution of random velocities relative to the local mean velocity of a circular orbit. The distribution of random velocities that is locally equivalent to eq. (1) is a triaxial Gaussian distribution in cylindrical coordinates,

$$f_0(z, \mathbf{v}) = \Omega\,(\sigma/m)(2\pi^2 c_r^2 c_z^2)^{-1}\exp[-(v_r^2+4v_\theta^2)/2c_r^2 - (v_z^2+\Omega^2 z^2)/2c_z^2]. \quad (2)$$

where $2c_r^2 = \langle e^2\rangle v_K^2$, $2c_z^2 = \langle i^2\rangle v_K^2$, v_θ is the azimuthal component of velocity relative to the local circular Keplerian velocity, $v_K = (GM_\star/r)^{1/2}$, and $\Omega = v_K/r$ is the orbit frequency.

Given the local random velocity distribution in eq. (2), one can write down a kinetic equation for the time evolution of the random velocity distribution caused by mutual planetesimal interactions and gas drag:

$$\partial f/\partial t + \mathbf{v} \cdot \partial f/\partial \mathbf{r} - [GM_\star \mathbf{r}/r^3 + (\mathbf{v} + \mathbf{v}_K) \cdot \partial \mathbf{v}_K/\partial \mathbf{r}] \cdot \partial f/\partial \mathbf{v}$$
$$= \delta f/\delta t|_{coll} + \delta f/\delta t|_{grav} + \delta f/\delta t|_{drag}. \quad (3)$$

The first two terms on the right hand side of eq. (3) denote the two kinds of planetesimal interactions: 1) physical collisions which dissipate some or all of the relative kinetic energy of the colliding bodies, and 2) gravitational scattering which conserves the relative kinetic energy but results in a rotation of the relative velocity vector. Physical collisions are usually modeled with a Boltzmann collision operator for hard spheres that has been modified to allow for inelastic collisions (see Trulsen 1971, Hornung *et al.* 1985). Most published calculations of planetesimal accumulation assume that collisions result in accretion because physically plausible rebound velocities rarely exceed the mutual escape velocity. Large protoplanets can stir random velocities to well beyond the escape speed from small planetesimals, leading to disruptive collisions among these bodies. In order to account for the gravitational enhancement of the collision cross-section, a correction factor must be applied to the hard sphere collision operator. An accurate determination of this enhancement factor is rather difficult in general, owing to the tidal influence of the star; detailed discussions of its calculation is presented by Greenzweig and Lissauer (1990, 1992).

Gravitational scattering between planetesimals can be modeled with a Fokker-Planck operator similar to that used in stellar dynamics (cf. Binney and Tremaine 1987). Both gravitational scatterings and physical collisions can convert energy from the ordered differential rotation of the disk (Kepler shear) into random motions of the planetesimals. Inelastic collisions (including those resulting in accretion) convert some of the energy of random motions into heat. Gravitational scatterings tend to equipartition kinetic energy of random motion among planetesimals of different masses. Although equipartition is not actually achieved, this process, which is known as dynamical friction, can produce a distribution in which the small particles have a much greater velocity dispersion than do the large ones. Gas drag acts to damp eccentricities and inclinations of planetesimals, especially small ones. The mathematical formulas used to study the evolution of planetesimal velocities may be found in Stewart and Wetherill (1988) and Lissauer and Stewart (1992).

Most analyses of the evolution of the planetesimal velocity distribution assume that successive encounters between planetesimals are uncorrelated and that 3-body effects may be neglected. In the 2-body approximation, conservation of energy implies that two planetesimals must be moving at the same speed *relative to each other* after a gravitational encounter as prior to it. However, uncorrelated scattering increases the speed of a planetesimal *relative to mean circular motion* in a random walk fashion.

The 2-body uncorrelated encounter approximation is accurate for interactions among the bulk of the planetesimals in any given accretion zone. However, as Safronov (1969) recognized, the largest body in a given zone is special. This body is the most effective scatterer, but if it is much more massive than the next largest body in its accretion zone, then its stirring ability is far less than in the 2-body case, especially if its orbit is circular. For a protoplanet on a circular orbit, a given planetesimal will always approach the protoplanet at the same velocity, apart from changes due to scatterings by other planetesimals. Thus, the magnitude of the secular increase in velocity dispersion is limited to the same order as that of the much smaller secondary scatterings. For this reason, Safronov (1969) approximated the stirring ability of a protoplanet by that of the second largest body in its accretion zone. This approximation is probably adequate if the largest body is only a few times as massive as the second largest in its zone, but must break down if one protoplanet becomes much more massive than any other nearby body.

The region in which 3-body effects are important scales with the radius of the protoplanet's Hill Sphere,

$$h = (M_1/3M_\star)^{1/3}\, a\,. \tag{4}$$

where M_1 and $M_\star$ are the masses of the protoplanet and the star and a is the semi-major axis of their mutual orbit. For fixed density, h scales as the protoplanet's radius. A rocky protoplanet orbiting 1 AU from a 1 $M_\odot$ star has a radius $\approx 0.005\ h$. It is convenient to introduce the following Hill-scaled units for planetesimal eccentricity, e, inclination, i, and separation in semi-major axis from the planet, Δa:

$$e_H \equiv ea/h\,, \quad i_H \equiv ia/h\,, \quad b_H \equiv \Delta a/h\,. \tag{5}$$

Ida (1990) has presented an extensive series of 3-body orbit integrations in order to directly determine the ability of a protoplanet on a circular orbit to alter the velocity distribution of the surrounding planetesimal swarm and the ability of a planetesimal swarm to damp the eccentricity and inclination of a protoplanet via dynamical friction. Ida finds that the energy equipartition rate agrees well with the 2-body results at high velocities. However, when $e_H < 2$, the rate of equipartition stops increasing with decreasing velocity (unlike the 2-body case, where the rate increases indefinitely). Thus, dynamical friction continues to operate at low velocities, but the energy equipartition rate attains a maximum asymptotic value that is given by setting $e_H = 2$ in the two-body formula.

Collision Cross-Sections and Growth Rates

The size distribution of planetesimals evolves principally via physical collisions among its members. Stresses caused by tidal forces during close encounters between planetesimals may also fragment very weak bodies, but such disruptive encounters require special circumstances (Sridhar and Tremaine 1992), and thus can be neglected or folded into the formalism of fragmentation via physical collisions. The evolution of the size distribution of planetesimals can be studied using coagulation theory, as discussed below. The inputs required in the coagulation calculations are the collision frequency and physical assumptions regarding the outcome of collisions.

The velocity at which two bodies of radii R_1 and R_2 and masses m_1 and m_2 collide is given by:

$$v_c = (v^2 + v_e^2)^{1/2} \tag{6}$$

where v is the relative velocity of the two bodies far from encounter and v_e is the escape velocity from the point of contact:

$$v_e = [G(m_1 + m_2)/(R_1 + R_2)]^{1/2} \tag{7}$$

The rebound velocity is equal to εv_c, where $\varepsilon \leq 1$ is the coefficient of restitution. If $\varepsilon v_c \leq v_e$, then the two bodies remain bound gravitationally and soon recollide and accrete. Net disruption requires both fragmentation, which depends on the internal strength of the bodies, and post-rebound velocities greater than the escape speed. As we saw above, relative velocities of planetesimals are generally less than the escape velocity from the largest typical bodies in the swarm. Thus, unless ε is very close to unity, the largest members of the swarm are likely to accrete the overwhelming bulk of material with which they collide. Fragmentation is most common for very small planetesimals. The largest bodies in the swarms, which eventually are to grow to planetary size, accrete at a rate essentially identical to the collision rate.

The simplest model for computing the collision rate of planetesimals ignores their motion about the central star entirely. This problem can be treated using methods of the kinetic theory of gases (Safronov 1969, Wetherill 1980). Collisions occur when the separation between the centers of two particles becomes less than the sum of their radii, R_s. The accretion rate of a body in a planetesimal swarm of density ρ_{sw} (not to be confused with the density or specific gravity of the individual bodies, ρ) at relative velocity v is given by:

$$dM/dt_{PiB} = \rho_{sw} v \pi R_s^2 (1 + (v_e/v)^2) \tag{8}$$

where the second term in the parentheses represents the gravitational enhancement of the accretion cross-section.

Note that in a disk of given surface mass density, σ, the volume density of the planetesimal swarm is inversely proportional to v, assuming the mean ratio of horizontal to vertical motions (eccentricities to inclinations) remains fixed:

$$\rho_{sw} \approx \sigma/(2a\sin i) \approx (\sqrt{3}/2)(\sigma\Omega/v) \qquad (9)$$

where the factor of $\sqrt{3}$ assumes the velocity dispersion is isotropic. Substituting eq. (9) into (8), we find that the mass accretion rate of a protoplanet depends on v only through the gravitational focussing factor:

$$dM/dt_{PiB} = (\sqrt{3}/2)\sigma\Omega\pi R_s^2\,[1 + (v_e/v)^2] \qquad (10)$$

The exact value of the constant in front of the r.h.s. of eq. (10) depends on the velocity distribution; thus, various values have been quoted in the literature and used in estimating growth times for planets.

When $(v_e/v) \gg 1$, the duration of an encounter becomes comparable to an orbital period, and the tidal influence of the star becomes important. Accretion cross-sections calculated for low v using numerical restricted 3-body integrations of planetesimal trajectories are presented by Ida and Nakazawa (1989) and by Greenzweig and Lissauer (1990, 1992).

Early Stages of Planetesimal Accumulation

The velocity dependence of the collision cross-section described above leads to a strong coupling between the evolution of the velocity and size distributions of planetesimals. The study of planetary accumulation therefore requires a simultaneous calculation of the velocity evolution and size evolution of the planetesimal swarm. The early stages of planetesimal evolution have been modeled by several authors (Greenberg *et al.* 1978, 1984; Nakagawa *et al.* 1983; Ohtsuki *et al.* 1988; Wetherill and Stewart 1989, Spaute *et al.* 1991). All of these papers use some variant of the particle-in-a-box approximation, but they differ in their treatment of the velocity evolution and their method of simulating the planetesimal size evolution. The most important result that has emerged from this work is that the evolving planetesimal size distribution can follow two qualitatively distinct paths that are characterized by very different time scales. The slower evolutionary path exhibits an orderly growth of the entire size distribution wherein all the planetesimals remain tied to the continuous size distribution throughout the early stages of accumulation. The term "runaway accretion" refers to a different evolutionary path where the largest planetesimal in the local region grows much more rapidly than the remainder of the population and therefore becomes detached from the continuous size distribution.

The origin of runaway accretion can be best understood in the context of various solutions to the discrete form of the coagulation equation,

$$dn_k/dt = (1/2) \sum_{i+j=k} A_{ij}\, n_i\, n_j - n_k \sum_i A_{ik}\, n_i\,, \tag{11}$$

which describes the time evolution of the number of bodies n_k with mass m_k. The merger of smaller bodies increases n_k, whereas the incorporation of bodies of mass m_k into larger bodies causes n_k to decrease. Analytical solutions to eq. (11) are known for a few simple forms of the collision probability, A_{ik} (Safronov 1969, Trubnikov 1971, Wetherill 1990b). Wetherill (1990b) and Spaute *et al.* (1991) have utilized some of these solutions to verify the accuracy of the numerical procedures used in their calculations.

These analytic solutions of the coagulation equation indicate a bifurcation between situations in which orderly growth occurs and those resulting in runaway growth of the largest body. The decisive factor that determines which branch the solution takes during planetesimal accumulation is the velocity dependence of the collision rate. In general, a distribution which exhibits smaller velocities for the larger bodies increases the effective mass dependence of the collision rate and therefore increases the likelihood of runaway accretion. Wetherill and Stewart (1989) presented a series of calculations which serve to delineate the conditions required for runaway accretion to occur. In those calculations which omitted the energy equipartition caused by gravitational scattering, orderly growth of the planetesimal size distribution was found. These results show qualitative agreement with the earlier results of Safronov (1969) and Nakagawa *et al.* (1983), who also neglected the energy equipartitioning.

When Wetherill and Stewart (1989) included the energy equipartition terms in their velocity evolution equations, they found runaway growth of the largest planetesimal to occur. Complementary calculations by Ohtsuki and Ida (1990) and Spaute *et al.* (1991) confirm the result that runaway accretion occurs when the random velocities decrease with increasing planetesimal mass.

Limits to Runaway Growth

Runaway accretion requires low random velocities, and thus small radial excursions, $2ae$, of planetesimals. This implies that a protoplanet's feeding zone is limited to the annulus of planetesimals which it can gravitationally perturb into intersecting orbits. Thus, rapid runaway growth must cease when a protoplanet has consumed most of the planetesimals within its gravitational reach (Lissauer 1987).

For the case of a protoplanet on a circular orbit, the standard theory of the restricted 3-body problem places an upper bound on the initial semimajor axis, b_H, that may lead to collision. Neglecting gas drag and interactions with other planetesimals, a planetesimal whose orbital elements satisfy the inequality:

$$(3/4)\, b_H^2 - e_H^2 - i_H^2 \geq 9 \tag{12}$$

cannot enter the protoplanet's Hill sphere (e.g., Artymowicz 1987). Unpublished numerical experiments by Kary and Lissauer show that most planetesimals with initial $e_H = i_H = 0$ and $1.3 < b_H < 3.2$ come within 0.1h of the protoplanet during 20 synodic periods. For nonzero "initial" eccentricity and/or inclination, the accretion zone expands slightly, but for e and i low enough for gravitational focussing to be large, planetesimals with $|b_H| > 4$ cannot be accreted. Thus, the accretion zone of a protoplanet embedded in a disk of low random velocity planetesimals extends over the region:

$$|b_H| < B, \tag{13}$$

where B depends on the magnitude of other perturbations on the planetesimals, and typically is ~ 3.5 - 4 in a quiescent disk.

The size of a protoplanet's Hill sphere expands as it accretes matter. The mass of a protoplanet which has accreted all of the planetesimals within an annulus of width $2\Delta r$ is:

$$m = \int_{r-\Delta r}^{r+\Delta r} 2\pi r'\sigma(r')dr' \approx 4\pi r\Delta r\sigma(r) \tag{14}$$

Setting $\Delta r = Bh = Br(m/3M_\star)^{1/3}$, wc obtain the isolation mass to which a protoplanet orbiting at a distance r from a star of mass $M_\star$ may grow:

$$m = (4\pi Br^2\sigma)^{3/2}/(3M_\star)^{1/2} = 2.10 \times 10^{-3}\,(Br^2\sigma/2\sqrt{3})^{3/2}(M_\odot/M_\star)^{1/2}\,M_\oplus \tag{15}$$

where the mass of the Earth, $M_\oplus = 5.98 \times 10^{27}$g, r is expressed in AU and σ in g/cm^2 (Lissauer 1987). For example, assuming $B = 2\sqrt{3}$, a "minimum mass solar nebula" with $\sigma = 10$ g/cm^2 at 1 AU implies protoplanet isolation at 0.066 $M_\oplus$; whereas $\sigma = 3$ g/cm^2 at 5 AU implies protoplanet isolation at 1.36 $M_\oplus$.

Runaway growth can persist beyond the isolation mass given by equation (15) only if additional mass can diffuse into the protoplanet's accretion zone. Three plausible mechanisms for such diffusion are scattering between planetesimals, perturbations by protoplanets in neighboring accretion zones and gas drag (Weidenschilling *et al.* 1992). Alternatively, radial motion of the protoplanet may bring it into zones not depleted of planetesimals. Gravitational torques due to excitation of spiral density waves in the gaseous component of the protoplanetary disk have the potential of inducing rapid radial migration of protoplanets (Goldreich and Tremaine 1980, Ward 1986).

Recently it has been suggested that gravitational focussing of gas could vastly increase the rate of inward drift of protoplanets due to gas drag (Takeda *et al.* 1985, Ohtsuki *et al.* 1988). This result would have profound implications for models of planetary accretion. However, the fluid calculations assumed low Reynolds Number, which may be appropriate for a turbulent protoplanetary disk, but not for a laminar one. The effects of the stellar gravitational field upon the flow pattern must also be examined.

Final Stages of Planetesimal Accumulation

The self-limiting nature of runaway growth strongly implies that massive protoplanets form at regular intervals in semimajor axis. The agglomeration of these protoplanets into a small number of widely spaced planets necessarily requires a stage characterized by large orbital eccentricities, significant radial mixing, and giant impacts. Mutual gravitational scattering can pump up the relative velocities of the protoplanets to values comparable to the surface escape velocity of the largest protoplanet, which is sufficient to ensure their mutual accumulation into planets. The large velocities imply small collision cross-sections and hence long accretion times. In the outer solar system the limits of runaway growth are less severe; it is feasible that runaway growth of Jupiter's core continued until it attained the necessary mass to rapidly capture its massive gas envelope (Lissauer 1987).

The transition from runaway growth in isolated accretion zones to the mutual accumulation of protoplanets in large eccentricity orbits has so far only been studied qualitatively. At the end of the runaway phase, most of the original mass is contained in the large protoplanets, so their random velocities are no longer strongly damped by energy equipartition with the smaller planetesimals. Even if the protoplanets form in circular orbits, mutual gravitational perturbations among several bodies can eventually induce eccentricities of order 0.01 ($e_H \approx 5$), which is sufficient to enable their orbits to cross so the protoplanets can suffer close gravitational encounters. Stagnation of protoplanets in isolated orbits is unlikely because the width of the accretion zone at the end of runaway growth roughly coincides with the maximum orbit separation from which neighboring pairs of protoplanets can significantly perturb each other. Multi-zoned accretion simulations of this process suggest that gravitational perturbations by neighboring protoplanets can excite sufficient random velocities to substantially slow the rate of runaway growth (Weidenschilling *et al.* 1992).

Once the protoplanets have perturbed one another into crossing orbits, their subsequent orbital evolution is governed by close gravitational encounters and violent, highly inelastic collisions. Wetherill (1980, 1985, 1990a) has described numerous simulations of this final stage of accretion, exploring a wide range of initial conditions. Surprisingly, the results of these simulations are not strongly affected by the earlier runaway growth stage. Regardless of whether the simulations start from a swarm of 500 bodies of mass 10^{25} g or just 30 protoplanets of mass several times 10^{26} g, the end result is the formation of 2 to 5 terrestrial planets on a timescale of about 10^8 years. An important feature of Wetherill's simulations is that planetesimal orbits execute a random walk in semimajor axis due to successive gravitational encounters. The resulting widespread mixing of material throughout the terrestrial planet region greatly diminishes any chemical gradients that may have existed during the early stages of planetesimal formation, although some correlations between the final heliocentric distance of a planet and the region where most of its constituents originated are preserved in the simulations (Wetherill 1988).

Nonetheless, Mercury's high iron abundance is unlikely to have arisen from chemical fractionation in the solar nebula.

The mutual accumulation of numerous protoplanets into a small number of planets must have entailed many collisions between protoplanets of comparable size. Mercury's silicate mantle may have been partially stripped off in such a giant impact, leaving behind an iron-rich core (Benz *et al.* 1988). Wetherill's simulations also lend support to the giant impact hypothesis for the origin of the Earth's Moon (e.g., Stevenson 1987); during the final stage of accumulation, an Earth-size planet is typically found to collide with several objects as large as large as the Moon and frequently one body as massive as Mars. The obliquities of the rotation axes of the planets provide independent evidence of the occurrence of giant impacts during the accretionary epoch (Safronov 1966, Lissauer and Safronov 1991).

Accretion Timescales

Growth of planet-sized bodies from kilometer-sized planetesimals involves an increase in radius of four to five orders of magnitude. The planetary accretion rates given by eqs. (8) and (10) all vary as the surface area of the planet, R_s^2. Thus, for all other parameters constant, a planet's radius grows at a statistically uniform rate, and most of the growth time is spent in the last decade of radial expansion. The late phases of planetary growth are therefore crucial to determining the overall length of the accretionary epoch. A stage of runaway growth which ends in isolated protoplanets which must pump up velocities in order to collide and continue their agglomeration to planetary size, leads to accretion timescales similar to those models in which runaway never occurs.

Of course, protoplanet radius is not the only parameter in the equations which varies with time. Runaway accretion starts slowly (small F_g) and accelerates with time; thus it is possible that the "initial" size distribution of planetesimals significantly influenced the growth times of the planets. For example, a "protoplanet" initially of radius 10 km, at 5.2 AU from a 1 $M_{\odot}$ star, in a disk of surface mass density $\sigma = 15$ g/cm^2 whose velocity dispersion is controlled by planetesimals 5 km in radius, would require ~ 2 x 10^5 years to grow large enough that a gravitational enhancement factor of 3000 (at which point $e_H = 2$ and protoplanet perturbations become an important factor in exciting non-circular motions of planetesimals, cf., Greenzweig and Lissauer 1990). This protoplanet would take a subsequent ~ 6 x 10^5 years to attain a mass of 15 $M_{\oplus}$ (Lissauer 1987). A protoplanet in an equivalently skewed initial distribution of larger planetesimals would take much longer to reach a gravitational enhancement factor of 3000, whereas the runaway would proceed much faster in a swarm of smaller planetesimals, because the radius doubling time for the protoplanet would be shorter.

CONCLUSIONS

The planetesimal hypothesis provides a viable theory of the growth of the terrestrial planets, the cores of the giant planets and the smaller bodies present in the solar system. The formation of solid bodies of planetary size should be a common event, at least around young stars which do not have binary companions orbiting at planetary distances. Planets could form by similar mechanisms within circumpuslar disks of solar nebula parameters. The formation of giant planets, which contain significant fractions of H_2 and He, requires rapid growth of planetary cores, so that gravitational trapping of gas can occur prior to the dispersal of the gas from the protoplanetary disk. According to the scenario which we have outlined, the largest bodies in any given zone are the most efficient accreters, in the sense that they double in mass the fastest. Such runaway accretion of a few large solid protoplanets can lead to core formation in $\sim 10^6$ years, provided disk masses are a few times as large as those given by "minimum mass" models of the solar nebula. Thus, it appears possible that giant planets may also be common, although this conclusion must be regarded as much more tentative.

The ultimate sizes and spacings of solid planets are determined by their ability to gravitationally perturb each other into crossing orbits. Such perturbations often are caused by weak resonant forcing, and occur on timescales much longer than the bulk of planetesimal interactions discussed herein. These interactions are not yet fully understood, although a great deal of progress has been made in recent years (Wisdom 1983, Duncan and Quinn 1992). Although quantitative formulas for scaling planetary sizes and spacings will require a better grasp of these processes, a few qualitative remarks can be made. Firstly, a more massive protoplanetary disk will probably produce larger but fewer planets. Secondly, stochastic processes are important in planetary accretion, so nearly identical initial conditions could produce quite different outcomes, e.g., the fact that there are 4 terrestrial planets in our solar system as opposed to 3 or 5 or 6 is probably just the luck of the draw (cf. Wetherill 1988). Thirdly, migration of some planetesimals over significant distances within the protoplanetary disk probably occurred, leading to a radial mixing of material which condensed in differing regions of the solar nebula. Fourthly, although the spacing of giant planets is probably determined by similar processes to that of solid planets, their ultimate sizes may depend more on such factors as how fast they grew relative to the dispersal of the nebula and their potential ability to halt their own growth by gravitationally truncating the gaseous disk (Lin and Papaloizou 1979).

Many aspects of planetary growth remain poorly understood. The asteroid belt currently contains much less than a planetary mass of material, and that material is spread over countless bodies which move at high velocities relative to each other. Models of the protoplanetary disk suggest this is unlikely to be due to an initial absence of condensed material. Perturbations by Jupiter and/or Jupiter-scattered protoplanets have been invoked for preventing planetary growth in the asteroid zone by increasing relative velocities of planetesimals, and for clearing material from that region; however, many problems remain with these scenarios (cf. Wetherill 1989, Lissauer and Stewart 1992). Planetesimal dynamics in the Uranus-Neptune region are complicated by the fact that bodies in this region are not tightly bound to the solar system: The difference between circular orbit velocity and escape speed from the solar system at 30 AU is less than the escape speed from the Moon. No detailed model of accretion in such an environment has yet been produced.

ACKNOWLEDGEMENTS

We thank D. Kary and S. Weidenschilling for their valuable comments. This work was supported in part by the NASA Planetary Geology and Geophysics Program under grants NAGW-1107 at SUNYSB and NAGW-769 at CU.

REFERENCES

Adachi, I., Hayashi, C. and Nakazawa, K. 1976, *Prog. Theor. Phys.* **56**, 1756.
Artymowicz, P. 1987, *Icarus,* **70,** 303.
Barshay, S.S. and Lewis, J.S. 1976, *Ann. Rev. Astron. Astrophys.,* **14**, 81.
Beckwith, S.V.W. and Sargent, A.I. 1992, in *Protostars and Planets III,* ed., E.H. Levy, J.I. Lunine and M.S. Matthews, Univ. of Arizona Press, in press.
Benz, W., Slattery, W.L. and Cameron, A.G.W. 1988, *Icarus*, **74**, 516.
Binney, J., and Tremaine, S.D. 1987, *Galactic Dynamics,* Princeton Univ. Press.
Bodenheimer, P. and Pollack, J.B. 1986, *Icarus*, **67**, 391.
Cameron, A.G.W. 1962, *Icarus,* **1**, 13.
Duncan, M.J. and Quinn, T.R. 1992, in *Protostars and Planets III,* ed. E.H. Levy, J.I. Lunine and M.S. Matthews, Univ. of Arizona Press, in press.
Forrest, S.R. and Witten, T.A., Jr. 1979, *J. Phys.,* **A12**, L109.
Goldreich, P. and Tremaine, S., 1980, *Ap. J.,* **241**, 425.
Goldreich, P. and Ward, W.R. 1973, *Ap. J.,* **183**, 1051.
Greenberg, R., Wacker, J.F., Hartmann, W.L. and Chapman, C.R. 1978, *Icarus*, **35**, 1.
Greenberg, R., Weidenschilling, S.J., Chapman, C.R. and Davis, D.R. 1984, *Icarus*, **59**, 87.
Greenzweig, Y. and Lissauer, J.J. 1990, *Icarus*, **87**, 40.
Greenzweig, Y. and Lissauer, J.J. 1992, *Icarus*, submitted.
Grossman, J.N. 1988, in *Meteorites and the Early Solar System,* ed. J.F. Kerridge and M.S. Matthews, Univ. of Arizona Press, p. 680.
Grossman, J.N., Rubin, A.E., Hagahara, H. and King, E.A. 1988, in *Meteorites and the Early Solar System,* ed. J.F. Kerridge and M.S. Matthews, Univ. of Arizona Press, p. 619.
Hayashi, C., Nakazawa, K. and Adachi, I. 1977, *Publ. Astron. Soc. Japan,* **29**, 163.
Hayashi, C., Nakazawa, K. and Nakagawa, Y. 1985, in *Protostars and Planets II,* ed., D.C. Black and M.S. Matthews, Univ. of Arizona Press, p. 1100.
Herbert, F. 1989, *Icarus*, **78**, 402.
Hornung, P., Pellat, R. and Barge, P. 1985, *Icarus*, **64**, 295.
Hoyle, F. 1960, *Quart. J. R. Astr. Soc.,* **1**, 28.
Ida, S. 1990, *Icarus*, **88**, 129.
Ida, S. and Nakazawa, K. 1989, *Astron. Astrophys.,* **224**, 303.
Kant, I. 1745. *Allegmeine Naturgeschichte und Theorie des Himmels.*
Laplace, P.S. 1796, *Exposition du Système du Monde,* English translation:

H.H. Harte 1830, *The System of the World*, Dublin University Press.
Levy, E.H. 1988, in *Meteorites and the Early Solar System,* ed. J.F. Kerridge and M.S. Matthews, Univ. of Arizona Press, p. 697.
Lin, D.N.C. and Papaloizou, J. 1979, *Mon. Not. R. Astr. Soc.* **186**, 799.
Lin, D.N.C., Woosley, S.E. and Bodenheimer, P.H. 1991, *Nature,* **353**, 827.
Lissauer, J.J. 1987, *Icarus*, **69**, 249.
Lissauer, J.J. and Safronov, V.S. 1991, *Icarus*, **93**, 288.
Lissauer, J.J. and Stewart, G.R. 1992, in *Protostars and Planets III,* ed., E.H. Levy, J.I. Lunine and M.S. Matthews, Univ. of Arizona Press, in press.
Meakin, P. and Donn, B. 1988, *Ap. J. (Letters),* **329**, L39.
Mizuno, H. 1980, *Prog. Theor. Phys.,* **64**, 544.
Nakagawa, Y., Hayashi, C. and Nakazawa, K. 1983, *Icarus*, **54**, 361.
Nakamura, T. and Piran, T. 1991, *Ap. J. (Letters),* **382**, L81.
Ohtsuki, K., Nakagawa, Y. and Nakazawa, K. 1988, *Icarus*, **75**, 552.
Ohtsuki, K. and Ida, S. 1990, *Icarus*, **85**, 499.
Ohtsuki, K., Ida, S., Nakagawa, Y. and Nakazawa, K. 1992, in *Protostars and Planets III,* ed., E.H. Levy, J.I. Lunine and M.S. Matthews, Univ. of Arizona Press, in press.
Podolak, M., Hubbard, W.B. and Pollack, J.B. 1992, in *Protostars and Planets III,* ed., E.H. Levy, J.I. Lunine and M.S. Matthews, Univ. of Arizona Press, in press.
Podsiadlowski, P., Pringle, J.E., and Rees, M.J. 1991, *Nature,* **352**, 783.
Prinn, R.G. and Fegley, B. Jr. 1989, in *Origin and Evolution of Planetary and Satellite Atmospheres,* ed. S.K. Atreya, J.B. Pollack and M.S. Matthews, Univ of Arizona Press, p. 78.
Safronov, V.S. 1966, *Sov. Astron. AJ.,* **9**, 987.
Safronov, V.S. 1969, *Evolution of the Protoplanetary Cloud and Formation of the Earth and Planets*, Nauka Press. English translation: NASA TTF-677, 1972.
Spaute, D., Weidenschilling, S.J., Davis, D.R. and Marzari, F. 1991, *Icarus*, **92**, 147.
Sridhar, S. and Tremaine, S. 1992, *Icarus*, **95**, 86.
Stevenson, D.J. 1987, *Ann. Rev. Earth Planet. Sci.* **15**, 271.
Stewart, G.R. and Wetherill, G.W. 1988, *Icarus*, **74**, 542.
Takeda, H., Matsuda, T., Sawada, K. and Hayashi, C. 1985, *Prog. Theor. Phys.* **74**, 272.
Tavani, M. and Brookshaw, L. 1992, *Nature,* **356**, 320.
Terebey, S., Shu, F.H. and Cassen, P. 1984, *Ap. J.,* **286**, 529.
Trubnikov, B.A. 1971, *Soviet Physics - Doklady,* **16**, 124.
Trulsen, J. 1971, *Astrophys. Space Sci.,* **12**, 329.
Ward, W.R. 1986, *Icarus*, **67**, 164.
Weidenschilling, S.J. 1977, *Mon. Not. R. Astr. Soc.* **180**, 57.
Weidenschilling, S.J. 1980, *Icarus,* **44**, 172.
Weidenschilling, S.J. and Cuzzi J.N. 1992, in *Protostars and Planets III,* ed., E.H. Levy, J.I. Lunine and M.S. Matthews, Univ. of Arizona Press, in press.
Weidenschilling, S.J., Davis, D.R. Marzari, F. and Spaute, D. 1992, *Icarus,* submitted.

Weidenschilling, S.J., Donn, B. and Meakin, P. 1989, in *The Formation and Evolution of Planetary Systems,* ed. H.A. Weaver and L. Danly, Space Telescope Science Institute Symposium Series **3**, p. 131.
Wetherill, G.W. 1980, *Ann. Rev. Astron. Astrophy.,* **18**, 77.
Wetherill, G.W. 1985, *Science,* **228**, 877.
Wetherill, G.W. 1988, in *Mercury*, ed. F. Vilas, C.R. Chapman and M.S. Matthews, Univ. of Arizona Press, p. 670.
Wetherill, G.W. 1989, in *Asteroids II,* ed. R. P. Binzel, T. Gehrels and M. S. Matthews, Univ. of Arizona Press, p. 661.
Wetherill, G.W. 1990a, *Ann. Rev. Earth Planet. Sci.* **18**, 205.
Wetherill, G.W. 1990b, *Icarus*, **88**, 336.
Wetherill, G.W. and Stewart, G.R. 1989, *Icarus*, **77**, 330.
Wisdom, J. 1983, *Icarus*, **56**, 51.
Wolszczan, A. and Frail, D.A. 1992, *Nature,* **355**, 145.

Planets Around Pulsars
ASP Conference Series, Vol. 36, 1993
J.A. Phillips, J.E. Thorsett, and S.R. Kulkarni

PHYSICAL PROPERTIES AND EVOLUTIONARY TIME SCALES OF DISKS AROUND SOLAR-TYPE AND INTERMEDIATE MASS STARS

STEPHEN E. STROM AND SUZAN EDWARDS[1]
Five College Astronomy Dept., University of Massachusetts, Amherst, MA 01003

ABSTRACT This paper summarizes the observational evidence for circumstellar disks, their basic physical properties and evolutionary timescales.

INTRODUCTION

The weight of current observational evidence provides strong support for the hypothesis that circumstellar disks form naturally in the course of the star-formation process (see Strom, Edwards and Skrutskie, 1992 for a recent review). In this contribution, we will summarize

• the observational signatures from which we infer the presence of circumstellar disks:

1. Infrared, sub-mm and mm-wave photometric measurements of emission arising from small dust grains embedded within the disks. The shape of the observed spectral energy distribution provides evidence that geometrically flat, optically thick circumstellar accretion disks surround young stars of all masses.

2. Optical, near-infrared and mm-wave spectroscopic measurements of the gaseous component of disks. Line profiles of unresolved disks and the velocity profiles derived for spatially-resolved disks provide kinematic evidence of material in Keplerian motion around the central star.

• the basic physical properties of disks (mass, size, accretion rate, radial distribution of matter) surrounding stars of all mass during their early evolutionary phases, as inferred from emission arising from the gas and dust components

• the evolutionary timescales for the solid and gaseous components circumstellar disks, as inferred from the fraction of stars of a given mass and age which show measurable gas and dust emission arising from distributed small dust grains and gas (as opposed to dust and gas assembled into planetesimals and/or planets). These timescales can in principle provide astronomical constraints on the times available to build solid planetesimals and for those planetesimals to accumulate

[1]Also at Smith College, Northampton, MA 01065

remnant disk gas.

DIAGNOSING DISKS

Expectations regarding young disks

If we assume that the dust and gas components of disks surrounding young stars are similar to those in molecular "cores", then

- the dust component will be dominated by small dust grains with typical size $a = 0.1 - 1\mu m$. If the solid material in our solar system were all in the form of such grains, and if these grains were distributed in a disk of diameter ~ 100 AU, then the disk would be optically thick for wavelengths $\lambda \leq 100\mu m$.

- the gas component will dominate the total mass of the system. For interstellar conditions, $M_{gas}/M_{dust} = 100$.

Although these assumptions seem reasonable, we do not yet know how much grain growth takes place during the phase when interstellar grains are incorporated within the disk, or whether the gas chemistry or the gas/dust ratio in very dense disks is is in fact similar to that in protostellar "cores".

Expected emission from the dust component

Small dust grains embedded within the disk can be heated radiatively, via absorption of energy from the central star ("reprocessing disk"), and viscously, through dissipation of energy in an accretion disk ("accretion disk"); see Adams, Lada and Shu (1987) and Kenyon and Hartmann (1987). For a geometrically flat, optically thick disk, radiative and viscous heating each produce a radial temperature distribution of the form:

$$T(r) \sim r^{-3/4}.$$

When weighted by the emitting area in each disk annulus, this radial temperature distribution produces a spectral energy distribution of the form

$$\lambda F_\lambda \sim \lambda^{-4/3}$$

Although the spectral energy distributions for reprocessing and accretion disks are identical, the emergent disk luminosity can differ significantly. For flat reprocessing disks, the disk luminosity is a fixed fraction of the luminosity of the central star: $L_{disk} = 0.25 L_d$. For accretion disks, the disk luminosity depends on the accretion rate: $L_{disk} = GM_* \dot{M}_{acc}/2R_*$. For sufficiently large accretion rates $\dot{M}_{acc}$, disk emission arising from viscous dissipation of energy can exceed that from reprocessing. We illustrate in Figure 1 the computed composite (disk plus star) spectral energy distributions for a star of spectral type A0 surrounded by a set of model disks, assumed to begeometrically flat, optically thick and to extend continuously from the stellar surface to a radius $r = 3000$ AU. The model disks span a range of accretion rates $0 < \dot{M}_{acc} < 10 M_\odot/\text{yr}$.

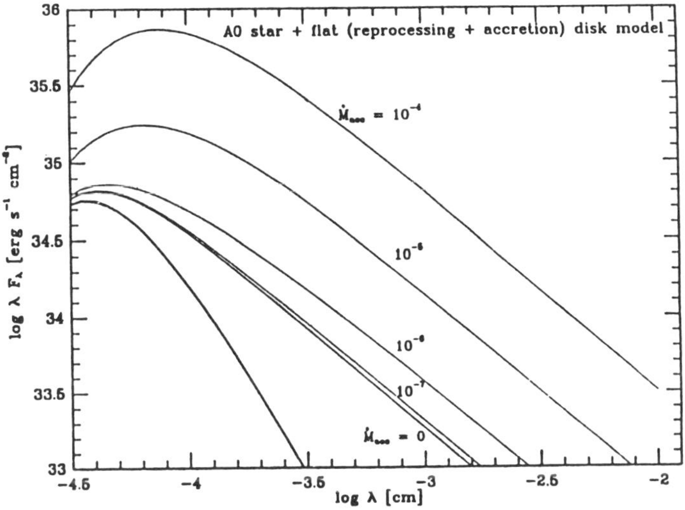

FIGURE I Composite (accretion plus reprocessing) models for geometrically flat, optically thick disks surrounding a central star characterized by parameters appropriate to a main sequence A0 star: $T_{eff} = 10^4$ K, $R_* = 3R_\odot$, and $M_* = 3M_\odot$. The dark solid line depicts the spectral energy distribution of the stellar photosphere. The assumed accretion rates are noted for each model.

Observations of the Dust Component

For wavelengths $\lambda \ll 60\mu$, the excess emission observed for young, optically visible solar-mass stars (T Tauri stars; K.M. Strom et al., 1989) and intermediate mass stars (Herbig Ae/Be stars; Herbig, 1960; Strom et al., 1972; Catala, 1989) has a spectral shape which compares reasonably well with the $\lambda F_\lambda \sim \lambda^{-4/3}$ distribution expected for flat, optically thick disks. This is illustrated in Figure 2, where we plot two histograms depicting the frequency distributions of the infrared color indices (L - N) and (L - Q); these indices reflect the spectral slopes $2.5\log[F(3.5\mu m)/F(10.6\mu m)]$ and $2.5\log[F(3.5\mu m)/F(20\mu m)]$, respectively. Superposed on this figure are lines denoting the spectral slopes predicted for perfectly flat optically thick disks, and slightly flared optically thick disks whose scale height increases slowly with disk radius, $h \sim r^{9/8}$ (see Kenyon and Hartmann, 1987). Because the observed spectrum for $\lambda \leq 60\mu$ is consistent with emission arising in nearly flat, optically thick disks, we infer that the mass in small particles in these disks must be at least as great as the mass of solid material contained within the present-day solar system.

A significant fraction ($\sim 50\%$) of all disks are found to have infrared excess luminosities which exceed the luminosity expected from a pure reprocessing disk; indeed, in many cases the disk luminosity, $L_{disk} > L_*$. Hence, during early evolutionary phases, circumstellar disks are best modeled as accretion disks, perhaps fed by infall of gas and dust from a protostellar core (see Hillenbrand et al., 1992a).

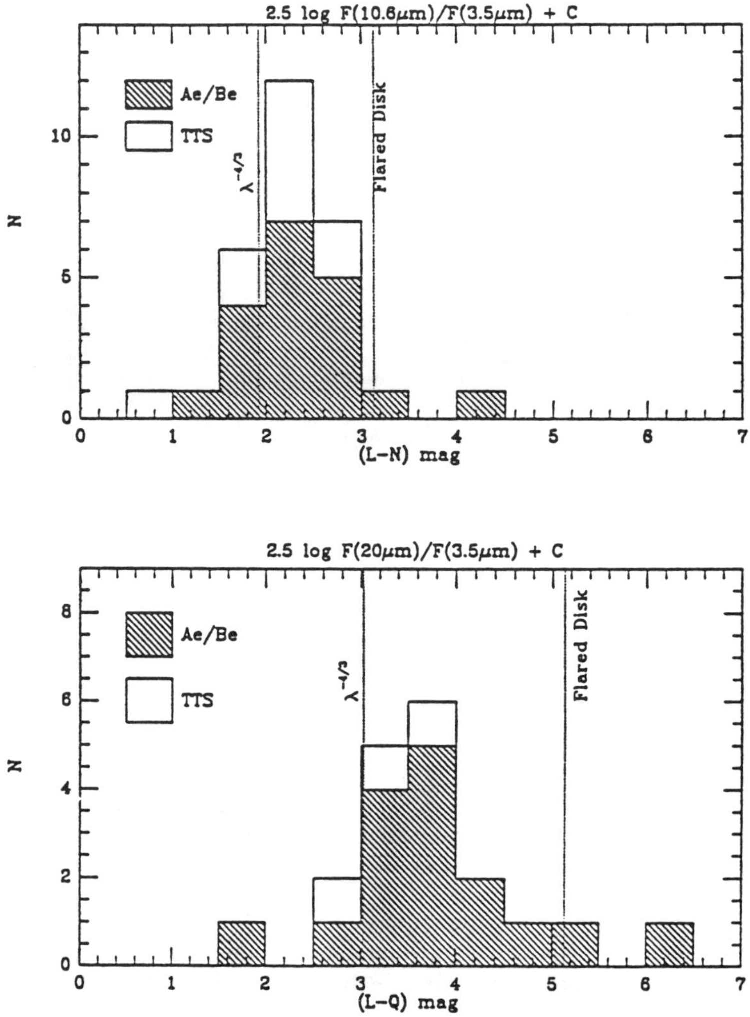

FIGURE II The frequency distribution of spectral slopes (L - Q) and (L - N) among Herbig Ae/Be stars and T Tauri stars; L, N and Q are ground-based 3.5μm, 10.6μm and 20μm broad-band magnitudes. The spectral slopes expected for flat and slightly flared optically thick accretion disks are indicated. Most Ae/Be stars and TTS have spectral slopes which are consistent with those expected for such simple disk models.

Emission from the Gas Component

Measurements of atomic or molecular lines arising in the gaseous component of a disk can provide direct kinematic evidence of material in Keplerian orbit around a central star, as well as independent estimates of disk size, mass, and radial structure. We expect circumstellar gas to be neutral or partially ionized within a few tens of stellar radii of the stellar surface, and largely molecular at $r \gg 10R_*$.

The signature of neutral or partially ionized disk gas in rotation around young, solar-type stars has been reported for a class of pre-main sequence stars known as FU Ori objects. These rare, photometrically eruptive stars are believed to be T Tauri stars surrounded by accretion disks undergoing a phase (duration $\sim$ 100 yr) of elevated accretion (see Hartmann and Kenyon, 1985). During the outburst phase, the disk spectrum dominates the spectrum arising from the central star by $\sim 100\times$. The accretion rates required to account for the inferred disk luminosity range from $M_{acc} \sim 10^{-3} - 10^{-4} M_\odot$/yr. The observations supporting this hypothesis are:

• infrared spectral energy distributions ($\lambda \leq 25\mu$m) which show the $\lambda F\lambda \sim \lambda^{-4/3}$ spectrum expected from a flat, steady accretion disk.

• absorption features which show the characteristic double-peaked structure expected for lines formed in a rotating circumstellar disk (Figure 3). The observation of absorption lines requires that FU Ori disks be hottest near the midplane and coolest at the disk surface The requirement of a negative vertical temperature gradient is consistent with the recent calculations of Calvet and Hartmann (1992) who show that in FU Ori stars, viscous heating near the disk midplane exceeds surface heating from the central stars.

• lines whose width decreases with wavelength (see Figure 4). This result is expected because absorption lines observed in the blue spectral region are formed predominantly in the hotter disk regions closer to the stellar surface, and thus exhibit higher rotational broadening widths than lines which are formed in the cooler regions further out in the disk.

The molecular gas exterior to $r \sim 10R_*$ is generally more difficult to study because young stars are usually still associated with their parent molecular clouds. In nearby ($d < 200$ pc) star-forming regions, the angular size of a disk is expected to be small ($\theta < 5''$) compared to the antenna beams of even the largest, single-dish mm-wave telescopes. As a result the emission arising from ambient molecular gas can overwhelm the emission from the disk. Nevertheless, some progress has been made recently:

• Millimeter-wave interferometric measurements are able to "resolve out" emission arising from the dense, optically-thick disk gas from the smoother, lower density background cloud emission. Measurements of two young, solar-type stars with the Owens Valley Radio Observatory interferometer reveal small-scale ($r < 1000$ AU) gaseous structures surrounding, and bound to these stars

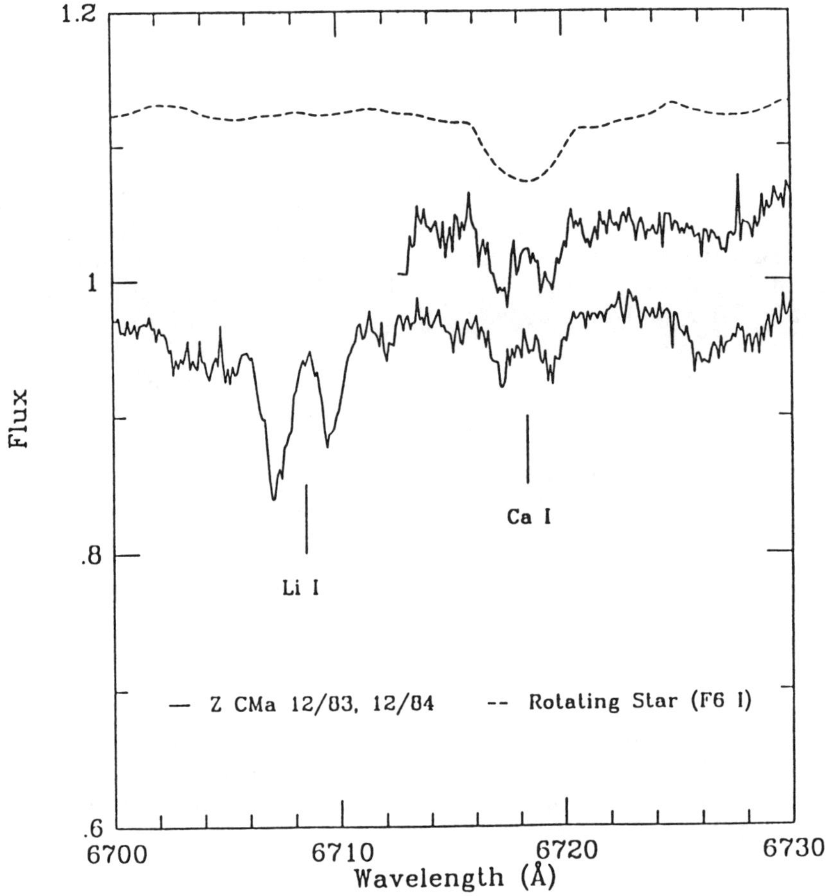

FIGURE III A plot of the region in the vicinity of the λ 6707 Å line of Li I for the FU Ori star, Z CMa (Hartmann et al., 1989). Note the characteristic double-peaked structure expected for absorption lines formed in a rotating circumstellar accretion disk.

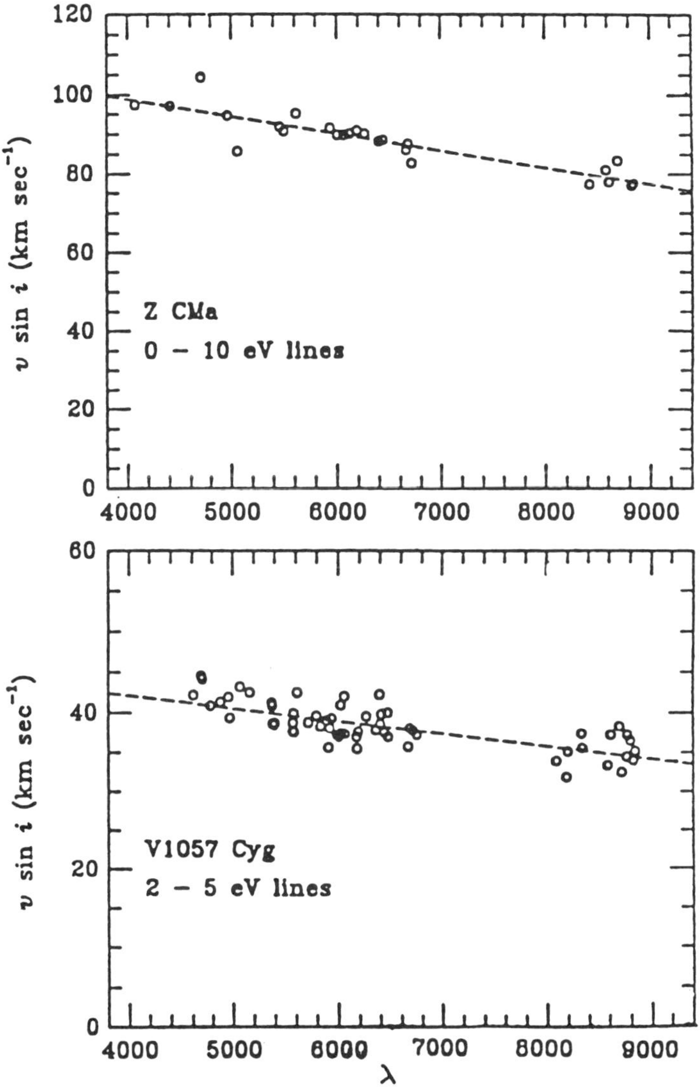

FIGURE IV A plot of measured line width against wavelength for the absorption lines formed in the accretion disks assumed to surround the FU Ori stars Z CMa and V1057 Cyg (Welty et al., 1992). Note that lines formed in the hotter, inner regions of the disk (blue spectral region) are broader than those formed in the cooler, outer disk regions. The observed correlation between line width and wavelength is presumed to reflect the higher rotational velocities in the inner regions of the disk.

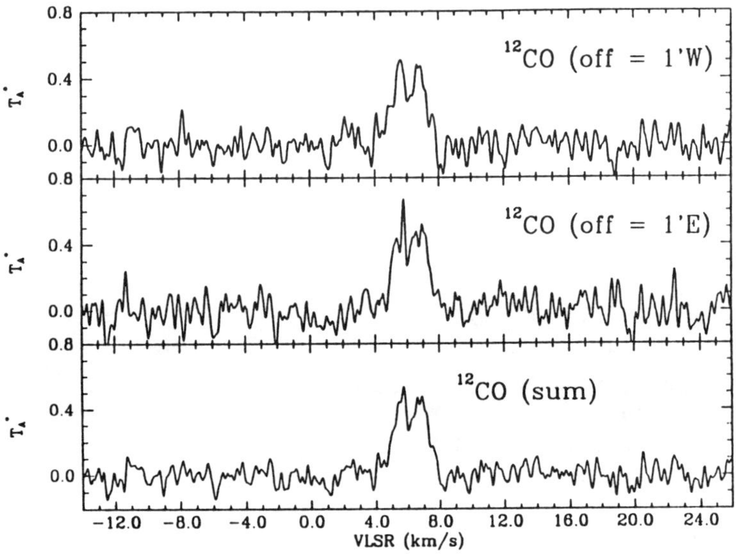

FIGURE V ^{12}CO (J=1-0) profiles of GG Tau obtained with the Nobeyama Radio Observatory 45-m telescope. The top two spectra are from independent 35 minute integrations on-source using different reference positions relative to GG Tau. The bottom profile is the average of the two spectra. Spectra have been boxcar smoothed over an 0.21 km s^{-1} interval. Note the double-peaked profile characteristic of gas in circular motion about the central source.

(e.g., Sargent and Beckwith, 1991; Koerner, Sargent and Beckwith, these proceedings). In the case of HL Tau, the molecular gas appears to show the kinematic signature of Keplerian motion. Unfortunately, the limited instantaneous uv-plane coverage and sensitivity of current mm-wave interferometers precludes surveys of large samples of young stars.

• Skrutskie et al. (unpublished) have recently discovered a sample of nearby solar-type PMS stars whose peculiar motions have apparently carried them just outside the boundaries of their parent molecular clouds. Absent emission from ambient gas, large, single-dish mm-wave telescopes (e.g. the IRAM 30-m and NRO 45-m telescopes) have sufficient sensitivity to enable sensitive searches for disk gas surrounding young stars located within $d \sim 200$ pc. Observations made recently with the Nobeyama Radio Observatory 45-m telescope of a T Tauri star (GG Tau) located in a "molecular hole" (Skrutskie et al., 1992) reveal a double-peaked ^{12}CO emission line profile which is best modeled by assuming that the CO emission is produced by gas in orbit around GG Tau and confined to a disk of radius $400 < r < 600$ AU (see Figure 5).

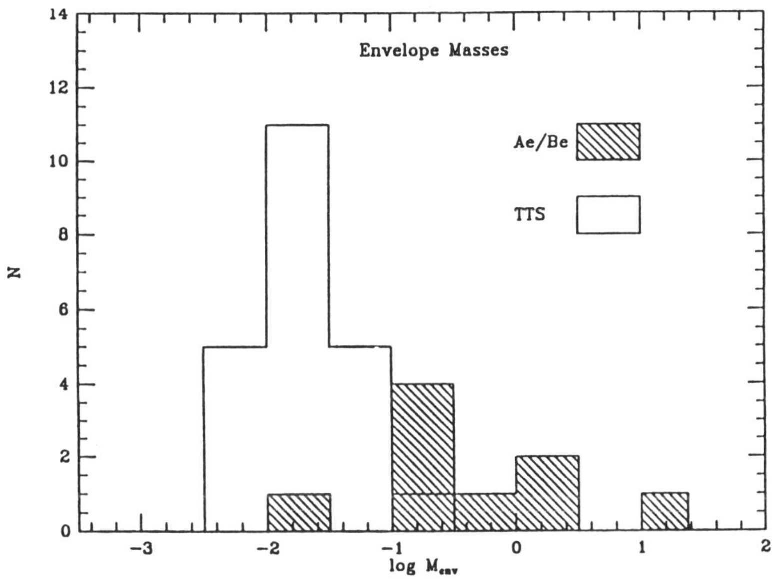

FIGURE VI The observed frequency distribution of disk masses derived from mm-continuum measurements for Herbig Ae/Be stars (hatched region) and T Tauri stars (open region). Note that we have assumed in all cases a *total* (gas plus dust) mass opacity coefficient $\kappa_\nu = 0.02$ cm^2 gm^{-1} at 1.3 mm. Also, the typical beam sizes used in making the measurements range from $15'' < \theta < 30''$ and thus could include emission from material outside the circumstellar disks surrounding these stars.

PHYSICAL PROPERTIES OF YOUNG DISKS

We summarize below the basic properties of disks inferred from measurements of the emission arising from circumstellar dust and gas.

Disk Masses

As we discussed earlier, disks with M > 0.01 $M_\odot$, r < 50 AU are optically thick at wavelengths $\lambda \leq 100\mu$m. Hence, flux measurements at $\lambda \leq 100\mu$m are not sensitive to mass. However, measurements at mm-wavelengths where disks are optically thin can in principle provide reliable mass determinations. An extensive mm-continuum survey of young *solar-type stars* is available from Beckwith et al. 1990. They find $0.01 < M/M_\odot < 0.2$, assuming $M_{dust}/M_{gas} = 100$. Sandell et al. (1992), Adams et al. (1991), and Hillenbrand et al. (1992a) have carried out similar mm-continuum measurements for a sample of young *intermediate mass stars* Herbig Ae/Be stars and find $0.01 < M/M_\odot < 6$. Extant results are summarized in Figure 6.

It is essential to note that disk mass estimates from dust emission are subject to large uncertainties. First, the emissivities for the grains contained within disks are poorly known and could differ significantly from typical interstellar grain emissivities. Second, emission arising from dust in regions exterior to the disk (e.g. in remnants of the protostellar core which gave birth to the star) may contribute significantly to the total flux density measured within a mm-wave antenna beam. The data summarized in Figure 6 have all been placed on a scale assuming (1) a total dust+gas opacity per gram of $\kappa_\nu = 0.02$ cm^2 gm^{-1} at $\lambda = 1.3$ mm, and (2) a grain emissivity which varies as λ^{-1} (Beckwith et al., 1990).

To date, estimates of the disk mass based on observations of circumstellar gas are available for only 2 optically visible young stars: HL Tau and GG Tau (both solar-type PMS stars surrounded by accretion disks). In the case of HL Tau, the inferred disk mass is ~ 0.1 $M_\odot$ (Sargent and Beckwith, 1991). For GG Tau, Skrutskie et al., 1992 infer a mass of $\sim 10^{-4}$ $M_\odot$ for the *outer disk regions* ($r >> 100$ AU), but note that a large amount of mass could be hidden within $r << 100$ AU and escape detection owing to beam dilution. By combining their NRO single-dish measurements of ^{12}CO emission for GG Tau with interferometric mm-continuum measurements made recently at IRAM (Simon and Gilloteau, 1992), Skrutskie et al. suggest that the gas/dust ratio in the outer disk regions of GG Tau may be lower than interstellar values by $\sim 10 - 100\times$.

Disk Radii

As noted earlier, disks with a mass in dust grains comparable to the mass of solid bodies in the solar system should be optically thick at $\lambda \leq 100\mu$m. If so, it is possible to estimate *minimum* disk sizes from the IRAS 60μm fluxes which are available for large numbers of young stars, since

$$A(\text{disk})B_\lambda(T_{\text{disk}})/4\,d^2 = F(60\mu\text{m})$$

and

$$r(\text{disk}) \sim A(\text{disk})^{1/2}$$

In Figure 7 we plot the minimum disk sizes derived from 60μm fluxes for solar-type and intermediate mass stars. Note that in assembling this figure, we have included only those objects whose infrared spectral energy distribution is well approximated by $\lambda F_\lambda \sim \lambda^{-4/3}$ (appropriate for a flat, optically thick reprocessing disk). Minimum disk radii range from $5 < r < 200$ AU for *solar-type* stars, and $30 < r < 3000$ AU for *intermediate mass* stars. It is worth remarking that the minimum radii for disks associated with nearly all intermediate mass stars and some solar-type stars can be considerably larger than the nominal radius of the solar system, and more comparable to the dust disks inferred from IRAS observations of objects similar to β Pictoris (e.g. Backman and Gillett, 1988).

Estimates of the size of gaseous disks surrounding young stars are again available only for HL Tau and GG Tau. For HL Tau disk gas extends to $r \sim 1000$ AU (Sargent and Beckwith, 1991), while for GG Tau, $400 < r < 600$ (Skrutskie et al. 1992; see also Koerner et al. in these proceedings).

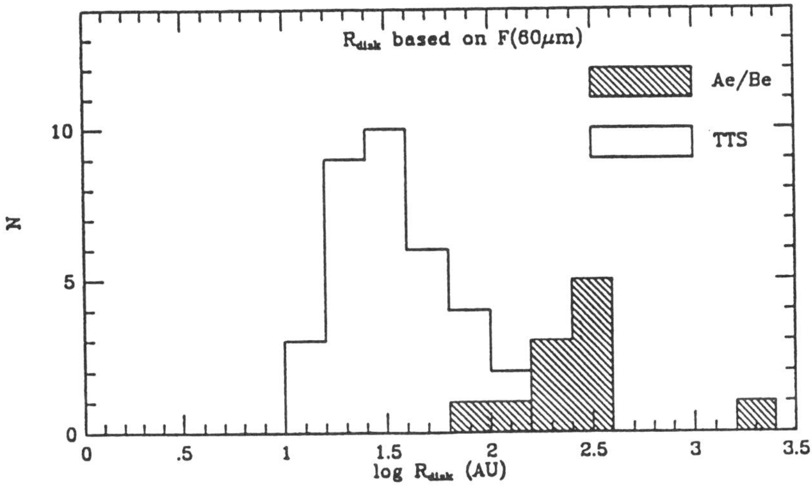

FIGURE VII A plot of the minimum disk radii derived from IRAS 60μm flux measurements for Herbig Ae/Be stars (hatched region) and for T Tauri stars (open region). Only those stars showing $\lambda F_\lambda \sim \lambda^{4/3}$ at wavelengths $2.2\mu m \leq \lambda \leq 60\mu m$ are included in this plot.

Accretion Rates

Many disks have excess infrared emission which lies well above that expected from disks heated by radiation from the central stars (Strom et al., 1988; Hillenbrand et al., 1992a). Estimating accretion rates from the observed infrared excess emission, we find

- for *solar-type* PMS stars:

$$10^{-8} < \dot{M}_{acc} < 10^{-6} M_\odot/yr$$

Independent evidence of these accretion rates is provided by observation of optical and near ultraviolet excess emission produced in a *boundary layer;* $L_{BL} \sim L_{IR}$ (Bertout, Basri, and Bouvier, 1988; Hartmann and Kenyon 1990; Hartigan et al., 1991).

- for *intermediate mass* stars:

$$10^{-6} < \dot{M}_{acc} < 10^{-4} \dot{M}_\odot/yr$$

At present, estimates of accretion rates for intermediate mass stars rest entirely on the magnitude of observed infrared excess emission (Hillenbrand et al., 1992a), and thus are subject to uncertainties regarding the fraction of such emission which arises from circumstellar disks and from more spherically-distributed "envelope" material (see Natta et al., 1992; Hartmann, Kenyon and Calvet, 1992).

However, if these accretion rates are correct the total amount of material passing through the disk and perhaps onto the star is in the range

$$0.1\ M_\odot < [\dot{M}_{acc} \times t_{disk}] < 1\ M_\odot,$$

where t_{disk} is the disk lifetime, discussed below. *Hence, a significant amount of material comprising a star may pass through a circumstellar disk.*

Disk Structure

The shape of the infrared spectral energy distribution can diagnose the radial distribution of material within the disk. For example

- were a disk optically thin in regions within a few R_* of the stellar surface, the excess emission at short infrared wavelengths ($\lambda \leq 2.2\mu m$) would lie below that predicted for an optically thick disk which extends inward to the stellar surface.

- were a disk optically thin at distances $r > 1$ AU, the observed excess emission at wavelengths $\lambda > 10\mu m$ would fall below that expected for an optically thick disk extending outward from the stellar surface to $r \sim 100$ AU

Examination of the spectral energy distributions observed for all Herbig Ae/Be stars and some T Tauri stars reveals an inflection in their near-infrared spectral energy distributions between $1.2\mu m$ and $2.2\mu m$ (see Figure 8). Hillenbrand et al. (1992a) suggest that this inflection signals the presence of optically thin "hole" of dimension $3 < r_{hole}/R_* < 20$ (see also Beckwith et al. 1990; Bertout, Basri and Bouvier, 1988). Such holes may indicate the combined effects of dust destruction at disk temperatures $T > 2000$ K, and a gap created by a stellar magnetosphere. In the latter case, the stellar magnetic field must be strong enough to invade the disk, and to force accreting material to lift off the disk surface and continue its path toward the stellar surface along magnetic field lines (thus effectively terminating the disk at r_{hole}). The linkage between the stellar magnetic field and partially ionized material within the disk can also regulate the star's angular momentum (e.g. Konigl 1991).

In this regard, Hartmann and Kenyon (1990) have noted the conundrum presented by the accretion of disk material at the rates we discussed above: addition of high angular momentum material at such rates will spin up a solar-type pre-main sequence star to 1/2 breakup velocity on timescales $t < 1$ Myr. Nevertheless, accreting TTS show projected rotational velocities $v \sin i < 0.1 v_{breakup}$. Ghosh and Lamb (1978; 1979 a,b) have suggested (in the context of accreting neutron stars) that if a magnetic field anchored in a star can invade the disk at radii large compared to the "natural co-rotation radius", r_{COR} (the disk radius at which the Keplerian angular velocity equals that of the star), the "spin up torque" applied to the star interior to r_{COR} will be balanced by a "spin down torque" exterior to r_{COR} via linkage between the stellar magnetic field and the disk. The inner holes diagnosed by near-infrared spectral energy distributions are expected to be located in a region interior to r_{COR} at the "forced co-rotation radius", where the stellar magnetic field terminates the disk. In the context of this model, the size of the inner hole correlates directly with the angular rotation

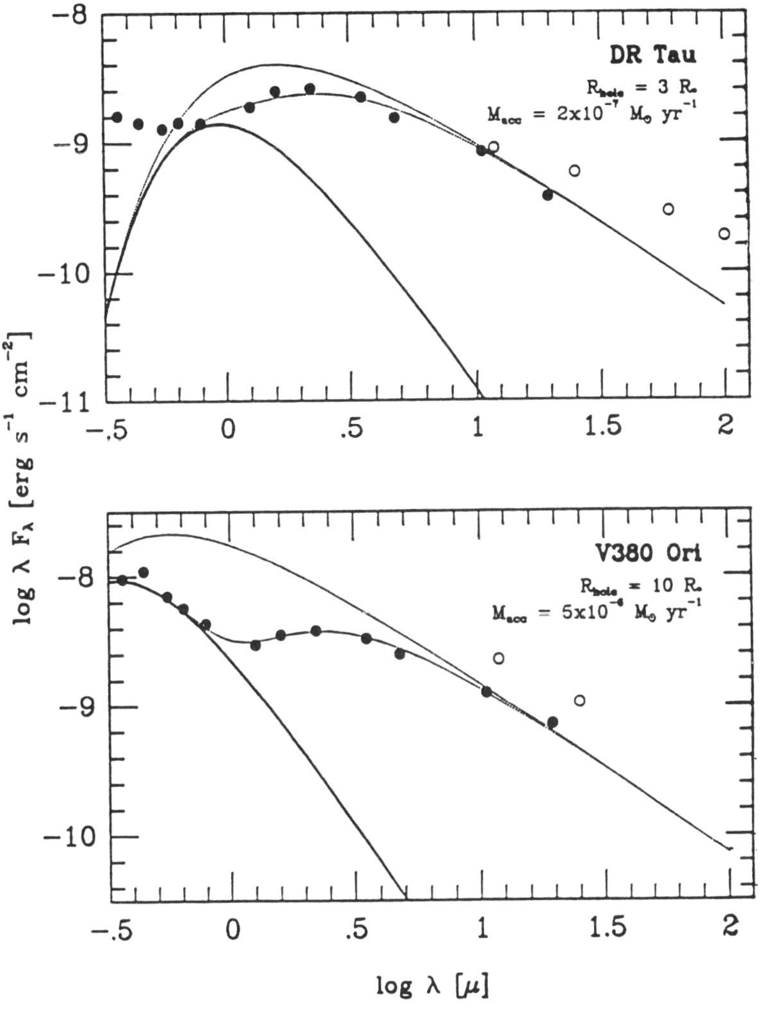

FIGURE VIII A plot of the spectral energy distribution derived for DR Tau (a solar-type PMS star; above) and V 380 Ori (an intermediate mass PMS star; below). The expected photospheric spectral energy distribution for each star is indicated with a heavy solid line. The dashed lines represent best fit disk models in which we posit the existence of "inner holes": optically thin regions between the stellar surface and r = 3 R_* (DR Tau) and r = 10 R_* (V 380 Ori) which do not contribute to the observed disk flux. The "inner holes" apparently required by the observed spectral energy distributions for DR Tau and V 380 Ori may reflect the combined effects of (1) reduced disk opacity owing to dust destruction, and (2) the truncation of the inner disk by a stellar magnetosphere.

rate of the star, and reflects the balance between magnetic pressure provided by the the field anchored in the star, and dynamical pressure of accreting material: stars with larger accretion rates will have disks with smaller holes for a given field strength.

Contracting pre-main sequence stars surrounded by accretion disks should be "held" at a fixed angular velocity determined by the balance between stellar magnetic field strength and disk accretion rate. By inference, when an accretion disk is no longer present, a contracting star will then be free to respond to changes in its moment of inertia, and should "spin up" as it approaches the main sequence, absent angular momentum loss mechanisms. By observing the distribution of inner hole sizes and rotational velocities for large enough samples of young stars it should be possible to search for two important effects predicted by this picture:

• a correlation between hole size and rotation; for stars of equivalent mass and internal structure, stars surrounded by disks with smaller magnetospheric holes should rotate more rapidly;

• a difference between the distribution of rotational velocities of stars (of identical mass and age) surrounded by disks and those which lack disks. If disks regulate stellar angular momentum, then stars no longer surrounded by disks should reflect the increase of angular velocity expected between the time a contracting star was "freed" from its disk and the present; those still coupled magnetically to their disks should, on average, have lower angular velocities.

Quantitative tests of various aspects of these ideas are currently underway (see Edwards, 1992; Attridge and Herbst, 1992; Hillenbrand et al., unpublished).

Disk Evolutionary Timescales

The inferred presence of relatively massive, optically thick disks surrounding young stars of all mass, combined with the absence of readily observable disk signatures for older main sequence stars requires that disks evolve. In order to discuss the timescales for disk evolution, we define two quantities:

• the *disk dust survival time:* the time it takes a disk to evolve from a massive, optically thick structure populated by small grains, to a structure which contains too few small grains to produce detectable infrared excess emission either because (1) such grains have been accreted by the star or dispersed; or (2) they have been assembled into larger bodies.

• the *disk gas survival time*: the time it takes the disk to evolve from a massive structure containing sufficient gas to account for the observed molecular emission associated with young stars, to a structure which contains insufficient gas to produce detectable emission, either (1) because the gas has been accreted by the star or dispersed; or (2) because the gas has been accreted by large rocky cores.

Note that many planet formation models require different survival times for the distributed (unassembled) dust and gas. Hence, it is important to establish the *relative* survival times for dust and gas.

Survival Time for the Small Dust Grain Component

In order to evaluate the dust survival time, we make use of the infrared excess emission above photospheric levels to determine whether disks are either (1) optically thick at wavelengths $\lambda \leq 100\mu m$, in which case the mass in distributed small dust grains is as least as large as the grain mass contained in the primitive solar nebula from which our planetary system was built; or (2) optically thin at $\lambda \leq 100\mu m$, or show no excess emission, in which case the disk mass is less than the mass required to build a planetary system similar to our own. We then measure infrared excess emission for stars spanning a wide range in ages and determine the fraction of young stars surrounded by optically thick disks as a function of age.

In Table 1, we summarize extant efforts to diagnose the presence of distributed small dust grains embedded within optically disks surrounding stars of differing ages and masses. These data provide the basis for the following conclusions regarding disk survival time:

- for *solar-type stars*, 50% of the stars of age $t \leq 3$ Myr are surrounded by disks, while only 10% of such stars with ages $t \sim 10$ Myr still exhibit the infrared signatures of massive optically thick disks. No stars with ages $t \geq 30$ Myr show evidence of such disks.

- for *intermediate mass stars*, we can only place upper limits on the disk survival times. The best limits come from the recent study of young intermediate mass stars in NGC 6611 (M 16) by Hillenbrand et al. (1992b). Among pre-main sequence B5-A0 stars, fewer than 10% show signatures of emission from optically thick disks. Hence, the survival time for stars of mass $2 - 5M_{\odot}$ appear to be $t < 0.3$ Myr.

A few disks surrounding solar-type pre-main sequence stars appear to exhibit large inner holes with $r \sim 1$ AU diagnosed by 1) the absence of significant excess infrared radiation $\lambda < 10\mu m$; and 2) the presence of strong excess emission at longer wavelengths (see Figure 9). These may represent "transition" cases (between optically thick disks and optically thin structures) in which assembly of small grains into larger bodies may have begun in the inner disk regions. The fraction of such transition cases is ~10% of the total number of stars apparently surrounded by massive, optically thick disks (as judged from their infrared excesses). If the average disk survives for a time $t \sim 3$ Myr, then the transition time for development of a large ($r > 1$ AU) inner hole is $t \sim 0.3$ Myr.

The presumed descendents of stars initially surrounded by optically thick disks are either (1) stars with no disks ($\tau \equiv 0$), or (2) stars with optically thin disks. Candidate optically thin disks have been identified from small excesses detected by IRAS (e.g. Vega, β Pic) around a large fraction (> 30%) B- through K- type main sequence stars for which IRAS sensitivities are sufficient to enable detection of excess emission (e.g. Backman and Gillett, 1988). Their significance derives from the fact that such optically thin disks must be populated by small dust grains in order to account for the observed IR emission. However, this dust cannot be leftover from the primordial disk because Poynting-Roberston drag will deplete small ($\alpha \sim 1\mu m$) grains on short timescales (10^3 to 10^4 yr). Because the ages of many of these stars exceeds 100 Myr, small grains must

Table 1 : SEARCHES FOR DUST EMISSION FROM CIRCUMSTELLAR DISKS

Study	Wavelengths	Clusters	Age Range	Sample	Detectable Disk Emission
K.M.Strom et al., 1989	$\lambda \leq 3.5\mu m$	Taurus-Auriga	1-10 Myr	83 c/wTTS	50% $t < 3$ Myr
					10% $t \sim 10$ Myr
Skrutskie et al., 1990	$\lambda \leq 10\mu m$	Taurus-Auriga	1-10 Myr	35 c/wTTS	50% $t < 3$ Myr
Beckwith et al., 1990	1.3 mm	Taurus-Auriga	1-10 Myr	80 c/wTTS	50% $t < 3$ Myr
Dutkevitch et al., 1992	$\lambda \leq 3.5\mu m$	T-A, α Per, Pleiades	30-100 Myr	93 solar-type	none; excesses < 10%
		UMa, Hyades, Praesape	300-600 Myr	83 solar-type	none; excesses < 10%
Hartigan et al., 1992	$\lambda \leq 10\mu m$	Sco-Cen; IC 2391/2602	7 - 30 Myr	41 B-F stars	none; excesses < 5%
Hillenbrand et al., 1992b	$\lambda \leq 2.2\mu m$	Ori; N2264; M8; M16	~ 3 Myr	>600 B-A stars	none; excesses <10%
Stapelfeldt et al., 1992	1.3mm	Sco-Cen; α Per	7 - 50 Myr	19 B-F stars	none; $F_{1.3} < 10$ mJy

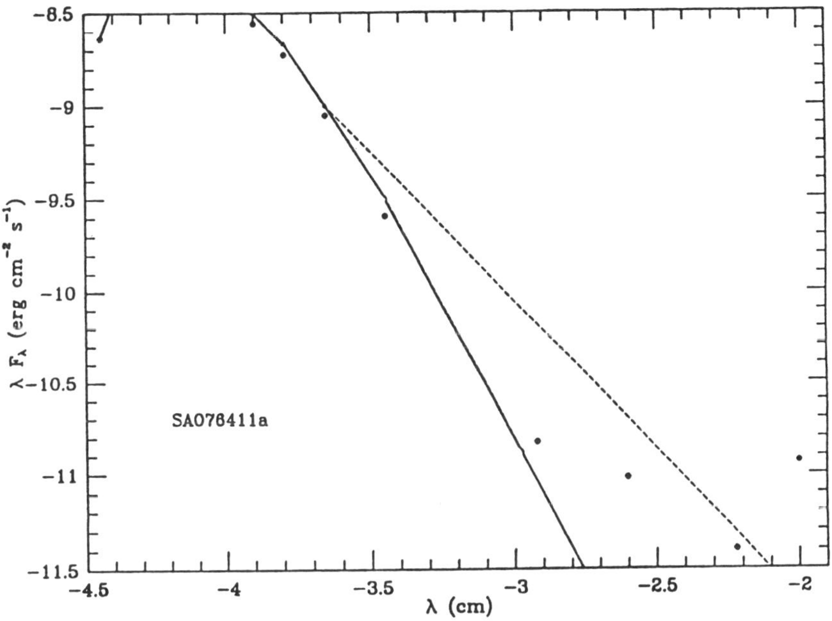

FIGURE IX A plot of the spectral energy distribution for the solar-type PMS star SAO 76411A. The heavy solid line represents the emission expected for the stellar photosphere, while the dashed line represents the emission expected for a flat, optically thick reprocessing disk viewed at an inclination of 60°. The solid points represent the observed ground-based and IRAS fluxes. Note the apparent absence of emitting material at disk radii r < 1 AU (diagnosed by the small or absent near-infrared excess emission), and the presence of significant emitting material at disk radii r > 1 AU (diagnosed by the large far infrared excesses). The structure surrounding SAO 76411A may be in "transition" between an optically thick disk which extends inward nearly to the stellar surface, and an optically thin disk similar to that surrounding β Pictoris.

be replenished via (1) introduction of small grains or cometesimals from a cold, outer disk region, or (2) *in situ* production via collisions among larger bodies.

In our view, the evidence gathered to date from analysis of the dust component of circumstellar disks suggests that

- all stars are surrounded initially by massive, optically thick accretion disks

- the solid components of these disks evolve on timescales $t < 10$ Myr for solar-type stars and $t < 1$ Myr for intermediate mass stars;

- the descendents of these disks are structures analogous to those surrounding Vega and β Pic: optically thin secondary disks populated by small dust grains perhaps produced via collisions among larger bodies assembled during the optically thick disk phases.

Survival Time for the Gas Component

Thus far, disk gas emission has been detected only among a few young stars still surrounded by circumstellar accretion disks. Little effort has been devoted to establishing the survival time for disk gas among their descendents. In large measure this reflects the difficulty in measuring emission arising in disks small compared with a typical antenna beam. To define the gas content for stars spanning the age range of relevance to planet-building ($1 < t < 100$ Myr) requires observation of objects located at distances between 100 and 200 pc. At 100 pc, a disk of solar system diameter has an angular size $\theta \sim 1''$. By comparison, the beam of the Nobeyama Radio Observatory 45-m diameter telescope has an angular size $\theta \sim 15''$ at 115 GHz ^{12}(CO (J= 1 – 0) transition). Nevertheless, the recent robust detection of disk gas associated with the accretion disks surrounding the solar-type pre-main sequence stars HL Tau and GG Tau provides some cause for optimism.

In Figure 10, we compare the strong double-peaked ^{12}CO profile observed for GG Tau (whose age is $t \sim 3$ Myr) using the NRO 45-m telescope to plots of antenna temperature against velocity (Skrutskie et al., unpublished) for 2 solar-type stars in the α Perseii cluster (age $t \sim 30$ - 50 Myr; distance $d \sim 150$ pc, or nearly identical to that of GG Tau). We note that there is no evidence for disk gas emission associated with the α Per stars. If there is remnant disk gas associated with the α Per stars, and if that gas is distributed over a structure comparable in size to that found around GG Tau, then the total disk gas mass cannot exceed 10^{-5} $M_{\odot}$. A search at similar sensitivity was made using the 14-m FCRAO telescope for 8 solar-type stars located in the Ursa Major cluster, which is closer ($d \sim 25$ pc), but much older ($t \sim 300$ Myr) than α Per. No emission was detected at a level exceeding that expected from gas with mass equal to $\sim 10^{-2}$ $M_{\odot}$ (Skrutskie et al., 1991).

These initial observations appear to rule out very long survival times ($t >> 100$ Myr) for disk gas, and imply that the giant planets must accumulate their gas on timescales $t << 100$ Myr. In the relatively near future, sensitive observations of much larger samples of stars should enable refined estimates of survival times for the gas component of circumstellar disks. Measurements of the likely immediate descendents of solar-type PMS stars surrounded by accretion disks –

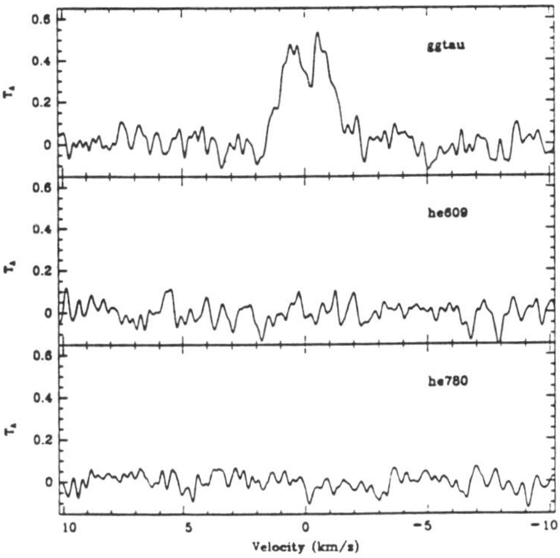

FIGURE X A plot of ^{12}CO antenna temperature against velocity for GG Tau (top panel) and two solar-type stars in the α Perseii cluster (bottom two panels); the age of GG Tau is t $\sim$ 3 Myr, while the age for the α Perseii members lies between 30 Myr and 50 Myr. These observations were made using the Nobeyama Radio Observatory 45-meter telescope. They suggest that the survival time for disk gas surrounding solar-type stars may be no longer than t $\sim$ 30 Myr.

the weak-line T Tauri stars (ages $1 < t < 20$ Myr; K.M. Strom et al., 1989) – will be of particular significance. The lack of near- and mid- infrared excess emission for these stars signals the end of the disk accretion phase. However, in at least some wTTS, cool dust appears to be present in the outer disk, as evidenced by far-IR and mm-continnum excesses (K.M. Strom et al., 1989; Beckwith et al., 1990). It will be of great interest to learn whether the gas component in such transition cases has declined significantly from the values which obtain among the accreting classical T Tauri stars. If so, this observation will place a stringent constraint on the time available for assembling gas-rich giant planets.

CONCLUSIONS

The current observation status regarding circumstellar disks and their evolutionary timescales can be summarized as follows:

• disks appear to be a natural outcome of the star-formation process;

• the disks which surround young stars initially are massive, optically thick structures comprised of gas and micron-sized grains;

• disk masses range from $0.01 < M/M_\odot < 0.2$ for *solar-type PMS stars*; $0.01 < M/M_\odot < 6$ for *young, intermediate mass stars*;

• minimum disk sizes range from $5 < r < 200$ AU for *solar-type* stars; $30 < r < 3000$ AU for *intermediate mass stars;*

• massive, optically thick disks are *accretion disks*; accretion rates are $10^{-8} < \dot{M}_{acc} < 10^{-6} M_{\odot}/yr$ for *solar-type* PMS stars; $10^{-6} < \dot{M}_{acc} < 10^{-4} M_{\odot}/yr$ for *intermediate mass* stars;

• disks evolve; survival times for the micron-size dust grains which initially populate massive, optically thick disks range from $t < 10$ Myr for *solar-type PMS stars;* $t < 1$ Myr for *intermediate mass stars;*

• these empirical disk survival times combined with the mass accretion rates quoted above, suggest that a significant fraction of the mass comprising the star may have passed through a circumstellar accretion disk;

• survival times for disk gas must be $t < 30$ Myr for *solar-type stars;*

• descendents of some massive, optically thick disks are optically thin structures analogous to β Pic in which small grains must be replenished either from outer disk regions or from collisions between larger bodies in the inner disk;

• large bodies must form in times $t < 10$ Myr for solar-type stars, and $t < 1$ Myr for intermediate mass stars;

• the transition time between the end of the disk accretion phase and the agglomeration of grains into planetesimals may be $t << 1$ Myr.

ACKNOWLEDGMENTS

We wish to thank Steve Beckwith, Nuria Calvet, Diane Dutkevitch, Lee Hartmann, Pat Hartigan, Lynne Hillenbrand, Scott Kenyon, Antonella Natta, Anneila Sargent, Michael Skrutskie and Karen Strom for their many valuable comments and for sharing their results prior to publication. This work was supported by grants from the National Science Foundation, the NASA Astrophysics Data Program, the NASA Origins of Solar Systems Program, and the NASA Planetary Astronomy Program.

REFERENCES

Adams, F.C., Lada, C.J. and Shu, F. 1987, *Ap.J.*, **312**, 788.

Adams, F.C., Emerson, J.P., and Fuller, G.A. 1990, *Ap.J.*, **357**, 606.

Attridge, J.M. and Herbst, W. 1992, *Ap.J. (Letters)* (submitted).

Backman, D.E. and Gillett, F.C. 1988, in *Cool Stars, Stellar Systems and the Sun*, ed. J. Linsky and R. Stencel, Springer Verlag: Berlin.

Beckwith, S.V.W., Sargent, A.I., Chini, R.S., Gusten, R. 1990, *Astron. J.*, **99**, 924.

Bertout, C., Basri, G. and Bouvier, J. 1988, *Ap.J.*, **330**, 350.

Calvet, N. and Hartmann, L. 1992 *Ap.J.* (submitted).

Catala, C. 1989 in *ESO Workshop on Low Mass Star Formation and Pre-Main Sequence Objects*, ed. B. Reipurth (ESO; Garching bei Munchen).

Dutkevitch, D., Skrutskie, M.F., Strom, S.E., Backman, D.E. and Stauffer, J.S. 1992 (in preparation).

Edwards, S. 1992 *A.J.* (in preparation).

Ghosh, P. and Lamb, F.K. 1978 *Ap.J. (Letters)*, **223**, L23.

Ghosh,P. and Lamb,F.K. 1979a, *Ap.J.*, **232**, 259.

Ghosh,P. and Lamb, F.K. 1979b, *Ap.J.*, **234**, 296.

Hartigan, P., Kenyon, S.J., Hartmann, L., Strom, S.E., Edwards, S., Welty, A., and Stauffer, J.S. 1991, *Ap.J.*, **382**, 617.

Hartigan, P., Hughes, J., Strom, S.E., Skrutskie, M.F., and Edwards, S. 1992 (in preparation).

Hartmann, L. and Kenyon, S.J. 1985, *Ap.J.*, **299**, 462.

Hartmann, L. and Kenyon, S.J. 1990, *Ap.J.*, **349**, 190.

Hartmann, L., Kenyon, S.J., Hewitt, R., Edwards, S., Strom, K.M., Strom, S.E. and Stauffer, J.S. 1989, *Ap. J.*, **338**, 1001.

Hartmann, L., Kenyon, S.J. and Calvet, N. 1992 *Ap.J.* (submitted).

Herbig, G.H. 1960, *Ap.J. Suppl.*, **4**, 337.

Hillenbrand, L.A., Strom, S.E., Vrba, F.J. and Keene, J. 1992a, *Ap.J.* (in press).

Hillenbrand, L.A., Strom, S.E., Massey, P., Merrill, K.M. and Gatley, I. 1992b, *PASP* (in press).

Kenyon, S.J. and Hartmann, L. 1987, *Ap.J.*, **323**, 714.

Koerner, D., Sargent, A. and Beckwith, S. 1992, *Ap.J. (Letters)* (submitted).

Konigl, A. 1991, *Ap.J. (Letters)*, **370**, L39.

Natta, A., Palla, F., Butner, H.M., Evans, N.J. and Harvey, P.M. 1992*Ap.J.*, **391**, 805.

Sandell, D., Weintraub, D., and Duncan, W.D. 1992 (preprint).

Sargent, A. and Beckwith, S. 1987, *Ap. J.*, **323**, 294.

Sargent, A. and Beckwith, S. 1991, *Ap. J. (Letters)*, **382**, L31.

Simon, M. and Gilloteau, S. 1992, *Astron. Ap.* (submitted).

Skrutskie, M.F., Dutkevitch, D., Edwards, S., Strom, S., and Strom, K. 1990, *Astron. J.*, **99**, 1187.

Skrutskie, M.F., Snell, R., Dutkevitch, D., Strom, S.E., Schloerb, F.P., and Dickman, R.L. 1991, *Astron. J.*, **102**, 1749.

Skrutskie, M.F., Snell, R.L., Strom, K.M., Strom, S.E., Edwards, S., Fukui, Y., Mizuno, A., Hayashi, M. and Ohashi, N. 1992, *Ap.J.* (submitted).

Stapelfeldt, K., Strom, S.E., Edwards, S. 1992 (in preparation).

Strom, S.E., Strom, K.M., Yost, J., Carrasco, L. and Grasdalen, G.L. 1972, *Ap.J.*, **173**, 353.

Strom, K.M., Strom, S.E., Kenyon, S.J. and Hartmann, L. 1988, *A.J.*, **95**, 534.

Strom, K.M., Strom, S., Edwards, S., Cabrit, S., and Skrutskie, M.F. 1989, *Astron. J.*, **97**, 1451.

Strom, S.E., Edwards, S. and Skrutskie, M.F. in *Protostars and Planets III*, ed. E.H. Levy and M.S. Matthews, (University of Arizona Press; Tucson, AZ).

Welty, A.D., Strom, S.E., Edwards, S., Kenyon, S. and Hartmann, L.W. 1992, *Ap.J.* (in press).

Planets Around Pulsars
ASP Conference Series, Vol. 36, 1993
J.A. Phillips, J.E. Thorsett, and S.R. Kulkarni

PLANETS AROUND PULSARS: IMPLICATIONS FOR PLANETARY FORMATION

PETER BODENHEIMER
UCO/Lick Observatory, Board of Studies in Astronomy and Astrophysics, University of California, Santa Cruz, CA 95064

ABSTRACT This paper summarizes some aspects of the data regarding the existence of planets around pulsars, discusses some of the models that have been proposed to explain the formation mechanism, and provides some comments on the implications of these results with regard to the formation of planets around "solar"-type stars.

DATA

The discovery (Wolszczan and Frail 1992) of two planets in orbit about the pulsar PSR 1257+12 is of great significance in its own right and may have further significance with regard to the existence of planetary systems in general in the galaxy. A brief summary of the relevant data here will serve as background information for a discussion of the possible wider implications of this discovery. The distance of the pulsar from the Sun is about 500 pc, its period is 6.2 ms, it has a surface magnetic field of 8.8 $\times 10^8$ gauss, its spindown age is estimated at 0.8 $\times 10^9$ year, and it has no stellar-mass companion. Typically, millisecond pulsars are found in binary systems; the high angular momentum of the pulsar is thought to occur because matter from a companion star has accreted onto it. In general only about 3% of radio pulsars are found in binaries (Bhattacharya and van den Heuvel 1991). The pulse timing measurements obtained with the Arecibo telescope indicate two periodicities, of 98.2 and 66.6 days. The observations are interpreted as evidence for two planets in orbits with eccentricities 0.022 and 0.020, distances of 0.36 and 0.47 AU, and masses of 3.4/sin i and 2.8/sin i earth masses, respectively (i is the angle of inclination between the orbital plane and the plane of the sky). This conclusion is based on approximately 2 years of data since the discovery of the pulsar in Feb. 1990. The detection is well above the observational limit, and confirming observations have been made at Green Bank by D. Backer (1992). The periods are in a near 3:2 resonance, and studies by Malhotra et al. (1992) and Rasio et al. (1992) indicate that orbital interactions resulting in changes in eccentricities, periods, and the longitude of periastron should be observable in a few years. They also find that the planets cannot be greater than about 2 Jovian masses, otherwise the system would be subject to orbital instability.

On the other hand, Bailes et al. (1991) report that several hundred other pulsars have been observed, and none shows a variation with periodicity and amplitude that would suggest the presence of one or more planets. Further surveys are under way based on archival timing data. For example, Thorsett and Phillips (1992) have ruled out earth-mass planets with periods between tens and thousands of days around PSR 1919 + 21 (period 1.337 s). Furthermore, they have examined PSR 1937 + 21, a millisecond pulsar not in a binary, and therefore a good candidate for possible planets. They rule out planets with periods between 10 and 100 days and masses down to 0.001 - .01 earth masses. Thus out of over 500 known pulsars, only one is definitely known to have planets, and that may once have been a binary pulsar.

Further evidence linking planets with binary pulsars is provided by the existence of two millisecond pulsars, PSR 1957 + 20 and PSR 1744 - 24A, which are in eclipsing binary systems. Both have very low mass companions, ≈ 0.02 and 0.09 $M_{\odot}$, respectively (Fruchter et al. 1988, Lyne et al. 1990). There is evidence from the eclipse data to suggest that the pulsars are in the process of ablating and vaporizing the companions, so that these systems are possible precursors of planetary systems. Overall, these data suggest the possibility that very special circumstances are required for the formation of planetary systems around pulsars, namely the prior presence of a millisecond pulsar with a close binary companion, probably a low-mass main-sequence star.

MODELS

Several different models have been suggested for the origin of planets around pulsars, and more details of their properties appear in other articles in this volume. Only a subset of these models is briefly discussed here; for example, models with pre-existing planets in the system that evolves to a supernova are not included. Some of the models were specifically designed to account for the planet around PSR 1829 - 10, which has since disappeared; these models have some difficulties in explaining the planets around PSR 1257 + 12, but they could be useful in the future.

(1) A planet could represent the remnant core of a companion star that was not quite completely vaporized by the action of the pulsar (Krolik 1991). This model is unlikely, it doesn't account for two planets, and it has difficulty explaining the fact that the present orbital distances of the companions in the systems with star-vaporizing pulsars are around 10^{11} cm, while the planets are at distances of $\sim 10^{13}$ cm. In fact the orbit of PSR 1957 + 20 is shrinking (Ryba and Taylor 1992), possibly because of angular momentum loss to the outflowing material (Banit and Shaham 1992).

(2) A planetary system could be formed from the fallback material in the Type II supernova explosion that produced the neutron star (Lin et al. 1991). The angular momentum of the planets is derived from the rotational velocity of the star that eventually became the supernova. This model has the problem of providing barely enough angular momentum, and it probably does not apply to a millisecond pulsar, because there is no companion to spin up the neutron star to a millisecond period. The fallback material initially forms a disk at a radius of about 10^8 cm. The transport of material and angular momentum to 10^{13} cm and the formation of planets there are treated in this model, and these features turn out to be useful in the context of other models. The composition of the material that forms the disk is primarily heavy elements, such as oxygen, silicon, and iron.

(3) A pre-existing solar-type star with a planetary system collides with a neutron star. The main-sequence star is vaporized, and the planets go into orbit around the neutron star. The problems are that the event is improbable except in globular clusters, and the remaining bound planets are likely to have elliptical orbits (Podsiadlowski et al. 1991). However the ejected envelope consisting of the original stellar material could aid in providing some degree of circularization; at the same time it expands and dissipates and also tends to cause the planets to spiral toward the center and to evaporate. All of these events occur on the same time scale, so it is not clear that circular orbits at the present observed distances from PSR 1257 + 12 would result. A system formed in this way is subject to later disruption, and the spinup of the neutron star to a millisecond period is unlikely.
(4) Two white dwarfs in orbit merge as the orbit shrinks (Podsiadlowski et al. 1991). The accretion-induced collapse of the more massive component produces the neutron star; the remains of the companion are left in orbit at about 10^9 cm. This disk then expands and forms planets in a manner similar to that of model (2). The advantages of the model are that the material would have sufficient angular momentum and would be composed primarily of heavy elements, which would promote rapid planetary formation. The difficulty is that it is not well understood under what circumstances the merger results in collapse rather than complete disruption of the massive white dwarf in a type I supernova explosion. The actual outcome depends on, for example, the composition of the more massive white dwarf and the accretion rate onto it. Detailed hydrodynamic calculations are needed to decide whether a millisecond pulsar would result and whether the orbiting material remains bound after the explosion.
(5) The neutron star has a low-mass, main-sequence companion in a close orbit, such as that observed around PSR 1957 + 20. The pulsar wind, driven by pulsar radiation, in combination with Roche-lobe overflow, results in complete evaporation of the companion and formation of a disk close to the pulsar (Ruderman et al. 1989, Stevens et al. 1992). The disk has sufficient angular momentum to form planets at their observed positions, but the details as to how the disk evolves and how the planets are formed are not specified. In a variation of this model, Tavani and Brookshaw (1992) make a numerical calculation of the evolution of material that is evaporated from the companion, and they show that a circumbinary disk is produced, in which planets have time to form. At a later stage the evaporation of the companion is completed. However it is not clear whether the parameters of the PSR 1257 + 12 system are appropriate for planetary formation to take place in that disk. The Tavani-Brookshaw model shows that for pulsar luminosities above 10^{34} erg s^{-1} the expanding disk is unbound, but for $L \approx 2 \times 10^{33}$ erg s^{-1} it is bound. Note however that the Shu et al. (1979) model of binary stars shows that the mass ratio of the system must be below 0.064, otherwise evaporated material is ejected from the system through the outer Lagrangian point, even without a wind. The mass ratio chosen in the calculations is actually slightly above this limit. At a smaller ratio, the ejected matter is bound; however further calculations are necessary to explore this region. Furthermore, the present luminosity of PSR 1257 + 12 is 5 $L_\odot$, a value which also suggests that a large fraction of the evaporated material will actually escape.
(6) A giant presupernova star has an orbiting companion of 1.5 $M_\odot$ at a sepa-

ration of 3×10^{12} cm (which actually must have been inside the primary at some stage). During the supernova explosion process some material is ablated from the side of the companion facing the supernova, and a small fraction of this material is captured into orbit around the neutron star at a distance comparable to that of the companion. The remaining ablated material, as well as the companion itself, become unbound and are lost from the system (Nakamura and Piran 1991). Because much of the captured material consists of heavy elements, the planet formation time is short. The parameters have to be fine-tuned; if there are significant deviations from the values given, an insufficient amount of material is captured. It is also unlikely that at the same time the pulsar can accrete enough matter to generate a millisecond period at the present time. This model has the advantage that it does not require substantial angular momentum transport in and evolution of the disk. The preplanetary material is already at the proper distance from the neutron star when it is released from the companion.

In fact, in general it seems that a very special kind of circumstance is required to produce a planet around a pulsar. The most likely theories involve the presence of a close binary companion, although other scenarios are also possible. Both theoretical and observational evidence strongly suggest that the situation is very different from that of the formation of the Solar System and of the presumed formation of other planetary systems around young stars.

IMPLICATIONS

What can we learn about planetary formation around solar-type stars from this discovery? There are a number of major unanswered questions. We first recall the statement by G. Herbig (1979) at a conference on Life in the Universe:

"It would be fine if astronomers would identify for us which stars could be accompanied by planets, and it would be staggering if through some technological *tour de force* they were able to demonstrate the positive presence of planets in a few such cases.... [T]he fact that this conference is being held is evidence of belief or hope that somehow or someday the subject may open up, probably in a way no one here can anticipate."

Certainly the subject has opened up in an unanticipated way. Nevertheless, based on the limited amount of existing data, a wide range of possible conclusions is still open. First, we could take the optimistic view and assert that firm data now exist for the presence of two planets in orbit around a pulsar. At least one reasonable theoretical model, that which involves the evaporation of a nearby companion, shows how such a system could form. Since it appears to be extremely difficult to form planets around pulsars, and unlikely circumstances are required, the fact that two indeed exist is a strong indication that it must be easy to form planets in general. The situation is made to look even more favorable because two pulsars have been found which convincingly appear to be in systems which are precursors to planet formation; a star is in the process of being ablated. As G. Marcy remarked (*San Jose Mercury*, Feb. 18, 1992) "if planets can form around pulsars, they could form anywhere."

However, we could also take a second, more skeptical point of view. One could note all of the models have great difficulties in explaining the existence of planets around pulsars. They require fine tuning and special values of parameters, they do not address all aspects of the problem, and they concentrate, for the most part, on forming a disk-like structure in the vicinity of a pulsar. Furthermore, they do not convincingly demonstrate either that the disk can expand to the appropriate radius or that, even if it does, the conditions are suitable for the formation of planets on a reasonable time scale. The observations, in this view, could be a figment of our imagination, a result of data reduction error, a product of timing noise, or a consequence of pulsar precession or neutron star oscillations. As Dave Black (1991) remarked, "the history of searches for other planetary systems is littered with published detections that vanish under further scrutiny." A few examples include the Crab pulsar PSR 0531 + 21, PSR 0329 + 54, Supernova 1987a, PSR 1829 - 10, HD 114762, Barnard's star, γ Cep , 61 Cyg A, ϵ Eridani, and Lalande 21185. In the case of the last 5 objects, observations of variable radial velocity and of astrometric perturbations have been reported, but the data are either inconsistent or simply insufficient to determine a firm period. In the case of HD 114762, radial velocity measurements indicated the presence of a companion of 0.02 $M_\odot$ (Latham et al. 1989). However, later analysis of the stellar rotational velocity of this F star (Cochran et al. 1991) indicated that most likely it was being observed pole-on, so the companion mass must be much greater. Of course it has also not been positively proved that any of these objects do *not* have planets.

A third point of view, more nearly neutral or intermediate between the first two, is also viable. We have data now that show the existence of two extrasolar planets. There is still a lot of pulsar timing data available which can be analyzed for the presence of further planets. However even when additional "pulsar" planets are discovered, one has to remember that the pulsar population is a tiny fraction of the total number of stars. What we are really interested in is the number of planets around "solar" (that is, G and K main-sequence) type stars. What the discovery seems to tell us is that the circumstances for their formation are very different from those near pulsars. If we are to reach any conclusions regarding the actual planetary formation process, planetary search techniques must be refined so that young stars, all of which could be forming planets, are scrutinized carefully for clues. Furthermore, for determination of the overall number of stars in the Galaxy which have "solar"-type planets, main-sequence stars must be systematically searched. Briefly, this point of view states that there could be many planets, but we don't know how many, and the search must be extended to other objects besides pulsars.

Going back to the question of what the planetary astrophysicist needs to know, it must be realized that the theory of planet formation is in a seriously incomplete and unsatisfactory state, and that there are several major unanswered questions, as reviewed by Safronov (1991). First, what were the initial conditions for formation of disks around young stars? What were the mechanisms for angular momentum transport in the disk and what was the time scale for evolution? In the solar nebula case the mechanisms include gravitational instability, turbulence due to convection, magnetic transport, and turbulence induced at the shear layer between the disk and the matter infalling onto it. Each one of these mechanisms may have operated at some stage, in some region of the nebula, and there are physical reasons for believing that the mechanisms were indeed operating. In the pulsar case none of these processes, except possibly the magnetic one, is likely to work. The study of the pulsar case is therefore not likely to lead to any great insights regarding the evolution of "pre-solar" nebulae. The nebula in the pulsar case evolves from inside out, that is from rather small radii out by at least 2 orders of magnitude to radii on the order of 10^{13} cm. Processes such as wind-driven hydrodynamics or generalized turbulence are invoked to accomplish this spreading. But the viscosity "is...a free parameter which can have values in accordance with the author's desire" (Safronov 1991). For the pulsar case, turbulent viscosity parameters are generally assumed which are taken from the solar nebula analog (α parameter $\approx 10^{-2}$), even though the appropriate mechanisms are not present. Observational constraints on the length of time needed for the pulsar disk to expand are not tight enough to give us much information on the time scale for angular momentum transport in more general situations. It is true, of course, that one particular solar nebula formation model involves expansion of the nebula from "inside-out" (Ruzmaikina 1981, 1985). In this model an interstellar cloud with relatively low angular momentum collapses to form a stellar core with radius of only ≈ 0.05 AU. Magnetic transport then results in the production of a disk, which expands outwards into the planet-formation zones. Again, the physical situation during this expansion is probably quite different from that near a pulsar.

The second question concerns what processes are involved in the coagulation of dust particles, how fast they grow, and what was the size distribution of protoplanetary bodies once the coagulation has taken place. At least in the case of disks around young objects, dust is present that is preserved from the interstellar cloud out of which the disk formed. In the pulsar case the dust is not originally present. Authors simply assume that once the material is transported out to radii where the temperature drops to 1500 K, the dust automatically forms. Actually the spectra of radiation and particles from the central object are completely different in the two cases. The initial growth processes leading to planetesimals in the "solar" case are not well duplicated in the environment of a pulsar.

The third important question concerns the buildup of protoplanetary objects into planets. It is still controversial to what extent runaway accumulation speeds up the formation times of terrestrial or giant planets, what the termination point of runaway accretion is, what the relative velocities of the objects of various sizes are during accretion, in what regions of the solar nebula runaway accretion occurs, and how long it takes to build up planets after runaway accretion stops. As an example we list a few estimates for the formation time of an approximately earth-mass planet at about 1 AU. Wetherill (1989) shows that runaway accretion stops at approximately lunar size and the earth does not reach its final mass until 10^8 yr. Safronov (1991) and Safronov and Ruzmaikina (1985) quote similar figures. Nakano (1987) uses a simple formula to obtain an estimate of 10^7 yr for formation of the earth in a minimum mass solar nebula in the gas-free state.

Regarding this question of planetary formation time scales, it appears that it could be possible to gain some information from the pulsar planets, given tighter observational and theoretical constraints. Using Nakano's formula, Nakamura and Piran (1991) obtain 2.6×10^5 years to form a planet near a pulsar at 0.7 AU in a disk with 1/8 the surface density of the minimum-mass solar nebula. The short formation time in this case comes from the fact that almost all the ablated material in their model is composed of heavy elements. The model by

Lin et al. (1991) also involves mainly heavy elements, and their time scale for planet formation is of the same order of magnitude. But for the case of PSR 1257 + 12, in the model involving the evaporation of a companion, the gas that forms the disk is probably near solar composition, so the formation time scale goes back up to more than 10^7 yr. The question is whether the circumstances, in the presence of the pulsar wind, allow that much time.

Other important questions concerning planetary formation include the interaction of the accretion of gas and solids in forming the giant planets, the dissipation time of the solar nebula in relation to the time required for formation of the giant planets, the process of formation of comets, the location of the origin of comets and its relation to the giant planets, and the role of the giant planets in ejecting them into the outer parts of the solar system. The important questions regarding the origin of giant planets are not likely to be answered by pulsar studies, although it would be of extreme interest if one were discovered near a pulsar. However, it is interesting to note that if the planets around pulsars are young, the comets that formed in their vicinity may not yet have been scattered out into the analog of an Oort cloud. Their presence may be detectable in emission from OH, CN, and C_2 (Chakrabarti and Krishna-Swamy 1992).

To summarize, the discovery of two planets around PSR 1257 + 12 opens up a wide range of interesting problems of dynamics and stellar evolution. Although some interesting constraints on terrestrial planet formation, as well as possibly on comet formation, could be obtained by further study of planets around pulsars, only a tiny fraction of all stars could evolve in such a way as to produce conditions favorable for their formation. At least some aspects of these conditions must differ from those necessary to form planets around young stars, and many of the outstanding questions regarding planet formation in the solar system or around other stars will not be answerable. I therefore conclude that the way to figure out how many planets there are in the galaxy and to learn about the more usual process of planetary formation is through (1) more intensive studies of the properties of disks near young objects, and (2) refinement and application of the techniques for detection of planets around main-sequence solar-type (G-K-M) stars.

ACKNOWLEDGMENTS

This work was supported in part by a special NASA astrophysics theory program which supports a joint Center for Star Formation Studies at NASA/Ames Research Center, UC Berkeley, and UC Santa Cruz. Further support was provided by National Science Foundation grant AST–8914173.

REFERENCES

Backer, D. 1992, this volume

Bailes, M., Lyne, A. G., and Shemar, S. L. 1991, *Nature*, **352**, 311

Banit, M., and Shaham, J. 1992, *ApJ*, **388**, L19

Bhattacharya, D., and van den Heuvel, E. 1991, *Physics Reports*, **203**, 1

Black, D. 1991, *Nature*, **352**, 278

Chakrabarti, S. K., and Krishna-Swamy, K. S. 1992, *A&A*, submitted

Cochran, W. D., Hatzes, A. P., and Hancock, T. J. 1991, *ApJ*, **380**, L35

Fruchter, A. S., Stinebring, D. R., and Taylor, J. H. 1988, *Nature*, **333**, 237

Herbig, G. H. 1979, in *Life in the Universe*, J. Billingham, Cambridge, MA, MIT Press, 77

Krolik, J. H. 1991, *Nature*, **353**, 829

Latham, D. W., Mazeh, T., Stefanik, R. P., Mayor, M., and Burki, G. 1989, *Nature*, **339**, 38

Lin, D. N. C., Woosley, S. E., and Bodenheimer, P. 1991, *Nature*, **353**, 827

Lyne, A. G., Manchester, R. N., D'Amico, N., Staveley-Smith, L., Johnston, S., Lim, J., Fruchter, A. S., Goss, W. M., and Frail, D. 1990, *Nature*, **347**, 650

Malhotra, R., Black, D., Eck, A., and Jackson, A. 1992, *Nature*, **356**, 583

Nakamura, T., and Piran, T. 1991, *ApJ*, **382**, L81

Nakano, T. 1987, *MNRAS*, **224**, 107

Podsiadlowski, P., Pringle, J. E., and Rees, M. J. 1991, *Nature*, **352**, 783

Rasio, F. A., Nicholson, P. D., Shapiro, S. L., and Teukolsky, S. A. 1992, *Nature*, **355**, 325

Ruderman, M., Shaham, J., and Tavani, M. 1989, *ApJ*, **336**, 507

Ruzmaikina, T. V. 1981, *Adv. Space Res.*, **1**, No. 7, 49

Ruzmaikina, T. V. 1985, *Astron. Vestn.*, **19**, 101

Ryba, M. F., and Taylor, J. H. 1992, *ApJ*, submitted

Safronov, V. S. 1991, *Icarus*, **94**, 260

Safronov, V. S., and Ruzmaikina, T. V. 1985, in *Protostars and Planets II*, D. C. Black and M. S. Matthews, Tucson, University of Arizona Press, 959

Shu, F. H., Lubow, S. H., and Anderson, L. 1979, *ApJ*, **229**, 223

Stevens, I. R., Rees, M. J., and Podsiadlowski, P. 1992, *MNRAS*, **254**, 19p

Tavani, M., and Brookshaw, L. 1992, *Nature*, **356**, 320

Thorsett, S. E., and Phillips, J. A. 1992, *ApJ*, **387**, L69

Wetherill, G. W. 1989, in *The Formation and Evolution of Planetary Systems*, H. A. Weaver and L. Danly, Cambridge, Cambridge University Press, 1

Wolszczan, A., and Frail, D. A. 1992, *Nature*, **355**, 145

Section 5:

Searches for Disks and Other Planets

Planets Around Pulsars
ASP Conference Series, Vol. 36, 1993
J.A. Phillips, J.E. Thorsett, and S.R. Kulkarni

MCDONALD OBSERVATORY PLANETARY SEARCH: A HIGH PRECISION STELLAR RADIAL VELOCITY SURVEY FOR OTHER PLANETARY SYSTEMS

WILLIAM D. COCHRAN AND ARTIE P. HATZES
McDonald Observatory, The University of Texas at Austin, Austin TX 78712

INTRODUCTION

There is increasing evidence that other stars in our galaxy may have sub-stellar mass companions. A major motivation for searching for sub-stellar companions is to investigate the process of star formation. Does the debris from star formation routinely accumulate into low mass objects? Is there a continuous function of stellar secondary masses extending from M stars through brown dwarfs to planets? Was the process that formed our sun and solar system typical of the general process of star formation? What properties of the central star (mass, age, metallicity, angular momentum, magnetic field strength, etc.) determine the characteristics of any planetary system which may be formed? The more examples of such systems we can find, the better chance we have of understanding the processes of star and planet formation. In addition to information on star formation, a systematic survey for brown dwarfs and planets will provide important new data on the minimum mass for hydrogen burning stars, the cooling sequence for brown dwarf stars, and the possibility that brown dwarfs provide the "missing mass" component of the galactic disk, or massive halos around galaxies in general.

THE MCDONALD OBSERVATORY PLANETARY SEARCH

In September 1987 we started a search for extra-solar planetary systems by obtaining very high precision measurements of radial velocity variations of the primary star to detect the orbital motion of the target star around the star-planet barycenter. Is it reasonable to think that our goal of searching for extra-solar planetary systems is observationally feasible? The velocity of the Sun around the Jupiter-Sun barycenter averages 12.3 m s^{-1}. Lacking any better knowledge, we consider our solar system to be typical, and set the observational goal as the detection of such systems around other stars. The problem of planet detection requires not only high precision, but also a long time baseline. Detection of planets more than a few AU from the primary star will require several years of effort. However, planetary systems with massive planets, or with Jupiter sized planets close to the star will be much easier to detect. In the course of this study, we will undoubtedly also come across many previously undetected double star systems. We would expect to find those systems which went undetected in previous surveys due to the low velocity amplitude of their radial velocity curves. This is a highly desirable side benefit of this program.

Our observing list, shown in Table , includes 33 nearby F, G, and K dwarfs. In monthly observing runs on the McDonald Observatory 2.7-m telescope coudé spectrograph we obtain high precision radial velocity observations of all stars on the program list which are available.

TABLE I List of Program Stars

HR	Star	Spec.	V	HR	Star	Spec.	V
219	η Cas	G0V	3.44	509	τ Cet	G8V	3.50
937	ι Per	G0V	4.04	963	α For	F8V	3.87
996	κ Cet	G5V	4.82	1084	ϵ Eri	K2V	3.73
1136	δ Eri	K0IV	3.54	1325	o^2 Eri	K1V	4.43
1543	π^3 Ori	F6V	3.19	2047	χ^1 Ori	G0V	4.41
2943	α CMi A	F5IV-V	0.38	3775	θ UMa	F6IV	3.17
4112	36 UMa	F8V	4.82	4540	β Vir	F8V	3.61
4983	β Com	G0V	4.28	–	HD114762	F9V	7.29
5019	61 Vir	G6V	4.75	5544	ξ Boo A	G8V	4.55
5933	γ Ser	F6V	3.85	6401	36 Oph B	K0V	5.33
6402	36 Oph A	K1V	5.29	6623	μ Her	G5IV	3.42
6752	70 Oph A	K0V	4.03	7462	σ Dra	K0V	4.68
7503	16 Cyg A	G1.5V	5.96	7504	16 Cyg B	G2.5V	6.20
7602	β Aql	G8IV	3.71	7948	γ^2 Del	K1IV	4.27
7957	η Cep	K0IV	3.43	8085	61 Cyg A	K5V	5.21
8086	61 Cyg B	K7V	6.03	8832	HR8832	K3V	5.56
8974	γ Cep	K1III-IV	3.21				

TECHNIQUE

The whole secret to making high precision stellar radial velocity variation measurements is to impose a stable set of reference absorption lines on the stellar spectrum before it enters the spectrograph. Thus, any instrumental wavelength drifts inherent to the spectrograph will affect both the stellar and comparison spectra equally. Very precise determinations of the variations in stellar radial velocity are made by measuring the apparent Doppler shift of the stellar lines with respect to the reference lines.

We started observations using the telluric O_2 band at 6300Å as its fiducial reference system (Griffin & Griffin, 1973) With this method, we obtained a precision of an individual observation of about 10-15 $\mathrm{m\,s^{-1}}$. We have recently shifted to use of a temperature stabilized I_2 absorption cell as the velocity reference system. A 30 cm long I_2 cell is placed in front of the entrance slit of the 2.7-m telescope coudé spectrograph. Starlight passes through the I_2 cell, which superimposes the fiducial I_2 spectrum on the stellar spectrum. This is illustrated in Figure I. The top panel shows a typical stellar spectrum alone. The middle panel shows the spectrum of the I_2 cell, and the lower panel shows the spectrum of the star through the I_2 cell. Radial velocity variations are then

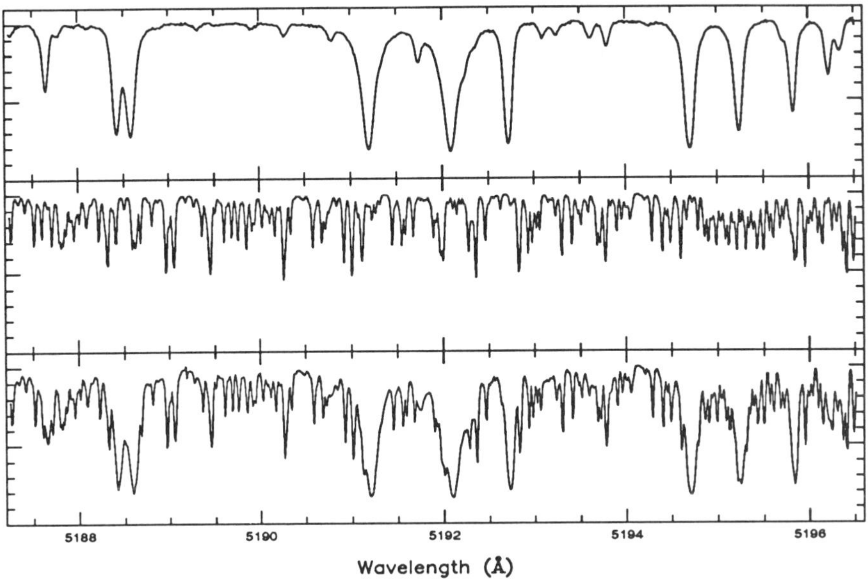

FIGURE I A sample of spectra for the McDonald Observatory Planetary Search. The top panel shows the spectrum of a program star τ Ceti. The middle panel shows the spectrum of the I_2 absorption cell in the same spectral region. The bottom panel shows the spectrum of τ Ceti taken through the I_2 cell.

determined by measuring the relative shift of the stellar spectrum and the I_2 spectrum, and correcting for the motion of the telescope with respect to the solar system barycenter. With the I_2 cell, we are able to achieve precision as good as $5\,\mathrm{m\,s^{-1}}$ under ideal conditions.

RESULTS

Moon

We observe a standard spot on the lunar surface as a calibration check. This is the best way to observe disk-integrated sunlight in a manner approximating as closely as possible actual stellar observations. The upper panel of Figure II shows data taken with the O_2 technique, and the lower panel shows the I_2 cell data. The O_2 data have an rms dispersion around the mean of $11.9\,\mathrm{m\,s^{-1}}$, and the I_2 data show an rms scatter of $6.4\,\mathrm{m\,s^{-1}}$. These results demonstrate that by using the I_2 cell, we have achieved our goal of long-term radial velocity precision in the 5-10 $\mathrm{m\,s^{-1}}$ range.

HD114762

The star HD114762 was recently found by Latham *et al.* (1989) to exhibit low amplitude periodic radial velocity variations, which they attributed to the reflex orbital motion of the star due to the presence of an unseen companion object. We have obtained 28 high precision relative radial velocity measurements on HD114762 as part of our McDonald Observatory Planetary Search program

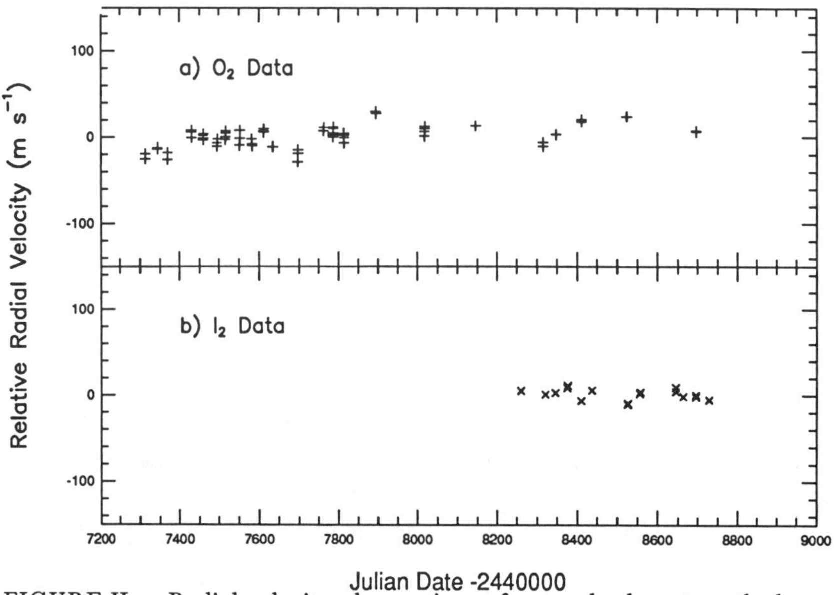

FIGURE II Radial velocity observations of a standard spot on the lunar surface. The upper panel shows data taken using the O_2 line technique, while the lower panel shows data taken with the I_2 cell.

(Cochran *et al.* 1991). The new data confirm Latham *et al.*'s discovery of the binary nature of the system. Our new orbital solution gives an eccentricity of 0.38 compared with the Latham *et al.* value of 0.25, but otherwise agrees well with the discovery orbit. Figure III shows the relative radial velocity curve calculated from our orbital solution. The large filled circles are the McDonald data and the small dots are the Latham *et al.* data. The mass function for the system implies a companion object with $M_c \sin i_{orb} = 0.011 M_\odot$. If the system is viewed edge on then the companion would be in the range of possible planetary masses. Intermediate values of the inclination angle would imply a sub-stellar "brown dwarf" companion. However if viewed nearly pole-on ($i_{orb} < 8°$), the companion could be a low mass star. A photometric investigation by Robinson *et al.* (1990) found no evidence for transits or eclipses in the system, effectively eliminating the nearly equator-on viewing geometry. However, they were only able to constrain the orbital inclination to a value less than 89°, and this does not place any tight constraints on the mass of the secondary object.

Fourier analysis of the shape of a photospheric absorption line of HD114762 gives a best estimate of the projected stellar rotational velocity $V \sin i_{rot} = 0\,\mathrm{km\,s^{-1}}$, and with an extreme upper limit on $V \sin i_{rot}$ of about $1\,\mathrm{km\,s^{-1}}$. The derived mass for the companion object then depends on the assumptions made for the rotational velocity of the star, and on the degree of alignment of the stellar rotational axis with the companion object orbital axis. If we assume that the actual rotation rate of HD114762 is similar to that of other F9 V stars studied by Soderblom, and that the stellar rotational axis is approximately parallel to the orbital axis, this gives a value for $\sin i_{orb}$ of the companion object of ~ 0.10 ($i_{orb} < 5.7°$). This results in the companion object having the mass of a late M dwarf star in a system viewed nearly pole-on. In the other extreme,

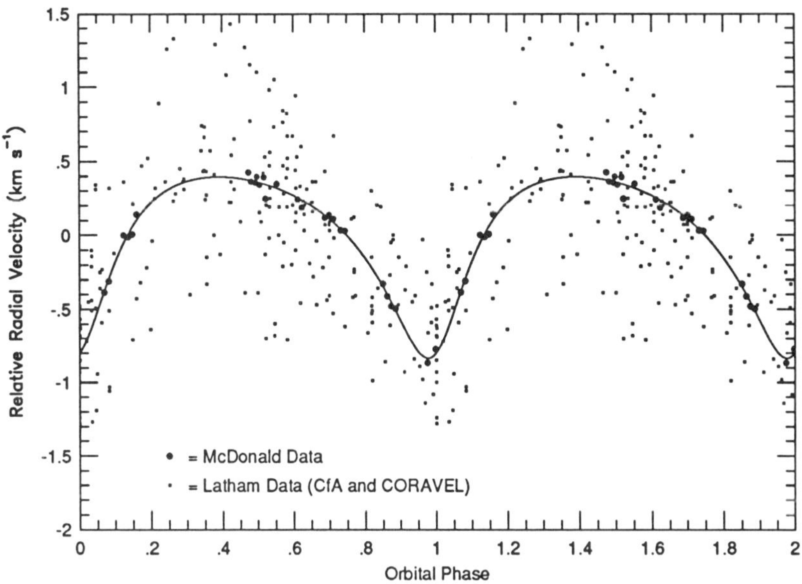

FIGURE III Radial velocity curve for HD114762. The McDonald data are shown as the large filled circles. The original discovery data of Latham *et al.* are shown as small dots. The solid curve is from the orbital solution derived from the McDonald data.

one can assume that HD114762 is a very slow rotator, that our hard upper limit on $V \sin i_{orb}$ is the actual value, and that there is a significant misalignment of the rotational and orbital axes, and derive a companion object mass as low as $0.02 M_{\odot}$.

α Canis Minoris A

Our measured relative radial velocities for Procyon are shown in Figure IV. Plus signs indicate data taken with the O_2 line technique, and the $\times$ symbols are data taken with the I_2 cell. The overall positive slope is due to the orbit of Procyon A and Procyon B. The solid line is the expected radial velocity variation from the published orbit solution. There is a substantial amount of scatter of the data around this line. This scatter is *not* noise. The rms variation of data from each individual night is about $10\,\mathrm{m\,s^{-1}}$, which is in line with the overall scatter in data from other stars. This scatter represents real radial velocity variability of Procyon A. It is quite possible that this star is a low-amplitude δ Scuti variable.

δ Eridani

Measured radial velocity variations for δ Eridani are shown in Figure V. The $\times$s represent data taken with the I_2 cell, and the plus signs are data taken with the O_2 line technique. No obvious systematic trends are evident in this star. These data are given as an example of most of the stars in the survey, which are constant to the precision of our measurements.

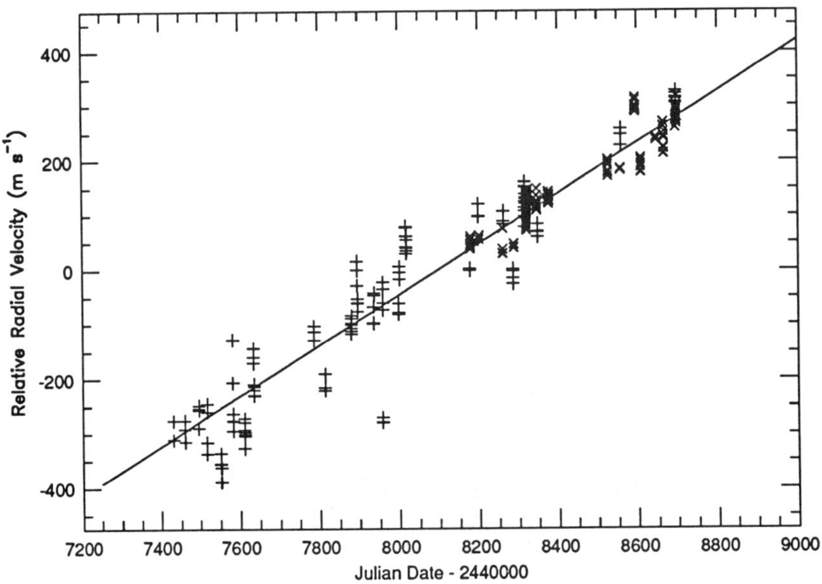

FIGURE IV Measured radial velocity variations of α Canis Minoris A. The solid line is the variation expected from the orbit of the white dwarf companion. The scatter in the data is due to intrinsic variability of the star.

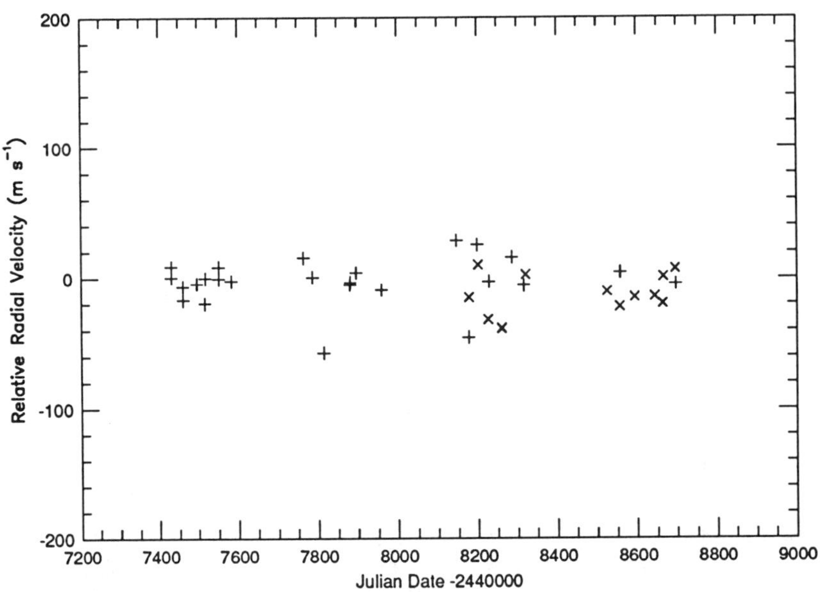

FIGURE V Measured radial velocity variations of δ Eridani.

CONCLUSIONS

- Measurement of stellar radial velocity variations to a precision of better than $10\,\mathrm{m\,s^{-1}}$ is now achievable in routine observations.
- This level of precision is capable of detecting Jovian planets in orbit around solar type stars.
- The McDonald Observatory Planetary Search program has been surveying a sample of 33 nearby F, G, and K stars since September 1987 to search for sub-stellar companion objects.
- We have confirmed the detection of a companion object to HD114762. The mass of the companion object is still somewhat uncertain. It is quite possible that we are looking at this system nearly pole-on, and the companion could well be a very low mass star.
- No other stars on the survey list show convincing evidence for planetary-mass companions at this time.

ACKNOWLEDGMENTS

This work has been supported by NASA Grants NAGW-2302 and NAGW-2465, and by NSF Grant AST-9017368.

REFERENCES

Cochran, W. D., Hatzes, A. P. & Hancock, T. J. 1991, *ApJ*, **380**, L35

Griffin, R. & Griffin, R. 1973, *MNRAS*, **162**, 243

Latham, D. W., Mazeh, T., Stefanik, R. P., Mayor, M., & Burki, G. 1989, *Nature*, **339**, 38

Robinson, E. L., Cochran, A. L., Cochran, W. D., Shafter, A. L., & Zhang, E.-H. 1990, *AJ*, **99**, 672

Planets Around Pulsars
ASP Conference Series, Vol. 36, 1993
J.A. Phillips, J.E. Thorsett, and S.R. Kulkarni

A DISK ABOUT PSR 1957+20?

A. S. FRUCHTER[1]
Department of Astronomy, University of California, Berkeley, CA 94720

ABSTRACT Although the PSR 1957+20 binary system displays a large orbital period derivative, radio observations of the ablated wind from the companion reveal a low free electron density. If the orbital period derivative provides a true indication of the mass loss rate, then the ablated material must be overwhelmingly neutral or largely confined to a narrow region about the orbital plane. However, the companion's high orbital velocity, the observed turbulence in the eclipsing wind, and the energy density of the pulsar's spin-down radiation conspire to make a neutral evaporate unlikely. Radio, optical and infrared observations are shown to suggest that the companion to PSR 1957+20 is bloated, perhaps approaching the size of its Roche lobe. The wind off the companion is most probably a low velocity flow which does not travel far from the orbital plane.

INTRODUCTION

Perhaps the most fundamental question facing those who would explain planets about pulsars is determining the source of the material for these orbiting bodies. If, as Wolszczan and Frail (1992) conjecture, the material is provided by the evaporation of a millisecond pulsar's companion, then PSR 1957+20 provides us with a superb opportunity to study the possible progenitor of a planetary system. Here I examine recent observations which strongly suggest that the ablated wind leaves the surface of PSR 1957+20's companion with a speed substantially smaller than that of the orbit and is restricted to the orbital plane. As shown by recent computer simulations (Tavani and Brookshaw 1992, Banit and Shaham 1992), such a flow may lead to the formation of a disk at a distance of ~ 1 A.U. from the pulsar binary.

DISCUSSION

Continuum observations of the eclipsing pulsar 1957+20 reveal that the evaporative wind off the companion is largely transparent to 20 cm radio waves (Fruchter and Goss 1992) (see Figure I). A plasma density of $\sim 2 \times 10^{10}$ cm^{-3} is capable of reflecting 20 cm radiation, and a density about two orders of magnitude less should cause an eclipse through refraction (Emmering and London

[1]Hubble Fellow

1989), therefore the density of the plasma throughout the region of 20 cm pulsed eclipse must generally be $\lesssim 10^9$ cm^{-3}.

However, the $\sim$ 9 hour orbit of PSR 1957+20 is apparently decaying on a timescale of 3×10^7 years (Ryba and Taylor 1991). In order to produce the observed orbital period derivative, a wind would require a density of

$$\rho_{wind} \approx \frac{10^{12} v_{esc}}{v_{wind}} \mathrm{cm}^{-3}.$$

Furthermore if $v_{wind} \gtrsim v_{esc}$, the escaping material has a ram pressure far in excess of the pulsar wind pressure. The mass loss can not be confined, and would be expected to cross our line of sight, even if the inclination of the orbit on the sky were large. Therefore, to agree with the continuum observations, the wind must either be largely neutral or the evaporating matter must be largely restricted to a region below our line of sight and be travelling relatively slowly. A neutral wind, however, is difficult to imagine given the large orbital velocity, the turbulence implied by rapidly variable pulsar timing delays (Fruchter *et al.* 1990), and the large Poynting flux of the pulsar wind at the distance of the companion (equivalent to a black body temperature of about 8000 K) most of which appears to be intercepted before it reaches the surface of the companion (Fruchter *et al.* 1990).

Many of the difficulties posed by these observations would disappear were the companion nearly as large as its Roche lobe. In this case the velocity necessary to escape the companion would be quite low. Mass loss could be gravitationally restricted to the plane of the orbit and need not interrupt the line of sight to the pulsar. The presence of an excretion disk (Rudnik and Paczynski 1981, Ryba and Taylor 1991), or the tidal interaction of a slow ablated wind with the companion (Banit and Shaham 1992), might then account for the negative orbital period derivative. The plasma we observe may be material entrained into the pulsar wind (Thompson *et al.* 1991) or merely the edge of a much denser wind.

Optical observations provide the color temperature and apparent magnitude of the companion's heated front side. The estimated size of the companion, therefore, scales approximately as the distance to the pulsar (Fruchter *et al.* 1988b, Djorgovski and Evans 1989). At the time of discovery, the dispersion measure of the pulsar, 29.1 cm^{-3} pc (Fruchter *et al.* 1988), was interpreted to imply a distance of 0.8 kpc. A recent reanalysis of pulsar dispersion measures leads to a revised distance to this pulsar of 1.6 kpc, albeit with an error of at least 50% (J. H. Taylor, personal communication). This changes the preferred size of the companion from very nearly equal to the expected size of a $0.025 M_\odot$ degenerate dwarf ($\sim 0.12 R_\odot$), to approximately the size of the companion's Roche lobe ($0.27 R_\odot$). Recent spectroscopic observations and kinematic modelling of the Hα nebula about PSR 1957 by Adlcroft, Romani and Cordes (preprint) appear to support this distance estimate and the suggestion of a bloated companion.

A companion this large, however, will be out of thermal equilibrium, and will collapse to degeneracy over the course of a Kelvin-Helmholtz timescale. However, ground based infrared (Eales *et al.* 1990) and space telescope I band observations (see Figure II) allow us to place a lower limit of on the thermal timescale of the companion equal to or greater than the $\sim 3 \times 10^7$ yr orbital

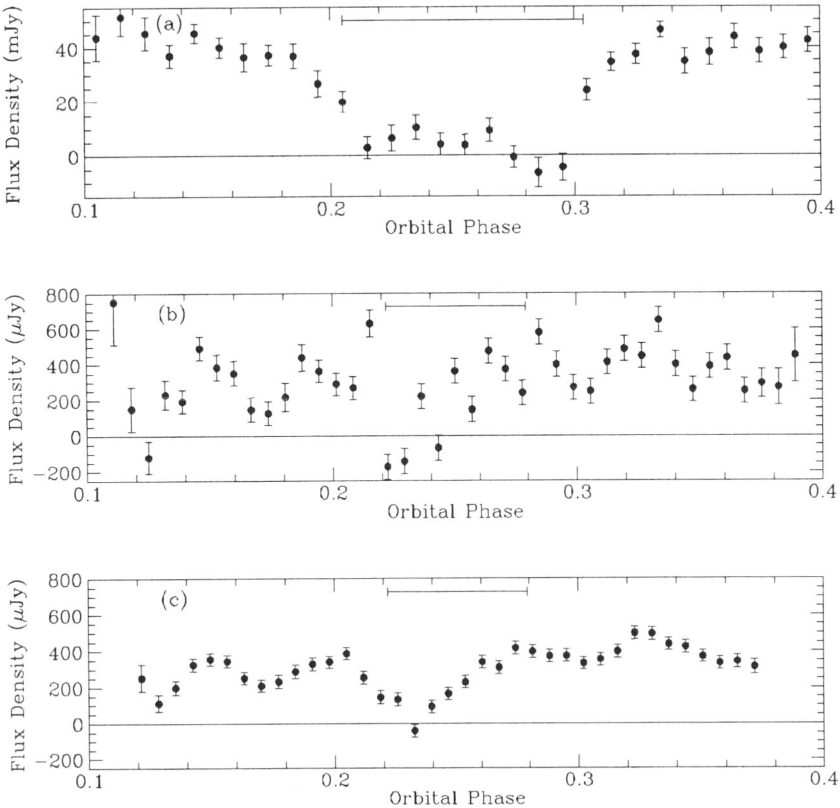

FIGURE I The continuum eclipse of PSR 1957+20 (work done in collaboration with W. M. Goss and T. J. Cornwell using the VLA.) Figure I(a) shows the 90 cm continuum eclipse of PSR 1957+20. The barred horizontal line indicates the typical length of pulsed eclipse at 90 cm as measured at Arecibo (Fruchter *et al.* 1990, Ryba and Taylor 1991). Note that throughout the 90 cm pulsed eclipse, the continuum flux density is considerably attenuated. Figures I(b) and I(c) show the 20 cm continuum eclipse of the pulsar. Figure I(b) contains the raw data, obtained by averaging observations of nine separate eclipses. Figure I(c) shows the same data as I(b),but smoothed with a boxcar 4 bins wide. Here the horizontal bar indicates the length of 20 cm pulsed eclipse (Ryba and Taylor 1991). Although a short asymmetric continuum eclipse may occur at orbital phase $\phi \approx [0.22 - 0.24]$, throughout most of the pulsed eclipse PSR 1957+20 is visible at 20 cm.

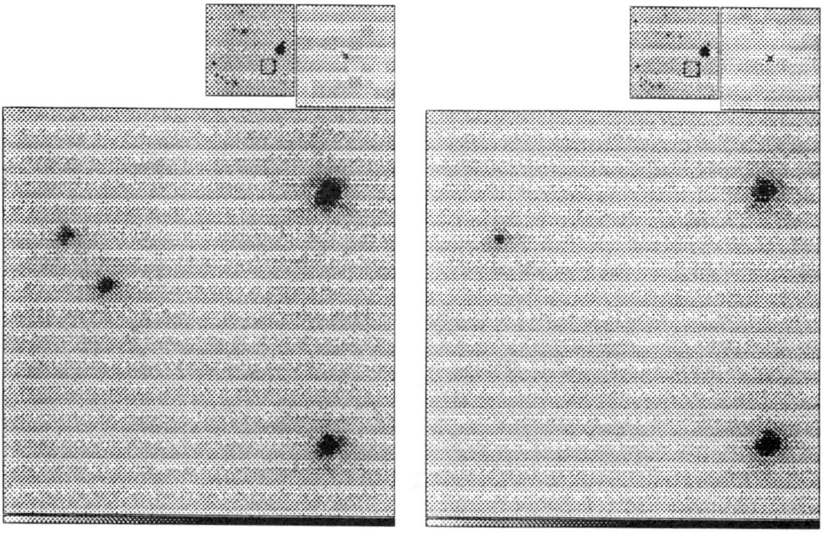

FIGURE II Images of the PSR 1957+20 system obtained with the Hubble Space Telescope in collaboration with Jay Bookbinder and Charles Bailyn. The V band image at left shows the system near optical maximum (orbital phase $\phi = 0.75$) and the I band image at right shows the same field at pulsar eclipse (orbital phase $\phi = 0.25$). While the nearby star to the north-east is a background main sequence dwarf, it has nearly identical colors to the pulsar companion at maximum, and therefore provides a basis for comparing the two images. At optical minimum, the I band magnitude of the companion is > 23.7, implying a surface temperature $\lesssim 2600$ K, and a Kelvin-Helmholtz timescale greater than 3×10^7 yrs.

decay time. Thus the orbital evolution may have been sufficiently rapid to prevent the secondary's collapse, though as Podsiadlowski (1991) has argued in another context, the pulsar spin-down radiation may also act to bloat the companion.

The arguments presented here assume that the measured orbital period derivative reflects the long term evolution of the system. However, we cannot exclude the possibility that the period derivative is non-secular. Large and erratic period derivatives are not unknown among close binaries (cf. Söderhjelm 1980, Hall & Kreiner 1980, Skillman & Patterson 1988). These systems typically display significant variability in orbital decay rate over time-scales of years, but, during the three years of data so far collected, the orbital decay rate of PSR 1957+20 has appeared constant (Ryba and Taylor 1991).

CONCLUSION

Observations of PSR 1957+20 suggest that the outflow off of the companion is massive and restricted to the orbital plane. This conclusion, however, relies upon the assumption that the orbital period derivative provides a true indication of the mass loss rate. Observations planned for the near future will substantially increase the time span over which the orbital evolution of this system has been measured, and thus should signifcantly improve our ability to distinguish between mass loss and non-secular changes in the binary system.

This article reflects the brevity of a poster. Readers interested in a more detailed discussion of some of the issues and observations mentioned here may wish to see Fruchter and Goss (1992).

ACKNOWLEDGMENTS

I thank my collaborators, Charles Bailyn, Jay Bookbinder, Tim Cornwell and Miller Goss for allowing me to present work in progress. Partial support for this work was provided by NASA through grant HF-1005.01-90A awarded by the Space Telescope Science Institute. The Very Large Array of the National Astronomy Observatory is operated by Associated Universities, Inc., under cooperative agreement with the National Science Foundation.

REFERENCES

Aldcroft, T. L, Romani, R. W. & Cordes, J.M., *preprint*

Banit, M. & Shaham, J. 1992, *Ap. J. (letters)*, **388**, L19

Djorgovski,S. & Evans, C. R. 1988, *Ap. J. (Letters)*,**335**, L61

Eales, S. A., Becklin, E. E., Zuckerman, B. & McLean, I. S. 1990, *MNRAS*, **242**, 17P

Emmering, R. T. & London, R. A. 1990, *Ap. J.*, **336**, 589

Fruchter, A. S. 1989, Ph.D. Thesis, Princeton University

Fruchter, A. S., Berman, G., Bower, G., Convery, M., Goss, W. M., Hankins, T. H., Klein, J. R., Nice, D. J., Ryba, M. F., Stinebring, D. R., Taylor, J. H., Thorsett, S. E., and Weisberg, J. M. 1990, *Ap. J.*, **351**, 642

Fruchter, A. S., Gunn, J. G., Lauer, T. R. & Dressler, A. 1988, *Nature*, **334**, 686

Fruchter, A. S., Stinebring, D. R. & Taylor, J. H. 1988, *Nature*, **333**, 237

Podsiadlowskii, Ph. 1991, *Nature* **350**, 136

Ryba, M .F. & Taylor, J. H. 1991, *Ap. J.* **380**, 557

Skillman, D.R. & Patterson, J. 1988, *A. J.* **96**, 976

Söderhjelm, S. 1980, *Ast. Astrophys.* **89**, 100

Tavani, M. & Brookshaw, L. 1992, *Nature*, **356**, 320

Thompson, C., Blandford, R. D, Evans, C. R. & Phinney, E. S., *Preprint*

Wolszczan, A. & Frail, D. A. 1992, *Nature*, **355**, 145

Planets Around Pulsars
ASP Conference Series, Vol. 36, 1993
J.A. Phillips, J.E. Thorsett, and S.R. Kulkarni

THE CIRCUMSTELLAR ENVIRONMENT OF THE GG TAURI MULTIPLE SYSTEM

D. W. KOERNER
Division of Geological and Planetary Sciences 170-25, California Institute of Technology, Pasadena, CA 91125

A. I. SARGENT
Division of Physics, Mathematics, and Astronomy 105-24, California Institute of Technology, Pasadena, CA 91125

S. V. W. BECKWITH
Max-Planck-Institut für Astronomie, Heidelberg, Germany and Department of Astronomy, Cornell University, Ithaca, NY 14853

ABSTRACT

We present $^{13}CO(2\rightarrow1)$ aperture synthesis maps of the circumstellar environment of GG Tau, a very young binary system. An extensive, approximately edge-on, gaseous disk in near-Keplerian rotation is detected. The disk emission is severely affected by the presence of GG Tau/c, a second binary system at a projected distance of 1500 AU from GG Tau and apparently in the plane of the disk.

INTRODUCTION

Most conventional theories of planet formation invoke a viscous circumstellar "accretion disk" of gas and dust as an initial condition (cf. Lynden-Bell & Pringle 1974). Among the many proposed mechanisms for forming planets around pulsars, however, only a few require such a disk (see Podsiadlowski, this volume). Evidence for the existence of a disk around a pulsar-planet system would thus be a powerful discriminant among theories of pulsar planet formation. It is perhaps timely that the feasibility of detecting disk-like structures around very young solar-type stars has now been demonstrated by millimeter-wave interferometry (Sargent 1989, and references therein; Beckwith & Sargent 1992a, and references therein; Sargent & Beckwith 1991). To illustrate the potential of the technique, we present 1.4 mm aperture synthesis observations of the pre-main sequence multiple system, GG Tauri.

GG Tau was originally classified as a classical T Tauri star with a companion, GG Tau/c, detected at an angular separation of $10''.3$ (corresponding to a projected distance of 1500 AU) in the direction PA = 184° (Cohen & Kuhi 1979; Moneti & Zinnecker 1991). Recent near-infrared speckle interferometric measurements have shown that both GG Tau and GG Tau/c are binary systems (Leinert et al. 1991; Ghez 1992). The separations are approximately $0''.3$ and

1″.4 for GG Tau and GG Tau/c, respectively.

The presence of a circumstellar disk around GG Tau was inferred from infrared and submillimeter continuum observations (Strom et al. 1989; Beckwith et al. 1990, hereafter BSCG). The large IRAS beam precluded an independent measurement of any mid-infrared signature of GG Tau/c, and it was not observed in the 1.3 mm continuum survey of BSCG. In the absence of high resolution photometry of the individual components of each binary, it is not clear how much of the excess infrared emission from the four-star system is attributable to the presence of a disk around GG Tau. By contrast, the stellar photospheres contribute negligibly at millimeter wavelengths. Thus, the conclusion that the 1.3 mm emission emanates from a disk around GG Tau rests on relatively solid ground. It is further supported by recent CO(1→0) observations with the Nobeyama Radio Observatory 45-meter telescope (Strom et al. 1992).

Due to limited sensitivity, detailed velocity analysis of the circumstellar disks around low-mass pre-main sequence stars has been carried out for only a few objects such as HL Tau (Sargent & Beckwith 1987; 1991), the strongest source of 1.3 mm continuum emission in the survey of BSCG. Here we present ^{13}CO (2→1) aperture synthesis maps of GG Tau – the second strongest 1.3 mm source in the BSCG sample. The observations suggest an inner disk of radius approximately 750-1000 AU in rotation about GG Tau. In addition, we find gas in near-Keplerian rotation outside the projected radius of GG Tau/c. The symmetric disk morphology and rotational signature are noticeably disrupted in the vicinity of the secondary system. The GG Tau environment thus affords the opportunity to observe both circumbinary gas kinematics and gas motions *interior* to two gravitational potential wells.

OBSERVATIONS

Observations in the ^{13}CO (2→1) transition and λ = 1.4 mm (220 GHz) continuum were carried out between 1990, December and 1991, March, using the Owens Valley millimeter-wave interferometer. Three configurations of the three 10.4 m telescopes, with baselines up to 55m E–W and 80m N–S resulted in a synthesized beam size of 5″.5 × 3″.2. Cryogenically-cooled SIS receivers on each of the telescopes gave system temperatures of 500–1000 K.

Two filter banks, 32 × 1 MHz and 32 × 50 kHz, provided spectral resolutions of 1.4 and 0.07 km s^{-1}, respectively. Continuum measurements were made simultaneously in a 375 MHz broad-band channel. Observations of the strong quasar, 0420–014, were used for phase calibration. Measurements of Uranus, Neptune, and 3C 273 provided absolute flux density calibration. Spectral line and continuum maps were made using the NRAO AIPS software package. Uncertainties in fluxes and positions are estimated to be of order 15% and 0″.5.

RESULTS AND ANALYSIS

Line Emission

Figure Ia shows a map of the ^{13}CO (2→1) line emission integrated over a velocity range -5.0 to 5.9 km s^{-1} relative to the systemic value, V_{lsr}=6.9 km s^{-1}. No

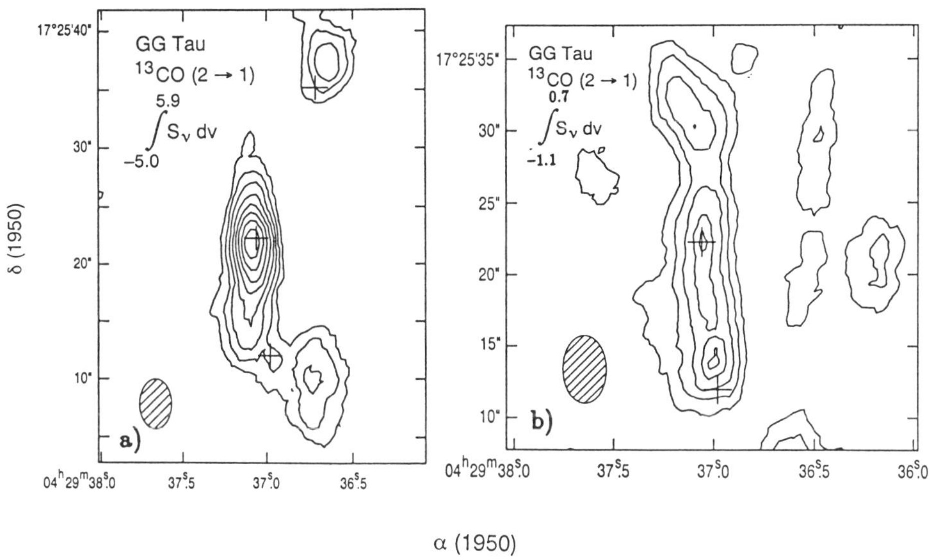

FIGURE I a) A map of the ^{13}CO (2→1) emission from GG Tau integrated over the velocity range -5.0 to 5.9 km s^{-1} (relative to V_{lsr}=6.9 km s^{-1}). The emission peak, 8.8 Jy/beam km s^{-1}, is centered on the star. Contour levels begin at the 2σ level and are spaced by 1σ. The synthesized beam is $5''.5 \times 3''.2$ at PA 0°. The positions of GG Tau and GG Tau/c are marked with crosses. b) A map of the ^{13}CO (2→1) emission from GG Tau integrated over the velocity range -1.1 to 0.7 km s^{-1}. Emission is again centered on the star. Contour levels begin at the 2σ level and are spaced by 1σ.

emission above a 3σ level of 600 mJy is detected outside these channels. The elongated structure seen is centered on the star. It is unresolved in the E-W direction; a Gaussian fit yields a nominal deconvolved source size of $8''.9 \times 1''.7$ at PA = 0.3° and an integrated line flux of 17.9 Jy/beam km s^{-1}. An upper limit on the disk inclination angle of $\sim$ 80° (with respect to the plane of the sky) is then derived assuming the emission morphology arises from a slightly-inclined infinitely-thin disk.

The ^{13}CO disk is evidently extended to the southeast in the vicinity of GG Tau/c. This extension is essentially unresolved; we believe it may be related to the gravitational influence of the GG Tau/c system on the GG Tau disk. A further concentration of gas is present southwest of GG Tau/c, but its relation to the overall system is not yet clear.

The ^{13}CO intensity distribution integrated only over velocities which are close to the systemic value (-0.7 to 1.1 km s^{-1} relative to V_{lsr}=6.9 km s^{-1}) is presented in Figure Ib. This map reveals the location of gas which is either rotating very slowly at large radii or is closer to the star and moving in a direction nearly perpendicular to the line of sight. Because the emission is relatively weak at these velocities, this extended structure is not readily visible in Figure Ia.

The structures seen in Figures Ia and Ib are elongated along slightly different position angles. The position angle of the low-velocity gas (Figure Ib) is in better accord with the direction of separation between GG Tau and GG Tau/c. The northern extension in Figure Ib is quite distinct from the central feature; to the south, the emission is sharply truncated just inside the radial distance of GG Tau/c. If the masses for GG Tau and GG Tau/c are 0.7 and 0.3 $M_{\odot}$, respectively (Strom et al. 1992), the emission "gap" to the north and the truncation point to the south are approximately equidistant from the center of mass of the system. One possible explanation for such a morphology is that an outer disk region has been cleared by GG Tau/c moving in a nearly-circular orbit. Gas should then also be observed in orbital trajectories exterior to the region, but in an asymmetric distribution.

Velocity Analysis

Figure IIa, a map of the mean gas velocity measured in the 1 MHz channels at low spectral resolution, confirms that gas closest to GG Tau is predominantly rotating about an east-west axis. A systematic velocity gradient across the edge-on axis of the disk extends from a mean velocity of nearly 9 km s^{-1} in the north to less than 5 km s^{-1} in the south. Further analysis of the data shows that the velocity dispersion also increases toward GG Tau as would be expected for gas in near-Keplerian rotation. Detailed position-velocity information is, however, available only for velocities within about 1 kms^{-1} of the systemic velocity. The high velocities of the gas orbiting the close binary were not accessible to the 50-kHz filterbank at $\lambda = 1.4$ mm.

A plot of the position of peak ^{13}CO emission relative to the star as a function of velocity is shown in Figure IIb for the *low* velocity gas only. Integrating the emission across a $5''.0$ strip width along the axis of the Figure 3 structure yields the axial intensity as a function of velocity. It should be emphasized that, while emission can be strong at particular velocities, the contribution to the integrated emission can be negligible. At radii of 2000 AU, for example, emission is barely detectable in Figures Ia and Ib. The most readily apparent trend is a velocity gradient indicating rotation in the same sense as in the map of Figure IIa. Emission maxima following this trend are marked by crosses; they are most readily attributable to gas at large radii orbiting the entire system. The systemic velocity is 6.9 km s^{-1} if the position of GG Tau is nearly identical with the barycenter of the system. The rotational center is probably slightly south of GG Tau due to the influence of GG Tau/c, implying a slightly lower systemic velocity (by a few tenths of a km s^{-1}).

Secondary emission peaks which do not fit the trend are also apparent in figure 5. Several, marked by filled circles, appear retrograde to the overall rotational trend. Most occur in the vicinity of GG Tau/c where orbital trajectories are likely to be complicated by the near-field gravitational influence of this secondary binary. The few retrograde emission peaks on the far side of GG Tau may indicate chaotic motion akin to that associated with the L3 Lagrange point in a two-mass potential. Triangles which occur near the GG Tau position over a wide

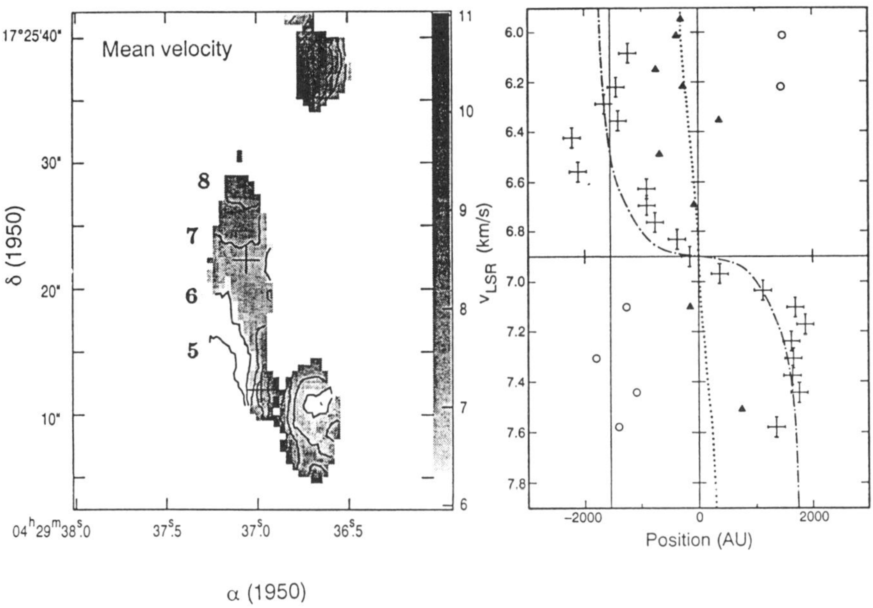

FIGURE II a) A map of the mean velocity as determined by a first moment velocity calculation from the emission in the ^{13}CO (2→1) 1 MHz channels. Contour levels are marked in km s^{-1}. A systematic velocity gradient is evident across the length of the disk. Gas motions in the proximity of GG Tau/c can be seen to disrupt this velocity structure. b) Distances from GG Tau of the peak ^{13}CO (2→1) emission plotted as a function of velocity. Observed peaks which follow a general rotational trend are represented by crosses. Circles mark the peak positions of gas not associated with this trend. Triangles mark the position of peaks which may be associated with an inner disk around the GG Tau system. The dotted and dashed line represents is the locus of peak positions predicted by a model of a Keplerian outer ring at about 2250 AU. The dotted line is that for an inner disk with 750 AU radius. The solid line at -1500 AU indicates the apparent position of GG Tau/c with respect to GG Tau.

velocity range are maxima which probably result from high-velocity, inner-disk gas moving orthogonally to the line of sight.

Frequently, two or more emission peaks occur at a specific velocity in Figure IIb, one close to GG Tau and another located on the axis of the ^{13}CO emission. This suggests two distinct gas distributions at different radii. Model rotation curves for an inner, circumbinary disk (the dotted line) around GG Tau and a circum-quadruple outer ring (dotted and dashed line) are also presented in Figure IIb; the system configuration for these models is illustrated in Figure III. The inner disk was modeled by convolving the emission predicted by the edge-on optically-thick model of Beckwith & Sargent (1992b) with the synthesized beamwidth and with a molecular linewidth of 0.3 km s^{-1}. Gas was assumed to have $\tau_\nu \geq 1$ out to a 750 AU radius and to be in Keplerian rotation about a 0.7 $M_\odot$ central mass. The outer ring model consists of a similarly-convolved distribution of the emission coming from rotating gas with a Gaussian radial distribution centered at 2250 AU. It is truncated at inner and outer radii of 1750 AU and 2750 AU, respectively.

While the peak positions in Figure IIb appear to fit the outer ring model fairly well, we caution that the actual emission intensities exhibit a fair degree of scatter. In addition, it can be seen that deviations from the model of an inner circumbinary disk require that conclusions about a signature of Keplerian rotation remain tentative at that scale. We thus caution that the system configuration illustrated in Figure III is primarily meant to represent the models shown in Figure IIB; the data also show additional complicated behaviour that is not yet understood. Details in the figure, such as the uniform nature of the outer ring or the sharp truncation at gap boundaries, are not meant to be interpreted literally.

DISCUSSION AND SUMMARY

The role of accretion disks in binary formation is currently the subject of much theoretical work (e.g. Mathieu 1991; Clarke & Pringle 1991; Artymowicz, et al. 1991). Only recently, however, has the distribution and kinematics of interstellar and circumstellar material in multiple systems been observable on relevant scales. A circumbinary disk has been imaged for T Tau (Weintraub, Masson, & Zuckerman 1989); chemical gradients between the two components of IRAS 16293-2422 have been observed (Mundy et al. 1991); and spectral energy distributions diagnostic of disks have been identified around many close binaries (Beckwith & Sargent 1991). However, the possibility of studying kinematic motions of circumstellar material both around close binaries and in the region *between* well-separated members of an apparent pair is thus far unique to the GG Tau multiple system.

The orientation of the GG Tau disk along the projected direction to GG Tau/c, and the disruption of symmetry and Keplerian rotation in that vicinity, strongly support the view that the apparent nearness of GG Tau/c to GG Tau is a consequence of it being gravitationally bound to the system rather than the result of chance projection. The association of GG Tau/c with the disk may be further construed as supporting evidence that the well-separated components of the system formed coevally rather than by capture. In such a picture, the

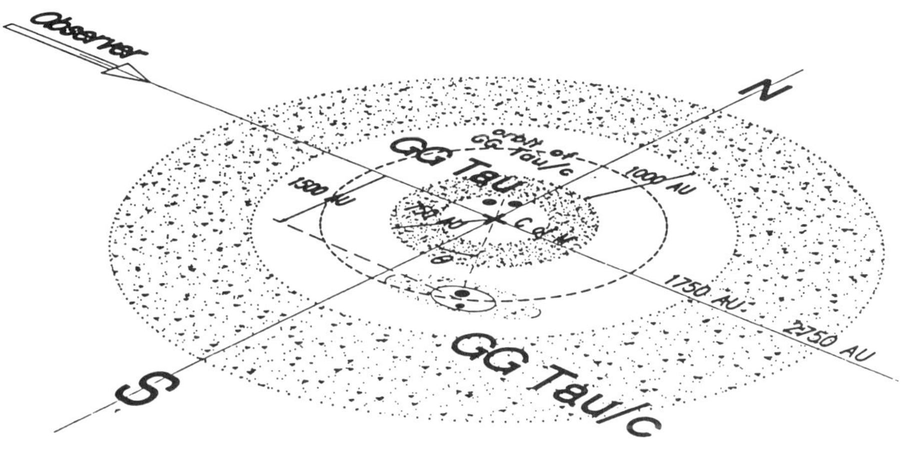

FIGURE III Schematic illustration of a model configuration for the GG Tau quadruple system. It was used to generate synthetic position-velocity curves in Figure IIb. The viewing angle is different from that of the maps in Figures I and II; comparison can be made by noting the "Observer" position.

planar extent of initially collapsing material must be seen to be an order of magnitude larger than is commonly imagined in the customary single-star formation scenario. The radial extent and influence of circumstellar disks has often been assumed to be confined to size scales like that of our own solar system, i.e., of order 100 AU; but the extensive distribution of material around the GG Tau system argues for consideration of a much larger scale in the early stages of star formation.

The motion of gas in such a many-body system is necessarily complex and chaotic. Differences from the commonly-accepted picture of a viscous accretion disk around a single star are to be expected and are apparent in the GG Tau system. Such differences include radial density and temperature distributions which do *not* vary smoothly as a power law, azimuthal asymmetry, and non-Keplerian rotation rates around the center of mass. The analysis of such deviations should help constrain the mass-values and orbital parameters of the multiple components and, in general, inform the effort now underway to understand the dynamical evolution of multiple star-disk systems.

ACKNOWLEDGMENTS

We are indebted to the staff of Owens Valley Radio Observatory for their unstinting assistance in acquiring the observations described here. We thank D. Stevenson for illuminating discussions. This work was supported by NSF Grant AST-9016404.

REFERENCES

Artymowicz, P., Clarke, C.J., Lubow, S. H., & Pringle, J. E. 1991, *ApJ*, **370**, L35.

Beckwith, S. V. W. & Sargent, A. I. 1991, *ApJ*, **381**, 250.

Beckwith, S. V. W. & Sargent, A. I. 1992b, *ApJ*, in press.

Beckwith, S. V. W. & Sargent, A. I. 1992a, in *Protostars and Planets III*, ed. G. Levy, J. Lunine, & M. S. Matthews (Tucson:University of Arizona Press), in press.

Beckwith, S. V. W., Sargent, A. I., Chini, R. S., & Güsten R. 1990, *AJ*, **99**, 924.

Clarke C. J., & Pringle J. E. 1991, *MNRAS*, **249**, 588.

Cohen, M., & Kuhi, L. V. 1979, *ApJS*, **41**, 743.

Ghez, A. 1992, Ph.D. Dissertation, California Institute of Technology, in preparation.

Leinert, Ch. Haas, M., Richichi, A., Zinnecker, H., & Mundt, R. 1991 *A&A*, **250**, 407.

Lynden-Bell, D., & Pringle, J. E. 1974, *MNRAS*, **168**, 603.

Mathieu, R. D. 1992, in *IAU Symposium 151, Interacting Binary Stars*, ed. Y.Kondo, R. Sistero, & R. Polidan (Kluwer, Dordrecht), in press.

Moneti, A., & Zinnecker, H. 1991 *A&A*, **242**, 428.

Mundy, L. G., Wootten, H. A., Wilking, B. A., Blake, G. A., Sargent, A. I. 1992, *ApJ*, **385**, 306.

Ohashi, N., Kawabe, R., Hayashi, M., & Ishiguro, M. 1991, *AJ*, **102**, 2054.

Sargent, A. I. 1989, in *The Formation & Evolution of Planetary Systems*, ed. H. A. Weaver & L. Danly (Cambridge:Cambridge University Press), 111.

Sargent, A. I. & Beckwith, S. V. W. 1987, *ApJ*, **323**, 294.

Sargent, A. I. & Beckwith, S. V. W. 1991, *ApJ*, **382**, L31.

Strom, K. M., Strom, S. E., Edwards, S., Cabrit, S., & Skrutskie, M. F. 1989, *AJ*, **97**, 1451.

Strom, S. E., Skrutskie, M., Snell, R., & Fukui Y. 1992, in preparation.

Weintraub, D. A., Masson C. R., & Zuckerman B. 1989, *ApJ*, **344**, 915.

Planets Around Pulsars
ASP Conference Series, Vol. 36, 1993
J.A. Phillips, J.E. Thorsett, and S.R. Kulkarni

ABLATING MILLISECOND BINARY PULSARS: PROGENITORS OF PULSAR PLANETARY SYSTEMS?

DAVID J. NICE
Joseph Henry Laboratories and Physics Department, Princeton University, Princeton, New Jersey 08544

STEPHEN E. THORSETT
Owens Valley Radio Observatory, 105-24, California Institute of Technology, Pasadena, California 91125

INTRODUCTION

Millisecond pulsars are formed when mass and angular momentum are transferred to a slowly spinning neutron star from a binary companion. Often this companion star or its remnant is still found in a binary system with the millisecond pulsar. In a few cases, millisecond pulsars are found alone, having lost their companions after spin-up to either supernova explosions or pulsar-wind-driven evaporation.

ECLIPSING PULSAR BINARIES

There are two known systems in which a millisecond pulsar is currently ablating its companion, PSRs 1957+20 and 1744−24A. Both are eclipsing binaries in which the eclipses last too long to be explained solely by a companion confined within its Roche surface. The eclipses are caused by material evaporated off of the companion's surface by the relativistic pulsar wind. In the case of PSR 1957+20, the heating of the companion's surface to several thousand degrees on the side nearer the pulsar can be directly measured by optical observations (Kulkarni *et al.* 1988). Because the ablated material, once free of the Roche surface, is no longer gravitationally bound to the companion star, there must be a continual flux of matter from the companion's surface. The orbital period of the PSR 1257+20 system is known to be decaying with a time scale of about 30 million years (Ryba & Taylor 1991). While uncertainties about the specific angular momentum carried away by the ablated material remain, it is likely that the ablation time scale is similar.

The pulsar observed to have a planetary system, PSR 1257+12 (Wolszczan & Frail 1992), is a millisecond pulsar, so it must haved been spun-up via mass transfer from a stellar companion which is no longer present. If this companion was destroyed by an evaporation process like that occuring in the ablating binaries, then a natural question arises: were the pulsar's planets formed as a consequence of the ablation of the companion? Tavani and Brookshaw (1992) have suggested the planets could indeed have formed during such a process. They argue that during ablation, some of the material may remain gravitation-

TABLE I Binary System Parameters. (All figures assume an edge-on orbit.)

Pulsar Name	1744−24A[a]	1957+20[b]	1257+12[c]	
Pulsar Period (ms)	11.563	1.607	6.218	
Pulsar Period Deriv.	[d]	1.61×10^{-20}	1.21×10^{-19}	
Spin-down Age (Myr)	[d]	1600	800	
Luminosity (erg/s)	[d]	1.5×10^{35}	2.0×10^{34}	
Psr. Orb. Radius (lt-s)	0.120	0.089	0.00131	0.00141
Orb. Period (days)	0.076	0.382	66.6	98.2
Orb. P. Deriv. (10^{-12})	0 ± 5	-40 ± 10		
Orb. Time Scale (Myr)	> 40	28		
Companion Mass [m_2]	$0.089 M_\odot$	$0.022 M_\odot$	$3.4 M_\oplus$	$2.9 M_\oplus$
Comp. Orb. Rad. (lt-s)	1.89	5.68	183	232
Ang. Mom. (erg s) [L]	5.5×10^{50}	2.4×10^{50}	6.6×10^{47}	6.1×10^{47}
L/m_2 (g^{-1}erg s)	3.1×10^{18}	5.5×10^{18}	3.3×10^{19}	3.6×10^{19}

[a]Parameters from Nice & Thorsett (1992). [b]Parameters from Ryba & Taylor (1991).
[c]Parameters from Wolszczan & Frail (1992).
[d]Intrinsic period derivative is not measurable (due to external acceleration effects), so spin-down time scale and luminosity are not known.

ally bound in an excretion disk, where dust could form and planetesimals could condense on a $\sim 10^6$ yr time scale, producing the observed planets. Rapid mass loss in the late stages of companion evaporation could sweep the remaining gas and dust out of the system.

Regardless of the detailed model, ablating-companion precursors to pulsar planetary systems are attractive. In table I we list parameters of the known ablating systems and PSR 1257+21 and its planets. Several facts encourage speculation about connections between the two types of systems: even at low conversion efficiency, the companions contain sufficient mass to form planets; while the planets have more specific angular momentum than the ablating companions, it is not so much more as to require exotic angular momentum transfer mechanisms; excretion disk friction can explain large negative period derivative of the PSR 1957+20 orbit (Ryba and Taylor 1991); and the ablation time scale is much shorter than the pulsar lifetime, so post-ablation systems should be relatively common.

TESTING FOR PLANETS

Only three millisecond pulsars without stellar mass companions are known outside of globular clusters. Of these, one has a planetary system. This suggests that planet formation is not rare, unless we have been extraordinarily lucky. Since the ablating systems may be precursors to isolated and planetary pulsars, we might hope to see planetary formation occuring around the ablating systems at the present time. Both known systems are sufficiently old that planets could have already formed.

To test this hypothesis, we have analyzed pulse arrival time series for each of these two systems, searching for periodicities indicative of planetary companions. The observations are discussed in Ryba & Taylor (1991) and Nice & Thorsett (1992). Details of the analysis method are given in Thorsett and Phillips (1992). Results are given in figures I and II. From the figures, we can deduce that the PSR 1957+20 system has no planetary companions with period under one year and mass greater than $0.015M_{\oplus}$ and the PSR 1744−24A system has no companions with period under one year and mass greater than $0.5M_{\oplus}$. The relatively short data sets available for these pulsars preclude strong limits at periods greater than a few years.

CONCLUSION

The data show that planets are not present in either of the known ablating systems. For PSR 1957+20, it is possible, as Tavani and Brookshaw have suggested, that the pulsar spin-down luminosity, an order of magnitude larger than that of PSR 1257+12, is sufficient to blow all the ablated material out of the system. Note however that this contradicts the suggestion that the orbital period decay of this system is due to braking by an excretion disk. In the case of PSR 1744−24, the pulsar spin-down energy flux is not known, but is unlikely to be higher than that of PSR 1257+12.

While we have found that planets have not already formed around the ablating binary systems, we cannot rule out a connection between these binaries and pulsar planetary systems. It is possible that ablating systems with lower pulsar luminosities form planets during an ablation process. It is also possible that undetectable low-mass planetesimals have formed around the ablating systems, and full-fledged planets will form in the future.

ACKNOWLEDGMENTS

The PSR 1744-24A data were collected at the VLA and the 140′ telescope at Green Bank; both are facilities of the National Radio Astronomy Observatory, operated by Associated Universities, Inc., under contract with the NSF. The PSR 1957+20 data were collected by J. H. Taylor and M. F. Ryba using the Arecibo Observatory, part of the National Astronomy and Ionosphere Center, operated by Cornell University under contract with the NSF.

REFERENCES

Kulkarni, S. R., Djorgovski, S. G., and Fruchter, A. S. 1988, *Nature*, **334**, 504.
Nice, D. J., and Thorsett, S. E. 1992, *ApJ*, in press.
Ryba, M. F., and Taylor, J. H. 1991, *ApJ*, **380**, 557.
Tavani, M., and Brookshaw, L. 1992, *Nature*, **356**, 320.
Thorsett, S. E., and Phillips, J. A. 1992, *ApJ*, **387**, L69.
Wolszczan, A., and Frail, D. A. 1992, *Nature*, **355**, 145.

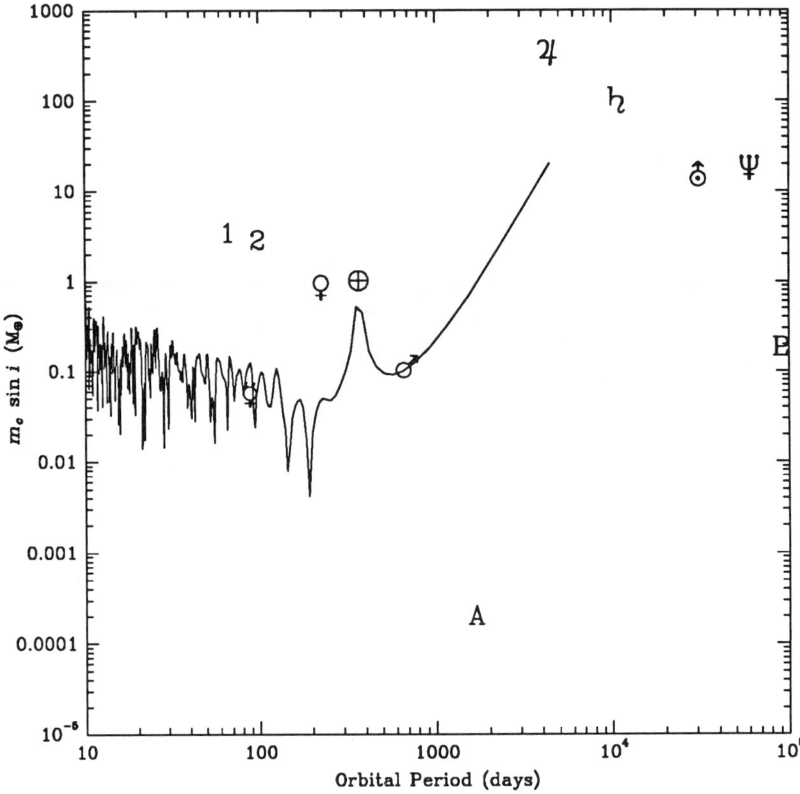

FIGURE I Limits on planetary companions of PSR 1744−24A. Allowed companion masses at various trial binary periods fall below the line; the region above the line is excluded by timing measurements. Indicated for comparison are the nine planets of the Solar System, the largest asteroid ("A"), and the two planets reported by Wolszczan and Frail (1992) ("1" and "2"). More discussion of the features of this plot (e.g., the loss of sensitivity near periods of one year) can be found in Thorsett and Phillips (1992).

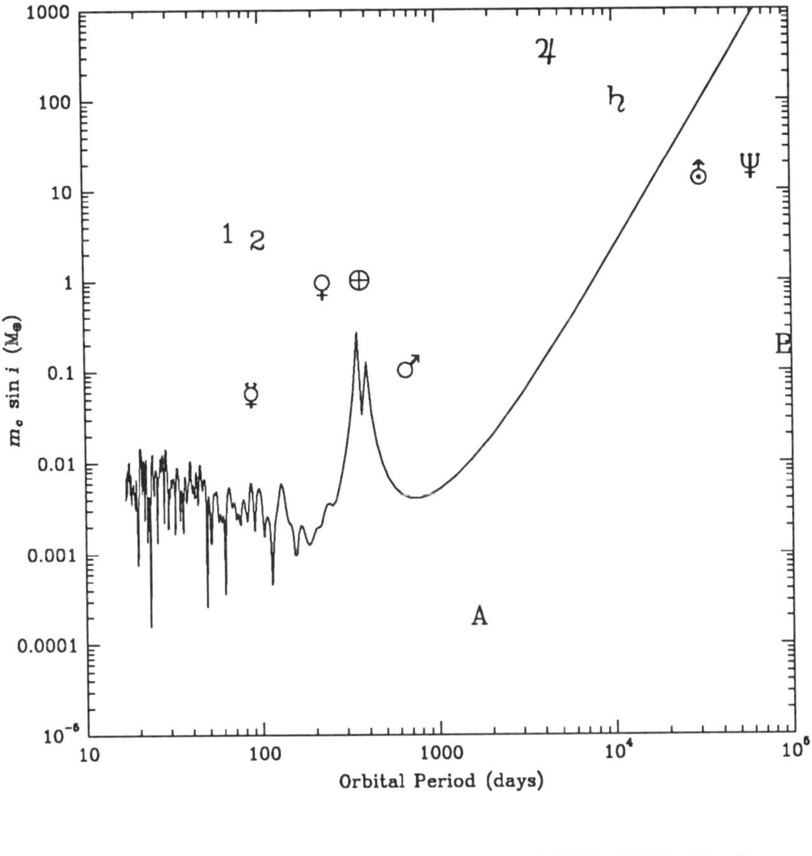

FIGURE II Limits on planetary companions of PSR 1957+20. See caption to figure I.

Planets Around Pulsars
ASP Conference Series, Vol. 36, 1993
J.A. Phillips, J.E. Thorsett, and S.R. Kulkarni

THE EFFECTS OF JOVIAN PLANETS ON SiO MASERS IN LATE-TYPE GIANT STARS

CURTIS STRUCK-MARCELL, Iowa State University, Dept. of Physics and Astronomy, Ames, IA 50011

ABSTRACT Jovian planets would have a significant effect on the winds in Mira-type AGB stars, and on SiO masers which are found in many of these stars. The masers in turn might serve as a beacon for discovering such planets.

INTRODUCTION

It seems clear from results presented at this conference that planets around pulsars are very rare creatures. However, we do not have a very complete understanding of the formation of planetary systems (but see the papers of Levy, Lissauer, and Strom at this conference) around pre-main sequence and main sequence stars, and these may be much more common. Once formed such systems may be retained, at least in part, until very near the end of the star's life. Several years ago I pointed out that the presence of Jovian planets, with orbits similar to those in our solar system, might have a significant effect on the strong stellar winds in Mira-type AGB stars (Struck-Marcell 1988). Moreover, the coincidence of the orbital radii of these hypothetical planets with the radii of SiO maser sources suggests that such planets might effect these sources, and possibly help understand their strong polarization. The masers might be used in turn as bright markers to help discover the planets. The purpose of this paper is both to bring this hypothesis to the attention of workers in other areas who might not be aware of it, and to review developments since the original paper.

THE STRUCTURE OF LPV WINDS

We begin with a little necessary background. All (isolated) stars of low to intermediate mass probably evolve to become pulsating, Mira-type long period variable stars (LPVs). There is a good deal of observational data available, allowing us to put quite good constraints on the basic properties of Mira winds. In addition, Bowen (1988a,b 1990) has constructed hydrodynamical models, based on these constraints and stellar evolutionary models, which show the structure of the spherically symmetric winds in detail.

In all of these models a nearly constant velocity wind is formed within a radius of about a few astronomical units (a.u.). For constant outflow velocity the continuity equation requires that the density in the wind decrease as $\rho \propto r^{-2}$. In the models that include an approximate treatment of the formation of dust grains, radiation pressure on grains is an important momentum source for the wind, which reaches terminal velocities of order 10 km/s. The grains also thermally couple the gas to the radiation field, so the gas temperature tends to be close to the local radiative equilibrium value. Typical gas densities at radii of 5-10 a.u. are of order 10^{-16} - 10^{-17} g/cm^3. Dust-free models, which may represent the "warmest" Miras, have lower wind velocities, i.e. about 1-2 km/s. An especially interesting feature of these dust-free models is that they have an acoustically heated "chromosphere" or "calorisphere" with temperatures of order 6000-8000 K at $r \approx 3$ a.u., and declining as roughly r^{-1} outwards. This quick summary is sufficient for our purposes; the reader is referred to Bowen's paper for further details.

SiO MASER PROPERTIES AND MYSTERIES

Many late-type giant stars show maser emission from some or all of the molecules: H_2O, OH, and SiO. Because of the higher dissociation energy of the SiO molecules these masers tend to be located closest to the star. Radio interferometry has been carried out on a few stars (McIntosh et al. 1989, Lane 1984, Colomer et al. 1992). The following basic properties have been derived from these and other observations.

- Typically, the sources are formed in only a few groups around each star, with a number of maser spots in each group.
- Sources are located within a volume of radius $\lesssim 10^{14}$ cm $\approx$ 5 a.u. $\approx$ 5 R_*, where R_* is the photospheric radius. An exciting new discovery is that the source size has now been observed to be much smaller than this radius, e.g. 10^{12} - 10^{13} cm. (Colomer et al. 1992). In order to achieve the observed maser gain these small clumps must either be very dense, or contain an anamolously high abundance of SiO (see discussion in Alcock and Ross 1986).
- Total line widths are of order 10-20 km/s, with typically several discrete features. Another recent discovery is that there is only one narrow feature, located at the stellar velocity, in spectra of ^{29}SiO and ^{30}SiO (Alcolea and Bujurral 1992). These line widths roughly equal to the circular velocities at these radii, but they are also comparable to the wind velocity. They are considerably larger than the shock jumps at these radii, according to the models.
- High linear polarization is common in the maser lines.
- Surprisingly high circular polarization has been discovered in some stars (Barvainis, McIntosh, and Predmore 1987).
- Maser intensities vary with stellar pulsation period, with little phase delay relative to the IR continuum.
- Velocity width and centroid do not vary with pulsation period, but the latter drifts over a longer timescale (Nyman and Olofsson 1986).

There are many mysteries about these masers. The excitation mechanism is not well understood. The densities in the extended stellar atmosphere are very low for collisional excitation. It would seem that the most probable sites for the masers are the post-shock regions, but the pulsational shocks are spherically symmetric, the masers are not.

The polarization properties have led to the suggestion that there are strong magnetic fields (≥ 1 Gauss) in the SiO maser zone. *How such strong fields could be generated or maintained in the extended stellar atmosphere is a mystery!*

JOVIAN PLANETS AND LPV WINDS

The presence of Jovian planets would have a significant effect on an LPV wind. Realistic simulations have not yet been carried out, but the magnitude of the effect can be seen with the aid of even very simple ballistic calculations. Figure 1 shows two simulations in which the wind "particles" move with a constant outward radial velocity, except for the perturbing force of a Jupiter mass planet. I.e., assuming that periodic shock or grain momentum sources reach a state of balance with gravity as in Bowen's simulations. It is further assumed that the planet is on a circular orbit of radius 5.0 a.u. around a 1.0 solar mass star. Each graph shows groups of wind particles at two times; before and after closest approach to the planet. The planet's motion between these two times is also shown. A clear spiral wake is formed in both the case of a slow (e.g. dust-free) wind and a fast (dusty) wind.

In a true hydrodynamical model pressure effects would be very important in the wake. It would not be a gap cleared of wind particles (like the gaps in Saturn's rings) with overdense edges. Rather, the wake would be a shock interaction zone. On a very small scale an MHD bow shock and Magneto-tail would be expected if the planet retained a magnetic field comparable to Jupiter's. The wake would still have a spiral shape with an opening angle that depends on the wind velocity relative to the planet's velocity. However, because of the dissipative forces, the density contrast between the wake region and its surroundings would decrease fairly rapidly with distance from the planet.

PLANETS AND SiO MASERS

If the solar system is typical, a Jovian planet system is likely to be found in the same region as the SiO masers. There are several ways that planetary perturbations acting on the wind might to understand the maser observations.

- An association with planets might explain why maser spots are so small. It is hard to think of many sources of inhomogeneity on the scales of the maser sources. Thermal instabilities would likely have a very small characteristic scale, rotational modulation of the flow would be on a larger scale. The wind might be perturbed on a comparable angular scale by large convective cells in the photosphere, but how this would lead to the development of maser spots at much larger radii is not clear.
- Density variations in a planetary wake could produce regions of relatively high density, especially in the vicinity of interacting shocks.

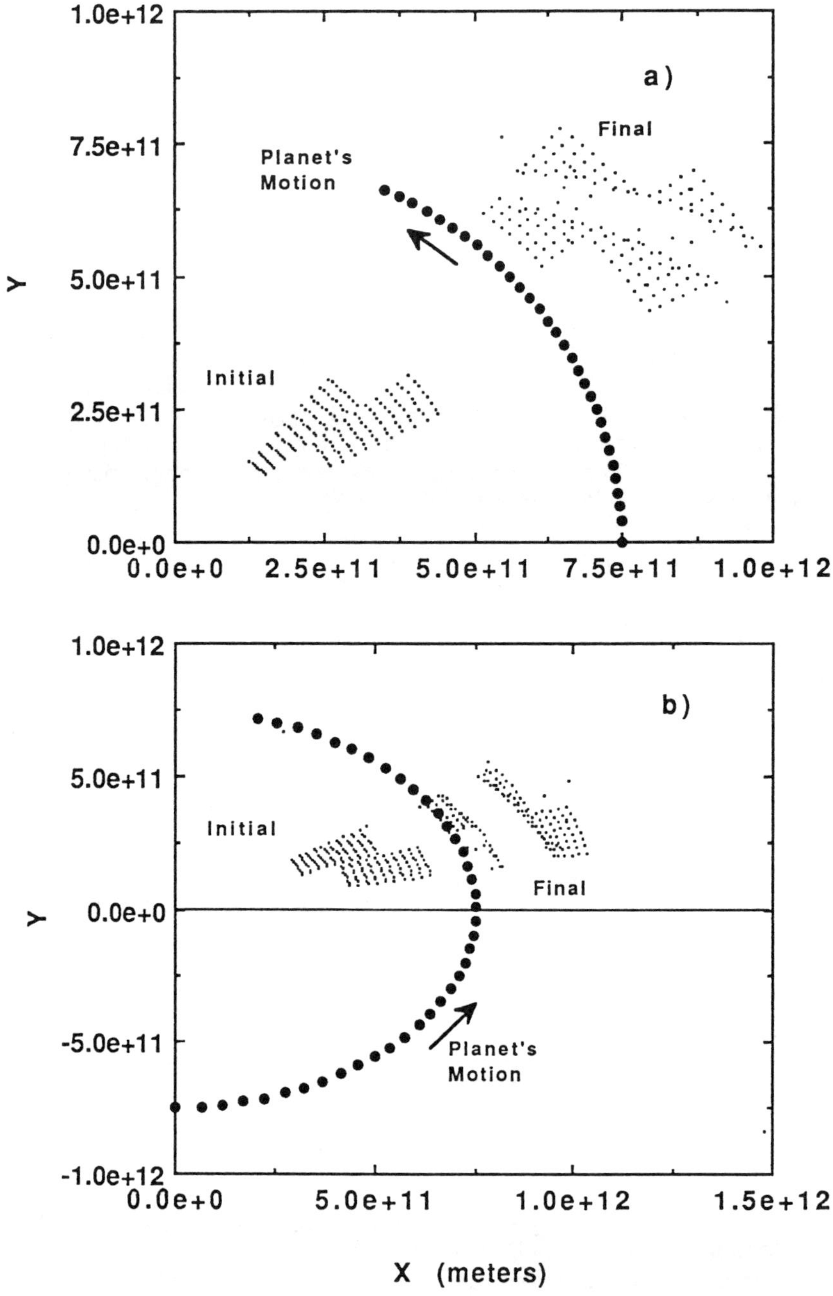

Fig. 1. Ballistic wind particles are shown before and after encountering the gravitational perturbation of a Jovian planet. In a) the particles have an outward radial velocity of 10 km/s, in b) the wind velocity is 2 km/s. See text for details.

- Anisotropies in the flow induced by the wake would help understand the observed polarization.
- Strong planetary magnetic fields and Magnetotails would help understand both the linear and the circular polarization.
- A more speculative possibility is that the vaporization of Galilean-type moons would dump large amounts of SiO into the magnetosphere and wake (see Struck-Marcell 1988).
- Another intriguing speculation is the suggestion of Sahai et al. (1991) that the equatorial torii inferred in some proto-planetary nebulae "may result from 'born-again' disks formed through the destruction of planetary systems at the end of the AGB evolutionary phase". These torii help collimate the bipolar outflows observed in such objects.

On the other hand, there are some apparent inconsistencies in the planet-maser interaction hypothesis.

- SiO masers are very common in Mira variables, it's hard to believe that all of them are associated with planets. But the maser stars with high circular polarizations seem to be less common, maybe they're the only ones associated with planets.
- Jovian planetary perturbations are not all that strong, so it is not clear that they could perturb the wind flow enough to account for the maser spots. On the other hand, the average "Jupiter" may be more massive than Jupiter, yielding a stronger perturbation.
- The two published maps of the star R Cas (see McIntosh et al. 1989 and refs. therein) have the same number of maser groups, and these groups moved between 1979 and 1985. However, the motions do not seem consistent with simple planetary orbits, unless they are not all in the same sense. Yet if the maser spots were transient features in a turbulent wake, or were the result of variable radiative transfer effects in the wake, they would not necessarily be tied to the orbital motions of the planets.

UNANSWERED QUESTIONS

At present there are very few answered questions in this subject! No detailed modeling has been carried out yet on planets in a Mira wind. The hydrodynamics is likely to be terribly complex. Pulsational shocks will interact with the bow shock and wake. The flows are transonic to mildly supersonic, so turbulence is likely. This makes the radiative transfer in the maser lines very complex. Moreover, if SiO emission plays an important role in cooling the wake, then the transfer and the hydrodynamics will be coupled by thermal effects.

Much observational work is needed to constrain all of the maser models. Continued VLBI monitoring is especially important for testing the planets hypothesis. Optical interferometry of nearby Miras also looks very promising (S. Kulkarni, private communication). *In any case, the possibility that SiO masers may prove to be a luminous tracer of Jovian systems is a strong motivation to persue these studies.*

REFERENCES

Alcock, C., and Ross, R. R. 1986, ApJ, 310, 838

Alcolea, J., and Bujarrabal, V. 1992, A&A, 253, 475

Barvainis, R., McIntosh, G., and Predmore, C. R. 1987, Nature, 329, 613

Bowen, G. H. 1988a, ApJ 329, 299

________ 1988b, in Pulsation and Mass Loss in Stars, ed. R. Stalio & L. A. Willson (Dordrecht: Kluwer), 3

________ 1990, in The Numerical Modelling of Nonlinear Stellar Pulsations, ed. J. R. Buchler (Dordrecht: Kluwer), 155

Colomer, F., Graham, D. A., Krichbaum, T. P., Ronnang, B. O., de Vicente, P., Witzel, A., Barcia, A., Baudry, A., Booth, R. S., Gomez-Gonzalez, J., Alcolea, J., and Daigne, G. 1992, A&A, 254, L17

Lane, A. P. 1984, in IAU Sym. 110, VLBI and Compact Radio Sources, ed. R. Fanti, K. Kellerman, and G. Setti (Dordrecht: Reidel), 329

McIntosh, G. C., Predmore, C. R., Moran, J. M., Greenhill, L. J., Rogers, A. E. E., and Barvainis, R. 1989, ApJ, 337, 934

Nyman, L.-A., and Olofsson, H. 1986, A&A, 158, 67

Sahai, R., Wootten, A., Schwarz, H. E., and Clegg, R. E. S. 1991, A&A, 251, 560

Struck-Marcell, C. 1988, ApJ, 330, 986

Section 6:

Comets, Planetesimals, and Debris Disks

Planets Around Pulsars
ASP Conference Series, Vol. 36, 1993
J.A. Phillips, J.E. Thorsett, and S.R. Kulkarni

DUSTY DEBRIS CLOUDS AROUND MAIN SEQUENCE AND POST-MAIN SEQUENCE STARS

B. ZUCKERMAN
Astronomy Department, UCLA, Los Angeles, CA 90024

ABSTRACT During the past decade, infrared observations by IRAS and from the ground have revealed that many stars are orbited by dusty debris disks. Somewhat more indirect arguments indicate that many of these systems also contain asteroids, comets, and/or planets, thereby suggesting that planetary systems may be quite common in the Milky Way. Dust clouds at the main sequence K5 star HD 98800 and the white dwarf Giclas 29-38 are particularly noteworthy and mysterious.

1. INTRODUCTION

This is a brief review of evidence for particulate matter in orbit around main sequence and post-main sequence stars and the implications of such matter for the existence of comets, asteroids and planets. Many of the objects that we shall meet comprise a zoo of strange creatures. The essentials of the review may be summarized as follows:

A) Vega-like main sequence stars: primordial or present day debris?
B) From red giants to planetary nebulae: too many binaries?!
C) Dust particles around white dwarfs: Giclas 29-38, still a mystery 5 years later.
D) Dust near PSR1257+12?

2. DUST TOWARD MAIN SEQUENCE STARS

Until 1983, stars were the only objects that were known with certainty to orbit stars other than the Sun. Then IRAS discovered that many nearby main sequence stars, for example, Vega, emit 25 to 100 μm radiation well in excess of that which would be expected from their photospheres (Aumann et al 1984). The excess emission is due to dust particles that orbit ~ 50 AU from the stars. At about the same time, ground-based infrared astronomers discovered dusty material orbiting pre-main sequence stars (discussed by Steve Strom in this volume). Subsequently, the dusty, Vega-like stars were investigated by many researchers, notably Backman, Gillett, and Aumann (e.g., Backman & Gillett 1987, Aumann 1988). The consensus seems to be that of order 15% or more of A, F, G, and K-type main sequence stars have orbiting debris disks. It is possible that more sensitive satellite systems, such as ISO, could reveal that most main sequence stars are orbited by significant quantities of particulate matter.

Beta Pictoris is the most studied of the Vega-like stars because, as of May 1992, it is the only one for which orbiting dust and gas have been detected at ultraviolet, visible, and near-infrared wavelengths (e.g., Smith & Terrile 1984, Lagrange-Henri et al 1992, Beust et al 1991, Backman et al 1992, Telesco et al 1988).

In this section we consider four questions:

i) How does the mass of orbiting particulate matter vary with time?
ii) Are the grains detected by IRAS "primordial" in the sense that they are remnants of the star formation process or are they replenished during the main sequence life span of a star?
iii) Are there debris disks around stars in binary systems?
iv) Are debris disks related to the "Kuiper cometary belt" in our solar system?

Concerning question (i), the submillimeter is probably the best spectral region to observe to obtain the relative mass of dust that orbits stars of various ages. The submillimeter flux is not very sensitive to the dust temperature and all grains, hot and cold, contribute to the optically thin emission (e.g., Sopka et al 1985). This is not the case near 10 μm wavelength or even at IRAS wavelengths. At 800 μm wavelength, probably optimum for ground-based measurements, a 15-m antenna, such as the JCMT at Mauna Kea Observatory, has a 17" beam (FWHP) which is much smaller than the IRAS beam sizes. So debris clouds at nearby stars, such as Vega and Fomalhaut, can be spatially resolved (Zuckerman and Becklin 1992a, hereafter ZB92a).

Figure 1 displays dust mass versus time based primarily on 800 μm fluxes. In comparison with the mass of dust particles near T Tauri and Herbig Ae stars at 3×10^6 years, Figure 1 indicates a decline in dust mass during the subsequent 3×10^8 years that is at least as rapid as $(\text{time})^{-2}$. A continued decline throughout a star's main sequence lifetime and beyond, at least for A-type stars, seems plausible based on the general absence in the IRAS data base of excess 60 μm emission toward G-type giant stars of luminosity class III (Jura 1990). An alternative possibility, very rapid decline in dust mass during the first few x 10^7 years or so followed by roughly constant dust mass thereafter, is consistent with Figure 1 but is difficult to reconcile with statistics summarized by Jura (1990). Such a rapid decline might occur, for example, when the dusty disks first become optically thin, turbulent convection ceases, and small dust particles are able to accumulate into larger objects that have decreased opacities per unit mass at 800 μm (Ruden & Pollack 1991).

Concerning question (ii), very small grains are blown away from a star by radiation pressure, while somewhat larger grains spiral in as a result of the Poynting-Robertson (PR) effect (e.g., Gillett 1986). The IRAS team assumed that the grains that they detected toward Vega were primordial and considered the PR effect to derive minimum particle sizes of about 1 mm (Aumann et al 1984). But there are at least three reasons to believe that the

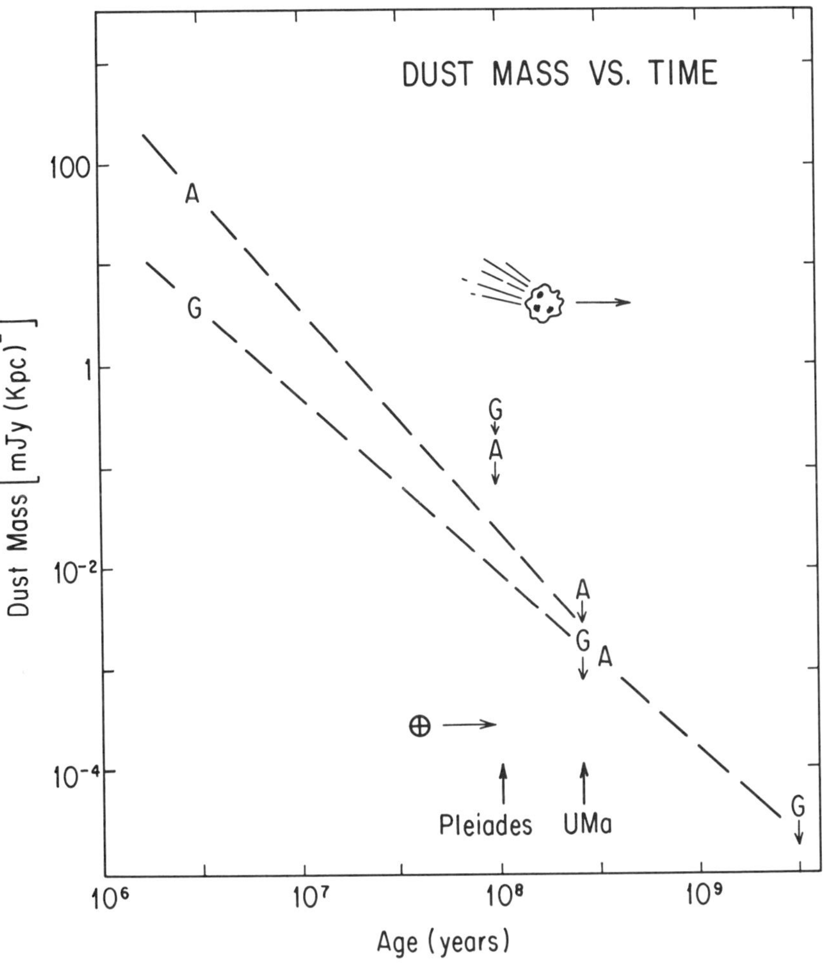

Fig. 1. Excess emission at 800 μm vs age for a variety of stellar sources (from ZB92a). The ordinate has been normalized so that all stars are at the same distance from Earth, and therefore to first order, location on the ordinate is proportional to dust mass. Data are plotted for G and A-type stars in the Pleiades cluster at 10^8 yrs., the nucleus of the UMa cluster at 2.7 x 10^8 yrs., representative A and G-type stars at 3 x 10^8 and 3 x 10^9 yrs., respectively, and for T Tauri stars and Herbig Ae stars, the G and A plotted at 3 x 10^6 yrs. The As and Gs above the downward pointing arrows indicate upper limits to dust masses. Also shown as the symbols $\oplus$ and a meteorite are the most commonly quoted times for the formation of the Earth and other terrestrial planets (~ 10^8 yrs.), and for the end of the period of heavy meteoritic bombardment in the inner solar system (at ~ 6 x 10^8 yrs.).

observed particles usually are not primordial, but rather are being replenished by comets and/or asteroids even at the present moment!

First, at β Pic there is substantial evidence from the temperature and spatial extent of both mid- and far-IR emission that the bulk of the radiating particles are micron and submicron in size (Gillett 1986, Telesco et al 1988, Backman et al 1992). Such particles are lost rapidly due to radiation pressure and the PR effect. The presence of a strong 10 μm silicate feature also suggests particle radii $\lesssim$ 10 μm (Telesco & Knacke 1991). Similar particle sizes are deduced from detailed models of the circumstellar disk that are based on optical and IRAS data (Artymowicz et al 1989). In a long, important series of papers, a French group (Beust et al 1991 & Lagrange-Henri et al 1992 and references therein) has analyzed time variable optical and ultraviolet absorption lines in the spectrum of β Pic. They conclude that infall of about 200 small, evaporating, comet-like bodies onto the star during the course of a year is the most plausible explanation of the data. Their models suggest that the cometary orbits are perturbed by a planet in an eccentric orbit about 3 AU from β Pic.

But β Pic is only one star. What of the debris disks around other Vega-like stars? Particles with radii between a few microns and a few millimeters which dominate the far-IR and submillimeter emission will be eroded and swept away due to collisions with atomic and molecular clouds at characteristic intervals of 4 x 10^7 and 4 x 10^8 years respectively according to calculations of Lissauer & Griffith (1989) and Stern (1990). These times are short compared to the main sequence lifetimes and/or estimated ages of almost all of the stars listed in Table II of Backman & Gillett (1987). When combined with the statistics in their Table I, it appears most unlikely that the typical particle disk can have survived without being replenished since the stars were formed.

Finally, dynamical considerations can help to constrain the properties of the debris disks (ZB92a). Define P to be the orbital period of a typical particle and $\tau = L_{IR}/L_*$, where L_{IR} is the total excess far-IR luminosity as measured by IRAS and L_* is the bolometric stellar luminosity. If the orbiting debris disks are initially vertically thick, then the component of motion that is perpendicular to the plane of the disk will be dissipated by frequent collisions -- on time scale P/τ -- of particles with orbits that have substantial inclination angles with respect to the disk plane. If the disks are thin and planar, as at Saturn's rings, then τ should never exceed $r_*/4r_i$ by any substantial amount (r_* is the stellar radius and r_i is the inner radius of the debris disk).

For seven of the 25 stars listed in Table II of Backman & Gillett (1987) τ exceeds $r_*/4r_i$ by about one order of magnitude and for HR 4796 (Jura 1991) by nearly two orders of magnitude. It is conceivable that the large values of τ at these eight stars are the results of warps in thin disks, although there is no hint of a warp in the β Pic disk. A more plausible explanation of these large τ's is that the disks are vertically thick and are sustained by an ongoing source of particulate matter, such as comets and/or asteroids which supply the zodiacal dust in our solar system. An additional or alternative mechanism that could keep the disks puffed up is the gravitational influence of massive, Jupiter-like planets.

In regard to question (iii), IRAS data indicate that debris clouds are found at members of multiple star systems, as well as at single stars, although a good statistical description of these clouds must await future satellites, such as ISO. Remarkably (?), two of the four main sequence stars with the largest known values of τ are members of multiple systems (see Table 1). At HR 4796 if the radiating dust particles are comparable in size to the zodiacal dust, then they orbit the primary star inside of the orbits of both apparent companions.

Walker and Wolstencroft (1988) published a list of potential Vega-like stars that included HD 98800 (SAO 179815). HD 98800 is one of the strangest objects in the entire IRAS data base. HD 98800 has a known visual companion ~20 AU away and apparently also a spectroscopic companion (period ~262 days; Stefanik & Latham 1992), that has a semi-major axis ~0.9 AU. If the dust particles at HD 98800 are comparable in size to the zodiacal dust, then they orbit typically 1 to 2 AU from HD 98800, i.e., at positions similar to those which characterize the zodiacal dust in the inner solar system. According to Good (see Backman & Gillett 1987), for the zodiacal dust, $\tau \sim 8 \times 10^{-8}$. So there is about 6 orders of magnitude more dust in orbit around HD 98800 than there is in orbit around the Sun! How can this much dust, ~ 0.1 lunar masses, be generated near HD 98800 and how can it be maintained in orbit in the presence of the PR effect? Zuckerman and Becklin (1992c) consider some models for the HD 98800 system, none of which are very satisfactory. Perhaps the most plausible involves gravitational perturbations due to the spectroscopic companion, that overwhelm the PR effect and inhibit dust from spiraling closer than

TABLE I Main Sequence Stars with Largest τ

Star	Spectral Class	$\tau = L_{IR}/L_*$	T_{dust}(K)	Remarks
β Pic	A5 V	2.4×10^{-3}	~ 100	single star; 10 μm silicate feature[1]
HR 4796	AO V	5×10^{-3}	~ 100	at least double, probably triple[2]
51 Oph	B9.5Ve	4×10^{-2}	100 - 1000	Waters et al (1988)
HD 98800	K5 V	9×10^{-2}	~ 160	at least double[3], very probably triple[4]; 10 μm silicate feature[5]

[1]Telesco & Knacke 1991.
[2]Jeffers et al 1963; Jura 1991; Jura et al 1992.
[3]Gliese & Jahreiss 1979; Heintz 1990; Star 8141 in Aitken's Catalogue of Double Stars.
[4]Stefanik & Latham 1992.
[5]Skinner et al (1992).

~1 AU from HD 98800. In this model, the gravity of the companion must also puff up the inclinations of the orbiting particles since a thin planar disk cannot account for the extraordinary value of τ at HD 98800.

Concerning question (iv), excepting HD 98800, the debris disks orbit roughly as far from the Vega-like stars (~ 50 AU) as the Kuiper cometary belt (Duncan et al 1988) is from the Sun. Using submillimeter opacities recently calculated by Pollack et al (1992), ZB92a estimated minimum dust masses of order one lunar mass at a star like Vega or Fomalhaut. These minimum masses are dominated by particles with radii, a ~ 300 μm. ZB92a use data for the Sun and the Vega-like stars to estimate particulate masses in the Kuiper belt. These masses, when combined with the cometary mass estimates of Duncan et al for the Kuiper belt, suggest that the power law spectrum of particle number density, $dn \sim a^{-3.5}da$, often assumed to result from the fragmentation of asteroids and comets, applies reasonably well between 300 μm and comet-size objects.

3. AFTER THE MAIN SEQUENCE; TOO MANY BINARIES?

What happens to orbiting, dusty debris after a star leaves the main sequence? Jura (1990) searched the IRAS data base for evidence of excess 60 μm emission toward G-type giant stars of luminosity class III. Only a tiny percentage of these stars, which are descendants of A-type main sequence stars, displays evidence of orbiting dust. Later, in the asymptotic giant branch (AGB) and post-AGB phases of stellar evolution, indications of debris disks reappear. Figure 2 locates on an HR diagram unusual evolved objects that are, or may be, orbited by dust and/or gas. Many of these are known or suspected to be binary systems.

Morris (1987, 1990) has championed a binary star model to explain the preponderance of nonspherically symmetric outflows from AGB stars and their descendants, the planetary nebulae (PN). In this model, the gravitational field of a comparison star deflects the spherically symmetric wind of its AGB star primary. Oftentimes some of the mass from the wind ends up in a long-lived disk, either around the companion or in a circumbinary orbit (see Figure 2 in Morris 1987).

IRAS discovered that a small group of carbon stars displays a 10 μm silicate feature (Little-Marenin 1986, Willems & de Jong 1986). The origin and nature of these stars, which are located in Figure 2, remain controversial. In one conceivable model the 10 μm emission is carried by a dusty, oxygen-rich disk that orbits a companion to the carbon star (Barnbaum et al 1991). In this picture the silicate dust was produced during an earlier time when the carbon star was an M-type, oxygen-rich giant.

An even more remarkable set of objects in Figure 2 is the dusty, high latitude, F-type "supergiants" (Waters 1992; see also Sasselov 1992). Their modest luminosities, ~ 4000 $L_\odot$, indicate that the F supergiants will evolve into white dwarfs. Their peculiar photospheric abundances may tie them also to the λ Boo and even some Vega-like stars (Venn & Lambert 1990). More germane in the present context is the establishment that many of the

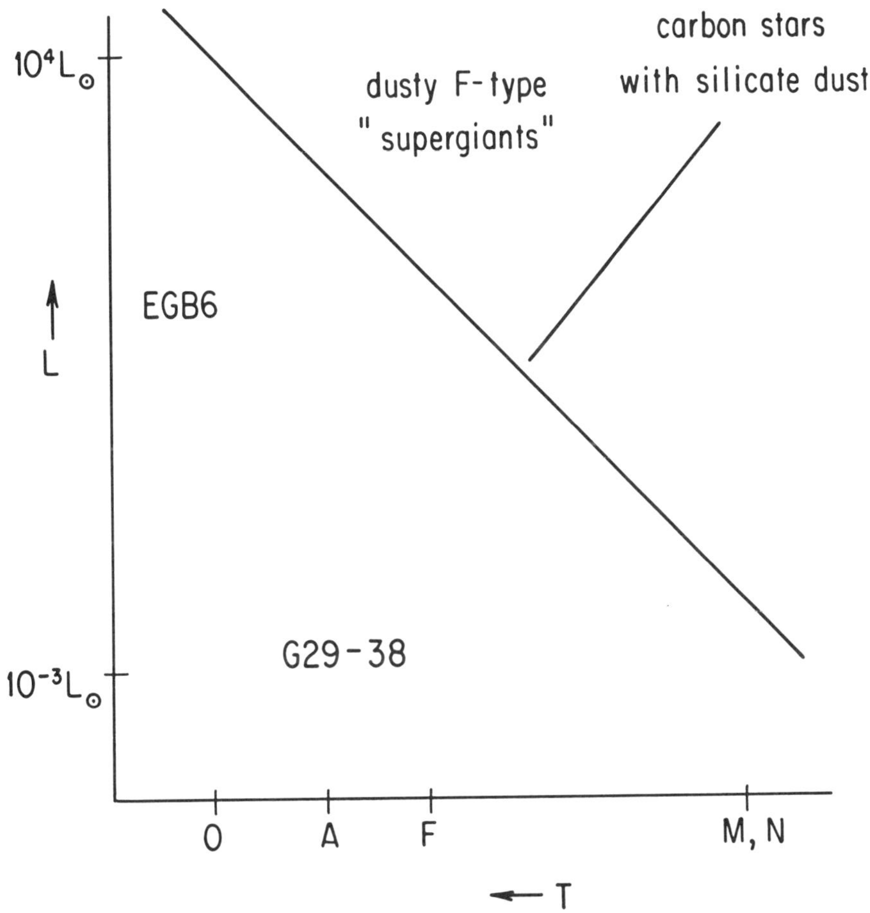

Fig. 2. A Hertzsprung-Russell diagram depicting some bizarre post main sequence stars. As described in the text, these stars are definitely, or possibly, surrounded by orbiting disks of dust and/or gas. In addition, many, possibly all, have orbiting companions.

F-type supergiants are members of binary systems with eccentric orbits and orbital periods ~ 1 year (Waters 1992). There is substantial evidence for orbiting gas and hot dust. The hot dust often (always?) displays 3 μm spectral features characteristic of carbon-rich material.

Also plotted in Figure 2 is EGB6, a large, old PN discovered by Howard Bond in 1978 during an examination of the Palomar Sky Survey prints. Liebert et al (1989) called special attention to this PN which has, coincident with its central star, a unique, ionized nebula that is spatially unresolved. Dopita & Liebert (1989) argued that the nebula might be due to the ablation of a Jupiter-like planet by the central star of the PN or an accretion disk about the (hypothetical) planet, or both. Infrared images (Zuckerman et al 1991) indicated that the central star has a late M-type companion. The presence of the companion vitiates any obvious need for an ablating planet. If the M star possesses an accretion disk ionized by the central star of the PN, then the unresolved nebula might result.

Bond, Liebert, and Renzini (1992) discovered, with the faint object camera on HST, that the nucleus of EGB6 appears double through narrow band filters at [OIII] λ5007 and at H β. The "secondary" is about 0.18" from the central star. The images are not well calibrated, so at present it is still not possible to say whether the secondary is the M star with an ionized accretion disk or alternatively if the secondary is merely a blob of nebulosity and the two stars are unresolved in the HST image.

So in each of the preceding examples there is evidence for orbiting dust and/or gas and, possibly or definitely, a companion star as well. If these were the only AGB and post-AGB stars to show evidence of binarity, then from a statistical point of view, no problems are encountered. However, a very large fraction, $\gtrsim$ 80% of post-AGB stars (so-called transition objects or protoplanetary nebulae) and PN show evidence for non-spherically symmetric outflow (e.g., Zuckerman & Aller 1986, Bujarrabal et al 1992). At present one of the major research questions in late stellar evolution is: What shapes planetary nebulae? Single star models have been considered (e.g., Soker 1990, Pascoli 1992, Soker & Harpaz 1992), but the current rage is definitely the binary star model, at least judging from remarks and papers at a recent conference on post-AGB evolution (e.g., Schwarz 1992). Zuckerman & Aller (1986) and Zuckerman & Gatley (1988) argued against the binary star model as an explanation of all of the asymmetrical shapes that are seen in PN. In particular, there simply do not seem to be enough sufficiently close binaries (Duquennoy & Mayor 1991) to account for the nearly ubiquitous presence of non-spherically symmetric outflows at post-AGB stars. The relevance of these considerations in the context of debris disks and planetary systems is that various authors (e.g., Struck-Marcell 1988, Sahai et al 1991) have suggested that single stars are often accompanied by close Jupiter-like planets that are evaporated in a common envelope post-AGB phase. The debris would form a disk that could collimate a wind from the primary. In any event, if Jupiter-like planets often exist near single stars, then that might help to explain the rarity of spherically symmetric outflows.

4. DUST AROUND WHITE DWARFS

Within an A.U. (or more) of an AGB star and its PN descendant, all solar system-like objects with masses up to and including that of Jupiter will be destroyed. Therefore, one anticipates that, excepting companion stars, the near vicinity of white dwarfs will be essentially devoid of matter. For this reason when Zuckerman and Becklin (1987) discovered excess 2 to 5 μm emission from the white dwarf Giclas 29-38, they argued that it was unlikely to be a white dwarf analog of the Vega-like stars. At that time a brown dwarf seemed to be the most natural carrier of the excess IR emission (see also Greenstein 1988). Subsequent observations have indicated that dust is present within a few solar radii of G29-38 and that the dust accounts for most and probably all of the excess IR emission (Graham et al 1990a, Telesco et al 1990, Tokunaga et al 1990, Patterson et al 1991). The dust is remarkable since Zuckerman and Becklin (1992b & private communication) have now searched ~ 400 white dwarfs without discovering a second example of the G29-38 phenomenon.

G29-38 is a ZZ Ceti variable with a cooling age since it first became a white dwarf of ~ 6 x 10^8 years. The optical pulsations of a few ZZ Ceti stars are sufficiently stable that Earth mass companions could be detected should they exist (Kepler et al 1990). Unfortunately, the pulsational spectrum of G29-38 is a mess! Timing of its optical pulsations by Winget et al (1990) initially led them to suggest that G29-38 is orbited by a neutron star or black hole. Graham et al (1990b) showed that variations in the radial velocity of G29-38 are too small to allow such a massive companion (see also Patterson et al 1991). But the data of Graham et al indicated that at the 5-10 km/s level the radial velocity of G29-38 might be variable. Follow-up observations by Barnbaum and collaborators (Barnbaum et al 1992, Barnbaum & Zuckerman 1992) confirmed that the velocity is probably variable with period ~ 11 months. Should observations during the next few years confirm the apparent variations, then G29-38 is orbited either by a high mass brown dwarf or low mass white dwarf with orbital plane nearly perpendicular to or nearly in the plane of the sky, respectively. Astrometric observations can resolve this ambiguity.

The origin, maintenance, and structure of the dust cloud at G29-38 are mysterious. The lifetime of micron-size dust particles is only of order 100 years against decay due to the PR effect. So Graham et al (1990a) suggested that the dust is produced by fragmentation of meter-size boulders. However, the dust absorbs ~ 3% of the radiation from G29-38 and arguments similar to those given in Section 2 imply that it cannot lie in a thin plane. Yet collision times for the dust particles and also the boulders in the Graham et al model are of order days (Tokunaga et al 1990). Conceivably, the disk could be vertically thin but warped. If however the disk is vertically thick, then because of the very short collision times, the puffiness cannot be due to dust generation by a myriad of comets with inclined orbits. Perhaps there is an Earth-like planet embedded in the disk that gravitationally scatters the dust particles. (The putative brown dwarf or white dwarf mentioned above would be nearly 1 AU from G29-38, probably too far from the dust to puff it up.)

In any event, the dust cannot originate from comets (e.g., Tokunaga et al 1990). Nor can it be from the tidal disruption of one white dwarf by another (e.g., Benz et al 1990). The narrow core of the Hα line that enables a moderately accurate measurement of the radial velocity of G29-38 also may be used to set an upper limit on its rotational velocity. All the 17 DA white dwarfs listed by Pilachowski & Milkey (1987) are slow rotators (vsini $\lesssim$ 80 km/s), as is G29-38. If ~ 0.1$M_\odot$ of orbiting debris were deposited onto a white dwarf, then its rotational velocity would be an order of magnitude larger than observed. Such rapid rotation might be present in some non-magnetic cataclysmic variables (Patterson 1992, private communication). So slow rotation probably rules out all scenarios whereby the debris of a stellar mass companion is accreted onto G29-38. The slow rotation also means that the gravitational quadruple moment of G29-38 probably is too small to warp an initially planar dust disk or unwarp one that somehow formed warped within a few solar radii of G29-38 (Goldreich 1991, private communication).

There exists no compelling model for G29-38. Perhaps it is a more evolved counterpart of the F-supergiants with orbiting hot dust and companions with orbital periods ~ 1 year mentioned in Section 3. In most cases, one would expect the companion to an F supergiant to be a low mass main sequence star that would swamp the infrared emission of any dust that remains in orbit around a white dwarf descendant. Only in those rare cases where the companion is another white dwarf or a brown dwarf could the dust be detected, as possibly in the case of G29-38. And if the arguments presented above regarding a puffy dust disk at G29-38 are germane, then even more specialized conditions that involve embedded planets may be required if the dust is to be detectable (see the following paragraph). In particular, Lynden-Bell (1991, private communication) suggests a system with two planets in close resonant orbits around G29-38, analogous to Io and Europa in orbit around Jupiter. The tidal stresses on the inner planet produce volcanos energetic enough to shoot dusty debris into orbit around the white dwarf. At least a partial test of this model and the F-supergiant analogy may be possible if 3 μm emission features can be detected at G29-38.

In summary, evidence for orbiting dust has been detected at only one white dwarf of about 400 searched at 2.2 μm wavelength (K-band). However, the contrast between dust and photospheric emission at K is too small for dust to be detected if it orbits in a thin planar disk. The K excess at G29-38 could be readily detected by Zuckerman & Becklin only because the disk seems to be vertically thick. The optimum wavelength for ground based detection of thin planar dust disks is 3.5 μm (L-band) where a total of only a few dozen white dwarfs have been examined to date (Kumar 1985, Zuckerman & Becklin, private communication).

5. DUST GRAINS NEAR PSR1257+12

If the planets orbiting ~ 0.4 AU from PSR1257+12 formed from a debris disk, as seems likely, then it is conceivable that substantial quantities of debris remain out beyond the planets. Assuming a spindown luminosity of 30 $L_{\odot}$ (Wolszczan & Frail 1992), dust grains as far as 10 AU from the pulsar could be sufficiently well-heated to radiate effectively at 10 μm wavelength. Zuckerman & Becklin (1992d) used the facility bolometer on the IRTF at Mauna Kea Observatory to measure a 10 μm flux of 7 ± 11 mJy from PSR1257+12. For an assumed distance of 500pc (Wolszczan & Frail 1992), this corresponds to a mass of well-heated, micron-size silicate particles of order $\lesssim 10^{23}$g ($\lesssim 10^{-3}$ lunar masses). Unfortunately, this limit may be meaningless; if the disk is thin and planar, then it will subtend such a tiny solid angle as seen from the pulsar that the dust will intercept and reradiate a negligible fraction of the spin down luminosity. Perhaps a pulsar wind and/or the gravitational field of the planets can warp or puff up the putative debris disk.

6. SUMMARY

During the past decade IRAS and ground-based infrared measurements have revealed dusty debris disks toward many main sequence and post-main sequence stars. Mostly indirect arguments imply that many of the disks also contain asteroids, comets and/or planets. Some objects, notably the F "supergiants" at high galactic latitude, the K5V star HD 98800, and the white dwarf Giclas 29-38 are very mysterious. Especially in the later stages of stellar evolution there are indications that binarity is often associated with the presence of dusty debris. A new generation of very large ground-based telescopes and the ISO satellite portend another decade of discovery.

ACKNOWLEDGEMENTS

Long term help from Eric Becklin, Mike Jura, Susan Kleinmann, and Mark Morris is gratefully acknowledged. We thank Robert Stefanik and Dave Latham for informing us of variations in the radial velocity of HD 98800. Esther Scoble kindly prepared the camera ready manuscript. The author's research on debris disks has been supported by grants from NASA and NSF.

REFERENCES

Artymowicz, P., Burrows, C., and Paresce, F. 1989, *Ap. J.*, **337**, 494.
Aumann, H. H. 1988, *AJ*, **96**, 1415.
Aumann, H. H. et al 1984, *Ap. J.*, **278**, L23.
Backman, D. E. and Gillett, F. C. 1987, in Cool Stars, Steller Systems and the Sun, ed. J. L. Linsky and R. E. Stencel, Springer-Verlag, Berlin, p. 340.
Backman, D. E., Gillett, F. C., and Witteborn, F. C. 1992, *Ap. J.*, **385**, 670.
Barnbaum, C., Morris, M., Likkel, L., and Kastner, J. H. 1991, *A & A*, **251**, 79.
Barnbaum, C., Misch, A., and Zuckerman, B. 1992, *BAAS*, **24**, 799.
Barnbaum, C., and Zuckerman, B. 1992, *Ap. J.*, in press.
Benz, W., Bowers, R. L., Cameron, A. G. W., and Press, W. H. 1990, *Ap. J.*, **348**, 647.
Beust, H., Vidal-Madjar, A., and Ferlet, R. 1991, *A & A*, **247**, 505.
Bond, H., Liebert, J. and Renzini, A. 1992, in preparation.
Bujarrabal, V., Alcolea, J., and Planesas, P. 1992, *A & A*, **257**, 701.
Dopita, M. A., and Liebert, J. 1989, *Ap. J.*, **347**, 910.
Duncan, M., Quinn, T., and Tremaine, S. 1988, *Ap. J.*, **328**, L69.
Duquennoy, A. and Mayor, M. 1991, *A. & A.*, **248**, 485.
Gillett, F. C. 1986, in Light on Dark Matter, ed. F. P. Israel, Reidel, Dordrecht, p. 61.
Gliese, W. and Jahreiss, H. 1979, *A & A Suppl.*, **38**, 423.
Graham, J. R., Matthews, K., Nengebauer, G., and Soifer, B. T. 1990a, *Ap. J.*, **357**, 216.

Graham, J. R., McCarthy, J. K., Reid, I. N., and Rich, R. M. 1990b, *Ap. J.*, **357**, L21.
Greenstein, J. L. 1988, *A. J.*, **95**, 1494.
Heintz, W. D. 1990, *Ap. J. Suppl.*, **74**, 275.
Jeffers, H. M., van den Bos, W. H. and Greeby, F. M. 1963, *Publ. Lick Obs. 21*, Part II.
Jura, M. 1990, *Ap. J.*, **365**, 317.
Jura, M. 1991, *Ap. J.*, **383**, L79.
Jura, M., Becklin, E. E., Zuckerman, B. and Smith, C. 1992, *Ap. J.*, submitted.
Kepler, S. O. et al 1990, *Ap. J.*, **357**, 204.
Kumar, C. K. 1985, *PASP*, **97**, 294.
Lagrange-Henri, A. M., Gosset, E., Beust, H., Ferlet, R., and Vidal-Madjar, A. 1992, *A. & A.*, in press.
Liebert, J. et al 1989, *Ap. J.*, **346**, 251.
Lissauer, J. J. and Griffith, C. A. 1989, *Ap. J.*, **340**, 468.
Little-Marenin, I. R. 1986, *Ap. J.*, **307**, L15.
Morris, M. 1987, *PASP*, **99**, 1115.
Morris, M. 1990, in From Miras to Planetary Nebulae: Which Path for Stellar Evolution?, eds. M. O. Menessier and A. Omont, Editions Frontieres, Montpellier.
Pascoli, G. 1992, *PASP*, **104**, 350.
Patterson, J., Zuckerman, B., Becklin, E. E., Tholen, D. J. and Hawarden, T. 1991, *Ap. J.*, **374**, 330.
Pilachowski, C. A. and Milkey, R. W. 1987, *PASP*, **99**, 836.
Pollack, J. B., Hollenbach, D., Simonelli, D. P., Beckwith S., Roush, T. and Fong, W. 1992, *Ap. J.*, submitted.
Ruden, S. P. and Pollack, J. B. 1991, *Ap. J.*, **375**. 740.
Sahai, R., Wooten, A., Schwarz, H. E., and Clegg, R. E. S. 1991, *A. & A.*, **251**, 560.
Sasselov, D. D. 1992, editor of Luminous High-Latitude Stars, ASP Conference Series, in press.
Schwarz, H. E. 1992, in Mass Loss on the AGB and Beyond, ed. H. E. Schwarz, ESO/CTIO Workshop, in press.
Skinner, C. J., Barlow, M. J. and Justtanont, K. 1992, *MNRAS*, **255**, 31 p.
Smith, B. A. and Terrile, R. J. 1984, *Science*, **226**, 1421.
Soker, N. 1990, *AJ*, **99**, 1869.
Soker, N. and Harpaz, A. 1992, preprint.
Sopka, R. J. et al 1985, *Ap. J.*, **294**, 242.
Stefanik, R. P. and Latham, D. W. 1992, private communication.
Stern, S. A. 1990, *Icarus*, **84**, 447.
Struck-Marcell, C. 1988, *Ap. J.*, **330**, 986.
Telesco, C. M., Becklin, E. E., Wolstencroft, R. D. and Decher, R. 1988, *Nature*, **335**, 51.
Telesco, C. M., Joy, M. and Sisk, C. 1990, *Ap. J.*, **358**, L17.
Telesco, C. M. and Knacke, R. F. 1991, *Ap. J.*, **372**, L29.
Tokunaga, A. T., Becklin, E. E. and Zuckerman, B. 1990, *Ap. J.*, **358**, L21.
Venn, K. and Lambert, D. L. 1990, *Ap. J.*, **363**, 234.
Walker, H.J. and Wolstencroft, R.D. 1988, *PASP*, **100**, 1509.

Waters, L. B. F. M. 1992, in Mass Loss on the AGB and Beyond, ed. H. E. Schwarz, ESO/CTIO Workshop, in press.
Waters, L. B. F. M., Cote, J. and Geballe, T. R. 1988, A & A, 203, 348.
Willems, F. J. and de Jong, T. 1986, *Ap. J.*, **309**, L39.
Winget, D. et al 1990, *Ap. J.*, **357**, 630.
Wolszczan, A. and Frail, D. A. 1992, *Nature*, **355**, 145.
Zuckerman, B. and Aller, L. H. 1986, *Ap. J.*, **301**, 772.
Zuckerman, B. and Becklin, E. E. 1987, *Nature*, **330**, 138.
______________________________ 1992a, *Ap. J.*, submitted (ZB92a).
______________________________ 1992b, *Ap. J.*, **386**, 260.
______________________________ 1992c, *Ap. J.*, submitted.
______________________________ 1992d, in preparation.
Zuckerman, B., Becklin, E. E. and Mclean, I. 1991, in Astrophysics with Infrared Arrays, ed. R. Elston, (ASP Conf. Series Vol. 14), p. 161.
Zuckerman, B. and Gatley, I. 1988, *Ap. J.*, **324**, 501.

Planets Around Pulsars
ASP Conference Series, Vol. 36, 1993
J.A. Phillips, J.E. Thorsett, and S.R. Kulkarni

PLANETESIMALS AROUND NEARBY MILLISECOND PULSARS

S.K. CHAKRABARTI[1,2]
Tata Institute of Fundamental Research, Homi Bhabha Road, Colaba, Bombay 400005, India

ABSTRACT We predict that it is possible to observe line emissions of OH, CN and C_2 from the planetesimals around some of the nearby millisecond pulsars, such as PSR1257+12. Observation of these lines will provide an independent test of either an existing planetary system or one which is in the process of formation.

Recently, Wolszczan and Frail (1992) reported the discovery of two planets around a millisecond pulsar PSR1257+12. The detection of planets in PSR1257+12 involves a crucial interpretation of the variation of the residuals. A similar variation is also seen in the system PSR1828+10 (Bailes, this volume). These observations have brought up the possibility that planetary system could exist around many other pulsars. In general, one would like to detect a planetary system, or even one which is in the process of formation in a manner free from interpretational hazards. It would also be helpful if the age of the planetary system (as well as the central pulsar, for that matter) were also determined at the same time. We propose that if the line emissions are detected from the planetesimal clouds around pulsars, then one can have these informations unambiguously. However, such a proposal is valid only when the pulsar is still very young and a negative detection does not rule out the possibility that an older planetary system could still exist.

The planets around PSR1257+12 are observed to be at distances of 0.36AU and 0.47AU away from the pulsar. The median masses of these planets are $3.9M_\oplus$ and $3.2M_\oplus$, and their time periods are 66.6d and 98.2d respectively. There is a possibility that a third planet also exists with a time period of around a year corresponding to a mean distance of 1.1AU from a $1.4M_\odot$ pulsar. Several models have been put forward to understand the genesis of the planetary system. According to Wijers et al. (1992), even before the supernovae explosion, the planets surrounded the binary stars one of which evolved to form the pulsar. The more likely possibility is that the planets form in a disk generated out of rapidly mass-losing low mass binary companion of the pulsar (Stevens, Rees & Podsiadlowski, SRP, 1992). A large number of other models include planets in disks from collisional break up of white dwarf companion or simply in circumbinary disks (this volume).

Though the disk around the pulsar is an excretion disk, rather than an ac-

---

[1]Visiting as Associate, ICTP, 11 Strada Costiera, 34100 Trieste, Italy

[2]Visitor, (130-33) California Institute of Technology, Pasadena, 91125, USA

cretion disk, the formation of a planetary system must be accompanied by the formation of a large number of planetesimals at the outer skirt of the circumpulsar disk exactly as in the case of solar nebula. If the planetary system is in its early stage of evolution, one might be able to observe its signature spectroscopically. These planetesimals might contain matter as well as angular momentum of comparable magnitude as those of the planets formed. Furthermore, since the binary companion need not be a highly evolved star (see, for example, SRP model), the composition of the planetesimals could be very similar to the comets in the solar system. In the solar type systems, a large fraction of the composition is water which is also very volatile. Thus, when the planets form in an excretion disk they may be completely devoid of water. The evaporated water could condensate farther from the pulsar, thereby increasing the water content in planetesimals formed in the outer disk. In case the composition is very much different for icy substances, some of the discussions presented below do not apply.

In the present analysis, we assume that even though, in general, a spun-up pulsar could be very old (say, up to 10^9 years), it is possible that presently some of the observed ones are very young ($\leq 10^{5-6}$ yr). In this case, the majority of the planetesimals are still in the cloud nearby to the pulsar, and are not scattered off to form a 'Oort cloud' far away (10^4AU) which might take as much as 10^{8-9} yr (Duncan, Quinn and Tremaine, 1987). In the latter case, the line emissions will be sufficiently weak to be observable, although it is still possible to observe through radio emission if the amount of matter contained in the planetesimal cloud is large enough (see below).

The angular momenta of the two planets are 7.4×10^{47} and 7×10^{47} g cm^2 s^{-1} respectively. One could imagine that a third and probably more planets might be present in the system. Following Chakrabarti and Krishna Swamy (1992), we assume that the mass M_{CPC} and the angular momentum J_{CPC} of the circumpulsar cloud (CPC) may be *at least* comparable to the corresponding quantities which went in forming the planets themselves (Marochnik, Mukhin, Sagdeev 1988, Weissman, 1991). If, for example, we assume that $M_{CPC} \sim 8M_\oplus$ and $J_{CPC} \sim 4 \times 10^{48}$ g cm^2 s^{-1} respectively, then the mean radius of CPC is $\sim$ 2.5AU. Assuming each planetesimal to be a sphere made of ice with radius $r_c \sim$ 10km and density $\rho \sim 1.0$ gm cm^{-3}, the number of planetesimals in the CPC is $\sim 1.2 \times 10^{10}$.

From the published data of I, ω and $\dot{\omega}$ for PSR1257+12 by Wolszczan and Frail, the luminosity of the pulsar is found to be about 2×10^{34} erg cm^{-2} s^{-1} $\sim 5.2 L_\odot$. Comparing with the properties of comets in solar system, it is noted that the vaporization rate at one astromonical unit (AU) around the sun should be comparable to the rate at one pulsar unit (PU) $\sim$ 2.3AU. The higher luminosity is clearly more efficient in vaporizing the planetesimals even if all were in circular orbits at $\sim$ 2.5AU. Given that the possible third planet is at a distance of 1.1AU, one could assume that the planetesimals are spread over a region of 1AU to around 2.5AU. Assuming the number density of the planetesimals $n_0(r)$ to be constant, we obtain, $n_0 \sim 2 \times 10^8$ pc^{-3}.

Following Delsemme (1982), we note that the rate of evaporation of water molecule is roughly $Z = Z_0/r^{2.15}$ where $Z_0 = 1.7 \times 10^{18}$ mol cm^{-2} s^{-1} is the evaporation rate at 1AU in the solar system i.e., at 1PU in our pulsar system. Using this rate we obtain the rate of vaporization of water molecules as $Q(H_2O) \sim 4 \times 10^{40}$ s^{-1}. This gives the *half-life* of CPC to be about 850 years.

This is too short compared to the lifetime of the pulsar, but by suitably choosing reasonable values for M_{CPC} and J_{CPC} one could increase the half-life by atleast a factor of 10^4, see below. From the cometary data in the solar system, it is easy to obtain the production rates of OH, C_2 and CN from $Q(H_2O)$ alone by comparing the corresponding production ratios (see, e.g., A'Hearn, Millis and Birch, 1981). Using the efficiency (see, e.g., Feldman and Brune, 1976) factor $g = 1.2 \times 10^{-3}$ s^{-1} and lifetime $\tau = 1.33 \times 10^5$s, we obtain the flux from OH emission at 3090A at earth as 0.1 photons cm^{-2} s^{-1}. Using the appropriate values of g and τ for C_2 and CN line emissions at 5165A and 3883A respectively (Festou et al. 1990), the flux obtained are respectively 0.086 and 0.037 photons cm^{-2} s^{-1}, values well above the limits of detectability.

One could imagine to observe radio emissions from the planetesimals also. For example, in the case of a system of planetesimals with above mentioned parameters, the brightness temperature turns out to be about $360mK$, however the flux turns out to be insignificant, since the solid angle subtended by the planetesimal cloud at earth is very small. One way to increase the flux would be to assume more material in the cloud and/or use a larger radial distance for the planetesimals; some typical examples of which are shown in Table below.

For a large optical telescope, such as Palomar, the limit of detection could be as low as 10^{-4} photons cm^{-2} s^{-1}. This allows one to observe the lines for a wider range of mass and angular momentum of the planetesimals. For example, Table 1 below shows a few of the possible combinations of the mass, angular momentum and the size of the planetesimals and the corresponding half-life of the planetesimal cloud around PSR1257+12. The columns respectively represent the mass of the planetesimal cloud in units of earth mass, the angular momentum of the cloud in units of 10^{48}gm cm^2 s^{-1}, the typical radius of the cloud in astronomical units, the size of each planetesimals in units of 10km, number of photons per square centimeter per second, (emission from OH, C_2 and CN) reaching on earth, and the *half-life* of the planetesimals in years. The last row indicates the case where the mass and angular momentum used are roughly similar as those in the present planetary system around the sun.

In our analysis we assumed that all the planetesimals are scattered in roughly circular orbits. It is quite possible that they are in general at the safely of a larger distance but have large eccentricities. In this case, only some of the planetesimals are evaporated when they pass close to the pulsar, and the half-life of the cloud could be much larger.

In the case of some other nearby millisecond pulsars, our consideration could also be applied to check if any planetary system is in the process of formation around them. Since, in general, the cloud half-life is found to be typically about a few million years, whereas millisecond pulsars could be very old, say 10^8 years of more, roughly speaking, one should therefore observe these lines one in every hundred systems or so, unless the telescopic limit is vastly stretched to detect much fainter lines. One could also reverse our arguments to obtain a rough estimate of the age of the system, provided one has some knowledge about the mass and angular momentum of the cloud.

TABLE I Examples of effects of mass and angular momentum parameters on the half-life of the planetesimals. See text for description of various columns.

M	J	R	*Size*	P_{OH}	P_{C_2}	P_{CN}	T
8	4	2.5	1.0	0.10e+00	0.86e-01	0.37e-01	8.5e+02
8	14	31	1.0	0.75e-03	0.64e-03	0.28e-03	1.1e+05
16	14	7.7	1.0	0.26e-01	0.22e-01	0.95e-02	6.6e+03
8	24	90	0.7	0.10e-03	0.94e-04	0.40e-04	7.6e+05
24	24	10	1.0	0.23e-01	0.19e-01	0.84e-02	1.1e+04
8	34	181	0.1	0.17e-03	0.15e-03	0.64e-04	4.9e+05
32	54	28.5	2.0	0.17e-02	0.15e-02	0.65e-03	1.9e+05
16	64	160	0.4	0.11e-03	0.96e-04	0.41e-04	1.5e+06
32	124	150	1.0	0.10e-03	0.88e-04	0.38e-04	3.3e+06
40	134	112	2.0	0.12e-03	0.10e-03	0.44e-04	3.5e+06
40	144	130	1.7	0.10e-03	0.88e-04	0.38e-04	4.1e+06
40	154	148	1.3	0.10e-03	0.87e-04	0.37e-04	4.2e+06
40	164	168	1.0	0.10e-03	0.86e-04	0.37e-04	4.2e+06
500	1000	40	1.0	0.27e-01	0.23e-01	0.99e-02	2.0e+05

ACKNOWLEDGMENTS

I am grateful to many participants of the conference for enlightening me on this subject in various ways. I also thank the Lunar and Planetary Institute, California Institute of Technology and International Center for Theoretical Physics for financial assistances which made my participation possible.

REFERENCES

A'Hearn, M.F., Millis, R.L., Birch, P.V., 1981, *AJ*, **88**, 1559

Chakrabarti, S.K., Krishna Swamy, K.S., 1992, *A&A*, (submitted)

Delsemme, A.H., in: *Comets*, L.L. Wilkening, Univ. of Arizona Press (Tucson), 1982

Duncan, M., Quinn, T. & Tremaine, S., 1987, *AJ*, **94**, 1330

Feldman, P.D. and Brune, W.H., 1976, *ApJ*, **209**, L45

Festou, M.C., Tozzi, G.P., Smallone, L.A., Felenbok, P., Falciani, R., Zucconi, J-M., 1990, *A&A*, **227**, 609

Marochnik, L.S., Mukhin, L.M., Sagdeev R.Z., 1988, *Science* **242**, 547

Stevens, I.R., Rees, M.J., Podsiadlowski, P., *MNRAS*254, 19p

Weissman, P.R., 1991, *Icarus* **89**, 190

Wolszczan, A. Frail, D.A., 1992, *Nature* **355**, 145

Wijers, R.A.M.J., van den Heuvel, E.P.J., van Kerkwijk, M.H., Bhattacharya, D., (1992), *Nature* **355**, 593

Planets Around Pulsars
ASP Conference Series, Vol. 36, 1993
J.A. Phillips, J.E. Thorsett, and S.R. Kulkarni

BISTATIC RADAR: REFLECTIONS FROM ASTEROIDS AROUND PULSARS

J.A. PHILLIPS
Owens Valley Radio Observatory, Caltech 105-24, Pasadena, CA 91125

ABSTRACT A pulsar can be thought of as a radar which illuminates objects in its neighborhood with an intense, rotating radio beam. The radar reflection from a planetary companion would be impossible to detect from Earth, but asteroids present a larger reflecting area per unit mass, and their reflection might be detectable. The reflection from a dense swarm of asteroids would be unpulsed, weakly polarized, and broadband. Observations of unpulsed flux from Vela – the strongest known pulsar – yield upper limits of $\sim 10^{-4} M_{\odot}$ of asteroid-like material within 1 AU of the neutron star.

INTRODUCTION

With the discovery of planets around PSR 1257+12 (Wolszczan & Frail 1992), it seems reasonable to suggest that pulsars might also be accompanied by planetesimals, comets, and/or asteroid belts. However, the timing methods used to detect planets around pulsars are not very sensitive to large numbers of orbital companions in a disk or belt. Each companion contributes a quasi-sinusoidal signal to the pulsar's timing residuals with a phase and amplitude determined by the phase and semi-major axis of the orbit. If there are a large number of objects with a random distribution of orbital elements, the net result would be an incoherent superposition of perturbations similar to timing noise (see, e.g., the paper by Cordes in these proceedings).

Timing observations are most effective for finding one to a few orbital companions. In the case of an asteroid belt one might hope to see the coherent sinusoidal signature of the largest asteroid in the timing residuals, while the bulk of the asteroids contribute to a background of timing noise. Unfortunately, the prospects for detecting a single asteroid are poor. Thorsett & Phillips (1992) showed that timing observations of millisecond pulsars can detect objects as small as $\sim 10^{-3} M_{\oplus}$. For comparison, the largest asteroid in the solar system has a mass of only $\sim 10^{-4} M_{\oplus}$, well below the calculated detection threshold.

Here I consider another way to detect asteroids and planetesimals around pulsars. Asteroids orbiting a pulsar would be periodically illuminated by the rotating radio beam – the effect is analogous to an airport radar "pinging" nearby aircraft. Millisecond pulsars are believed to possess fan-shaped beams that sweep out as much as 2π sr (Narayan & Vivekanand 1983). The wide beam would therefore illuminate a substantial fraction of any asteroid swarm. Solar-system asteroids have a relatively high albedo, $\sim 10\%$ at radio wavelengths. If asteroids around neutron stars exist, and are similarly reflective, one might expect to see a faint "reflection nebula" surrounding the pulsar.

The reflection nebula will have the following properties at radio wavelengths:

- *Unpulsed emission:* The reflection will be unpulsed if the delay depth of the belt (i.e., the radial width $\times c$) exceeds P, the pulsar period. The radial width of the asteroid belt in our solar system is $\sim$ 500 light-seconds (1 AU), a value that far exceeds the period of any known radio pulsar.

- *Low degree of polarization:* Pulsar signals are usually highly polarized but the reflection will be less so. One source of depolarization is the irregularity of asteroid surfaces, which are presumably pitted and scarred by collisions with other asteroids. Reflections from an interface that is rough on scales $\sim \lambda$ can be partially or totally depolarized. Radar experiments involving asteroids in our own solar system show that surface roughness depolarizes the reflected signal by 10 to 15% (Ostro, Campbell & Shapiro 1985).

  A more significant cause of depolarization is related to the pulsar beam itself: the polarization of the pulsar signal varies as a function of angular position within the radiation beam. Position angle gradients of 90° or more across the beam are not uncommon (see Fig. 10-5 of Manchester and Taylor 1977). Asteroids in different parts of the beam will therefore reflect signals of different polarization, and the net depolarization should be near 100%.

- *Broad bandwidth:* Pulsar emission is broadband (10 MHz $\lesssim \nu \lesssim$ 10 GHz) and steep spectrum ($S \sim \nu^{-2}$). The reflection should have the same spectral character as the pulsar.

- *Small Angular Size:* The nearest pulsars are $\sim$ 100 pc away, so a 1 AU asteroid belt would subtend an angle $\lesssim 10^{-2}$ arcseconds.

A SIMPLE MODEL FOR ASTEROID REFLECTIONS

I will assume that the pulsar is surrounded by a belt of rocky asteroids with mean orbital radius R. The mean flux density incident on any asteroid in the absence of shadowing is $S_p(D/R)^2$, where S_p is the time-averaged radio flux of the pulsar (as measured on Earth) and D is the Earth-pulsar distance. We can define the scattering cross-section as follows: $\sigma = A \cdot \pi a^2$, where A is the scattering efficiency and πa^2 is the true geometric cross-section. The total reflected power from a single asteroid is $\sigma S_p(D/R)^2$, and the flux received at Earth is

$$S(a) = \frac{S_p A}{4\pi R^2}\pi a^2. \tag{1}$$

Equation 1 is valid for isotropic scatterers – i.e., dielectric spheres with $A = (\sqrt{\epsilon} - 1/\sqrt{\epsilon} + 1)^2$. The dielectric constant, ϵ, ranges from 3 to 10 for Solar System asteroids.

Equation 1 can be used to compute the reflection from a planet around PSR 1257+12. The mean pulsed flux at 430 MHz is 20 mJy, so an Earth-sized planet 1 AU from the neutron star would reflect a signal with flux density $S \leq 8 \times 10^{-6}$ μJy. A single planet around Vela – the strongest known pulsar ($S_p = 5$ Jy at 400 MHz) – would reflect less than 2×10^{-3} μJy.

Planetary reflections are clearly impossible to detect, but an asteroid belt presents a much greater reflecting area per unit mass than a single planet. The sensitivity of radio observations for detecting such a belt can be estimated if we assume a power-law distribution of asteroid sizes.

$$N(a) = C_* a^{-\alpha} \qquad a_i \leq a \leq a_o, \tag{2}$$

where $N(a)$ is the number of asteroids with radii in the range $a \to a + da$; a_i and a_o are the smallest and largest radii, respectively; and C_* is a normalizing constant. The power-law index, α, is between 2.5 and 3.5 for a collection of collisionally-evolved rocky bodies (Gradie, Chapman & Tedesco 1989 and references therein).

If all the asteroids are assumed to have the same mass density ρ, then

$$C_* = \frac{3M}{4\pi\rho \int_{a_i}^{a_o} a^{3-\alpha} da}, \tag{3}$$

where M is the total mass in asteroids with sizes between a_i and a_o. The total reflection is given by the integral $S_u = \int_{a_i}^{a_o} N(a)S(a)da$, which when combined with equations (2) and (3) yields,

$$S_u/S_p = \frac{3MA}{16\pi R^2 \rho} \Gamma(\alpha, a_i, a_o), \tag{4}$$

where

$$\Gamma(\alpha, a_i, a_o) = \frac{\int_{a_i}^{a_o} a^{2-\alpha} da}{\int_{a_i}^{a_o} a^{3-\alpha} da}. \tag{5}$$

The maximum reflection results from a steep spectrum of asteroid radii. For $\alpha = 3.5$, I find $\Gamma = 1/\sqrt{a_i a_o}$ and,

$$S_u/S_p \approx 4 \times 10^{-6} \left[\frac{A}{0.1}\right] \left[\frac{M}{M_\oplus}\right] \left[\frac{R}{\text{AU}}\right]^{-2} \left[\frac{\rho}{3 \text{ gm/cc}}\right]^{-1}. \tag{6}$$

The values $A = 0.1$ and $\rho = 3$ gm/cc are typical of Solar System asteroids (e.g., Ostro 1989; Kowal 1988) In writing equation (6) I assumed $a_o \approx 100$ km – corresponding to the largest asteroid in the Solar system – and $a_i \approx \lambda$ ($\equiv 20$ cm in equation 6). This choice of a_i implicitly excludes particles in the Rayleigh limit ($a \ll \lambda$) where the scattering cross-section is reduced by $\sim (a/\lambda)^4$. Ignoring the small particles is a conservative approach that *underestimates* the ratio S_u/S_p.

OBSERVATIONAL LIMITS

Single dish measurements are not very useful for studies of unpulsed flux from pulsars due to confusion with background sources. The most sensitive interfometric measurements are those of Beitenholtz, Frail & Hankins (1991), who observed the Vela pulsar with the VLA. They gated the VLA correlator synchronously with the pulsar period to form ON- and OFF-pulse images of the pulsar field. There was no continuous emission at the pulsar position above a

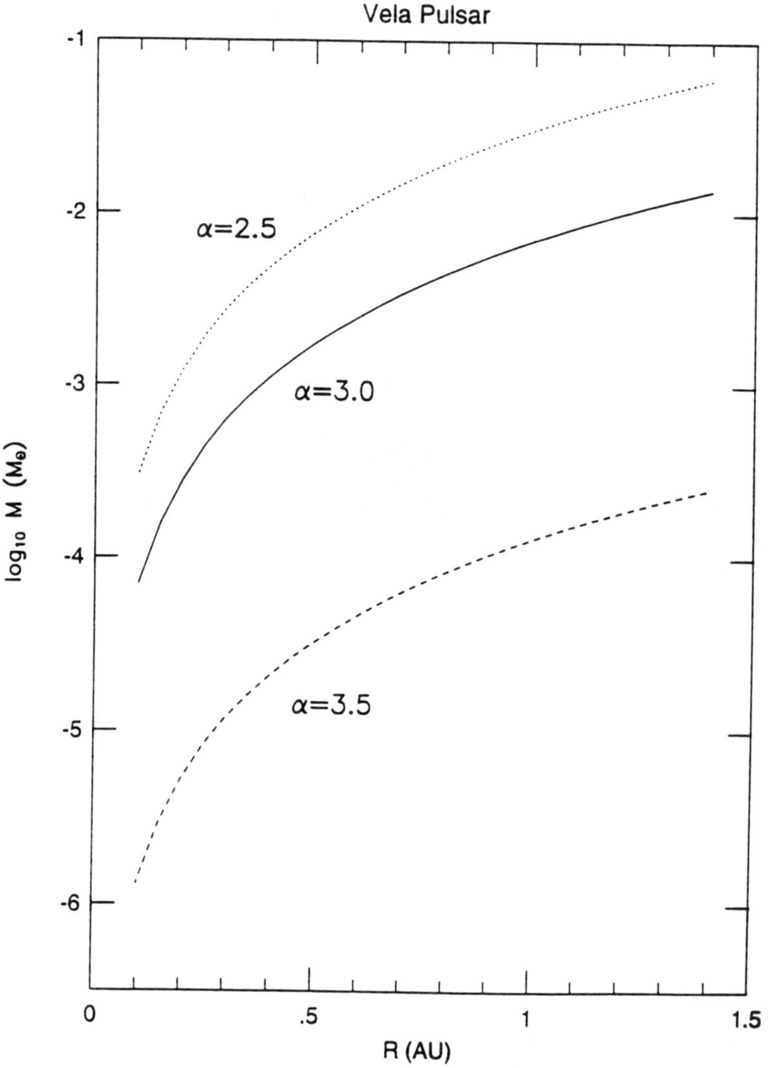

FIGURE I Asteroid mass limits for the Vela pulsar. At λ20 cm, there is less than 200μJy of unpulsed flux from the Vela pulsar (Beitenholtz, Frail & **Hankins** 1991). This limit constrains the total mass of asteroids illuminated by the pulsar emission beam. The plot shows the maximum total mass of asteroids within R AU of the neutron star. The limits are model-dependent and assume values for the scattering efficiency and mass density which are typical of Solar System asteroids ($A = 0.1$, $\rho = 3$ gm/cc). I also assume a power-law distribution of asteroid radii: $N(a) \propto a^{-\alpha}$ where $N(a)da$ is the number of asteroids with radii between a and $a + da$. The limits are plotted for three values of the power-law index in the range appropriate for a collection of collisionally-evolved rocky bodies.

threshold $S_u/S_p = 2 \times 10^{-4}$ at λ20 cm. Combining their flux limit with equation (4), I can compute limits on the total mass of illuminated asteroids orbiting the neutron star. Figure 1 shows the maximum mass of asteroids within orbital radius R as a function of the power-law index α. For $\alpha = 3.5$ the maximum mass within 1 AU of the neutron star is $M \lesssim 10^{-4} - 10^{-6} M_{\odot}$. These limits are smaller than the mass of circumstellar disks of gas and dust around some young stars (e.g., Sargent & Beckwith 1991) but they are orders of magnitude greater than the total mass of the asteroid belt in our own solar system, $M \sim 10^{-9} M_{\odot}$.

Perry & Lyne (1985) detected unpulsed emission from four pulsars using a two-element interferometer at 408 MHz. They obtained $S_u/S_u \sim 10^{-2}$ to 10^{-3} for PSR 1541+09, PSR 1604-00, PSR 1929+10 and PSR 2016+28. The unpulsed emission could be due to asteroids, but there is a less exotic explanation: Several of these objects are believed to be aligned rotators. The magnetic and spin axes of the pulsar are nearly parallel with the result that the observer never completely leaves the radio emission beam. Accurate polarimetry with a gated synthesis array is needed to distinguish between alternative explanations for the origin of the unpulsed flux.

ACKNOWLEDGMENTS

It is a pleasure to thank Leila Belkora, Dale Frail, and Steve Thorsett for stimulating discussions. Leila Belkora drew a beautiful picture of pulsar radio waves scattered by asteroids (in full color) for the display poster.

REFERENCES

Bietenholz, M.F., Frail, D.A. & Hankins, T.H., 1991, ApJ, 376, L41.

Gradie, J.C., Chapman, C.R. & Tedesco, E.F. 1989, in Asteroids II, Eds., R.P. Binzel, T. Gehrels & M. Shapley Matthews, University of Arizona Press, Tucson, 328.

Kowal, C.T. 1988, Asteroids: Their nature and utilization, Halsted Press, New York, 47.

Manchester, R.N. & Taylor, J.H. 1977, Pulsars, W.H. Freeman and Company, San Francisco, 220.

Nakamura, T. & Piran, T. 1991, ApJ, 382, L81.

Narayan, R. and Vivekanand, M. 1983, A&A, 122, 45.

Ostro, S.J. 1989, in Asteroids II, Eds., R.P. Binzel, T. Gehrels & M. Shapley Matthews, University of Arizona Press, Tucson, 192.

Ostro, S.J. Campbell, D.B. & Shapiro, I, 1985, Science, 229, 442.

Perry, T.E. and Lyne A.G. 1985, MNRAS, 212, 489.

Sargent, A.I. & Beckwith, S.V.W 1991, ApJ, 382, L31.

Thorsett, S.E. & Phillips, J.A., 1992, ApJ, 387, L69.

Wolszczan, A. & Frail, D.A., 1992, Nature, 355, 145.

Planets Around Pulsars
ASP Conference Series, Vol. 36, 1993
J.A. Phillips, J.E. Thorsett, and S.R. Kulkarni

IRAS SOURCES NEAR THE POSITIONS OF PULSARS

Dave Van Buren, Theoretical Astrophysics, California Institute of Technology, MS 130-33, Pasadena CA 91125

Susan Terebey, Infrared Processing and Analysis Center, California Institute of Technology and Jet Propulsion Laboratory, MS 100-22, Pasadena CA 91125

Abstract We searched the *IRAS* database at the positions of pulsars to see if they are associated with far-infrared sources, as might be expected if some pulsars possess preplanetary disks. There are 13 *IRAS* sources from the Point Source Catalog within 1 arcminute of nearly 500 radio pulsars. While this does not represent an enhancement over the number of chance coincidences, we argue that several are in fact associated with pulsars.

1. INTRODUCTION

Planets appear to be orbiting at least one pulsar, PSR1257+12 (Wolsczan and Frail, 1992). Many scenarios for their possible origin are discussed in these proceedings, some of which require protoplanetary disks around newly formed neutron stars. Such disks might be expected to be far-infrared sources.

To motivate this consider a young neutron star with a spin-down luminosity of $L_{\dot\omega}$ that has a protoplanetary disk of radius r_{disk} capable of absorbing the incident energy via any of a variety of mechanisms. For example, the pulsar's very low frequency radiation could be absorbed in a partially ionized but dense disk, with subsequent collisional heating of dust leading to infrared emission. In any case, let a fraction f of the pulsar luminosity be absorbed by the disk. Then the disk luminosity, assuming the incident energy is radiated, rather than drives mass flows, is

$$\begin{aligned} L_{disk} &= f\, L_{\dot\omega} \\ &= 10^{36}\, \frac{f}{0.01}\, \frac{L_{\dot\omega}}{10^{38}\ \mathrm{erg\ s^{-1}}}\ \mathrm{erg\ s^{-1}}. \end{aligned}$$

If the disk radiates like a blackbody, then its temperature is of order

$$\begin{aligned} T_{disk} &= \left(\frac{L_{disk}}{2\pi r_{disk}^2 \sigma}\right)^{1/4} \\ &= 594\, \left(\frac{f}{0.01}\right)^{1/4} \left(\frac{L_{\dot\omega}}{10^{38}\ \mathrm{ergs^{-1}}}\right)^{1/4} \left(\frac{r_{disk}}{10\ \mathrm{AU}}\right)^{-1/2}\ \mathrm{K}, \end{aligned}$$

and has a spectrum that peaks near 10 μm for our nominal parameter values. At a distance d, the flux density at the blackbody peak is

$$S = \pi r^2 B(\lambda = \lambda_{max}, T_{disk})/d^2$$
$$= 34 \left(\frac{r_{disk}}{10\ \mathrm{AU}}\right)^{1/2} \left(\frac{d}{1\ \mathrm{kpc}}\right)^{-2} \left(\frac{f}{0.01}\right)^{3/4} \left(\frac{L_{\dot{\omega}}}{10^{38}\ \mathrm{ergs^{-1}}}\right)^{3/4} \mathrm{Jy},$$

which lies in the regime surveyed by *IRAS*'s Point Source Catalog, $7 < \lambda < 120$ μm, $S > 0.5$ Jy, (*IRAS Point Source Catalog, Version 2 1988*). With these simple considerations in mind, we have studied the infrared emission near pulsars to look for protoplanetary disks.

TABLE 1. IRAS Fluxes of Pulsar-Coincident Sources[a]

Pulsar	PSC Name	F_{12} (Jy)	F_{25} (Jy)	F_{60} (Jy)	F_{100} (Jy)
0021-72	00218-7221	1.71	0.51	0.40u	1.00u
0531+21	05315+2158	1.25u	4.11	61.5	121.
1509-58	15099-5856	1.89u	19.9	37.0	138.u
1627-47	16269-4721	5.18u	2.14	68.1u	237.
1718-35	17182-3529	5.26	12.2u	404.u	3346.
1744-24	17450-2444	12.4	7.49	6.46u	78.7u
1745-20	17459-2020	1.17	0.69u	1.26u	11.4u
1806-21	18063-2110	3.03u	2.40	20.4u	371.u
1849+00	18498+0028	4.11	11.3	72.9	176.u
1902-01	19029-0101	0.52u	0.25u	0.71	33.8u
1907-03	19079-0315	2.27	12.0	4.21	16.6u
1933+16	19334+1609	0.39	0.26u	1.34	45.6u
2034+19	20349+1932	0.30	0.25u	0.40u	1.56u

a. Fluxes from the *IRAS* Point Source Catalog, Version 2, 1988. Upper limits denoted by lowercase u.

2. CATALOG SEARCHES

We obtained Taylor's pulsar catalog of 547 pulsars (Taylor, 1992). Because the *IRAS* positions are more accurate than 1 arcminute, we discarded pulsars with position uncertainties greater than this amount, leaving 478 pulsars in the trimmed catalog. We compared the pulsar positions with the 245,889 far-infrared sources listed in the *IRAS* Point Source Catalog and found the 13 coincidences listed in Table 1. The important systematics in this table are that young pulsars with supernova remnants (0531+21, 1509-58) and pulsars in globular clusters (0021-72, 1744-24) are overrepresented.

To assess whether the number of coincidences shows a significant enhancement of the far-infrared source density near pulsars, we compared the

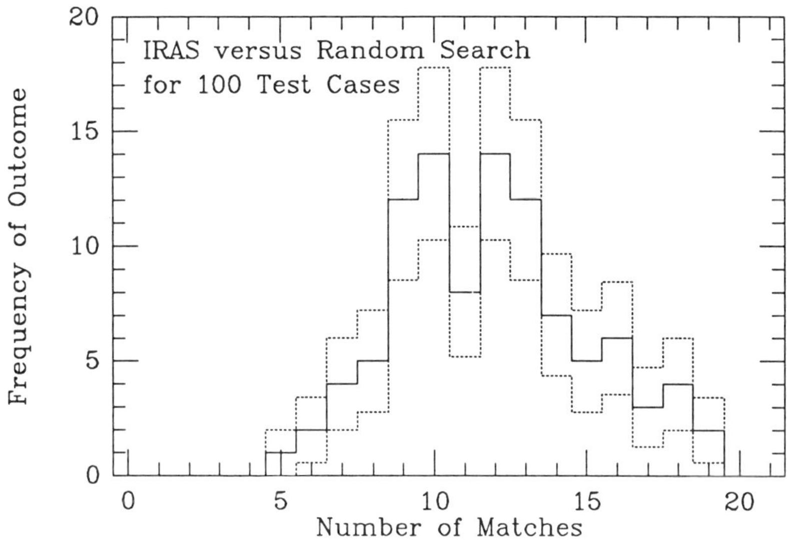

Figure 1. Histogram of the number of coincidences between perturbed pulsar catalogs and the PSC for 100 trials. Pulsar positions are perturbed by 30 to 60 arcminutes in a random direction. The number of coincidences with pulsars in the sky is 13 so no significant excess is seen.

PSC against 100 randomly generated pulsar catalogs. These catalogs were formed by perturbing the trimmed Taylor catalog's pulsar positions between 30 and 60 arcminutes in a random direction. The frequency of associations in this ensemble of random pulsar catalogs is shown in Figure 1. It is clear that the observed number of coincidences is consistent with chance associations of pulsars with far-infrared sources. In spite of this, the fact that some of the pulsars found coincident with PSC sources are special argues that under rare circumstances pulsars may be physically associated with far-infrared emission at flux levels > 0.5 Jy.

3. HIGH RESOLUTION 12 μm IMAGES

We used a variation of the standard *IRAS* HiRes processing (Aumann *et al.* 1990) to deconvolve the *IRAS* survey data at the positions of 12 of the 13 pulsars in Table 1. After 20 iterations the deconvolved images nearly achieve the diffraction limit. Gray scale representations of the 12 μm images are presented as Figure 2, where each field is 20 arcminutes on a side and crosses mark the pulsar radio positions. The interior gaps of the crosses are 2 arcminutes across. The *IRAS* morphology shows that some sources are unresolved while others are clearly extended.

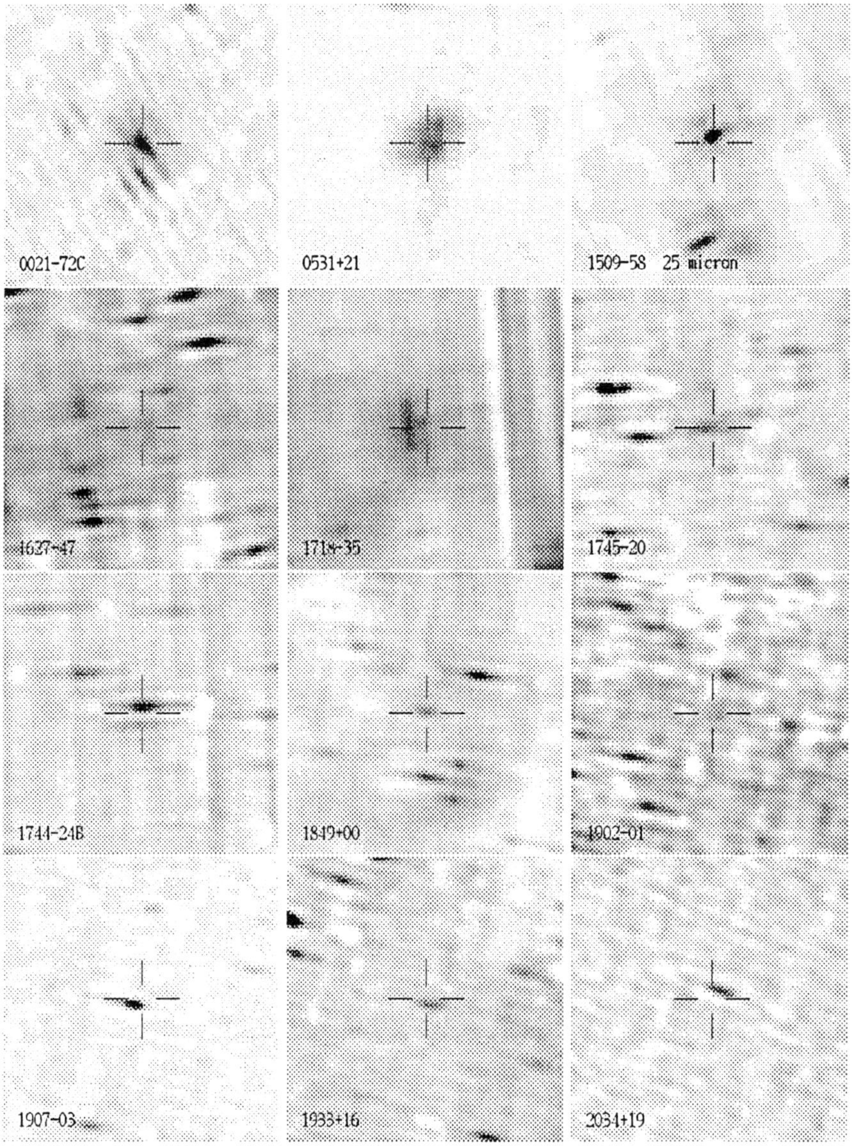

Figure 2. IRAS High Resolution images of pulsar fields. Displayed are the 12 micron images. Source 1509-58 is very weak at 12 microns so the 25 micron image is displayed for that source only. The images are 20′ on a side. The inner parts of the solid lines delimit a 2′ region centered on the pulsar position. Some sources are clearly extended while others are pointlike.

Figure 3 shows the relationship between the 1 arcminute search areas around each pulsar and the 95% confidence level error ellipses of the corresponding PSC positions. For those point sources inside extended objects, the PSC error ellipses underestimate the true positional uncertainties. When comparing Figures 2 and 3 bear in mind that the HiRes positions are less accurate that the PSC and radio positions and typically differ from the PSC positions by 20″.

3.1. 0021-72, THE 47 TUC PULSARS

There are at least 11 millisecond pulsars near the core of 47 Tucanae. A study of *IRAS* pointed observations of 47 Tuc by Gillett *et al.* (1988) found marginal excesses of 12, 25 and 100 μm radiation above that expected from stars, which could correspond to a small amount of intracluster dust. The far infrared excess, less than about 4 $L_\odot$, requires a color temperature of about 200K, rather higher than that expected for dust in radiative equilibrium at the cluster core. If these millisecond pulsars are similar to 1937+21, with $\dot{P} \approx 1 \times 10^{-19}$ s s^{-1}, then the spin down luminosity of one pulsar ($\approx 10^{36}$ erg s^{-1}) could provide the requisite power with $f \approx 0.001 - 0.01$. There are presumably a much larger number of undetected pulsars in 47 Tuc, whose beams do not sweep over the Earth as well, so only a small fraction of globular cluster pulsars need be of this class. Matching the color temperature requires r_{disk} of order 100 AU The existence of such objects seems rather implausible on a number of grounds however. For example, the dynamical environment at the cluster core is so violent it is not likely that a 100 AU debris cloud could survive from formation to the present day.

3.2. 0531+21, THE CRAB PULSAR

The 12 μm image of the Crab nebula, excepting the bright knots at the center, corresponds well with a map of the 20 cm radio continuum (Velusamy, 1985), suggesting that the extended infrared is synchrotron radiation. If the far infrared were from dust in the filaments, the morphology should instead reflect the optical line radiation's, which it does not. The position of the eastern knot is consistent with the both the pulsar's position and the peak of the radio map. The origin of the brighter western knot is unclear, but it does correspond to one of the more prominent optical filaments.

IRAS emission from the Crab nebula was discussed by Marsden *et al.* (1984) on the basis of some of the first data returned from the satellite. They concluded that there is an excess luminosity of $\approx 1000\ L_\odot$ in the *IRAS* bands above the synchrotron power law, explainable as due to $< 0.03\ M_\odot$ of dust with a color temperature of ≈ 70 K. The poor correlation of the far infrared with the optical filaments, where the bulk of the mass resides is then puzzling.

The protoplanetary disk model requires $f \approx 0.01$ but $r_{disk} \approx 1000$ AU, which is rather large to yield planets in 1 AU orbits without large angular momentum loss. Nonetheless, the 12 *IRAS* emission from the Crab nebula is interesting enough to warrant ground based 10 μm observations at the highest possible spatial resolution.

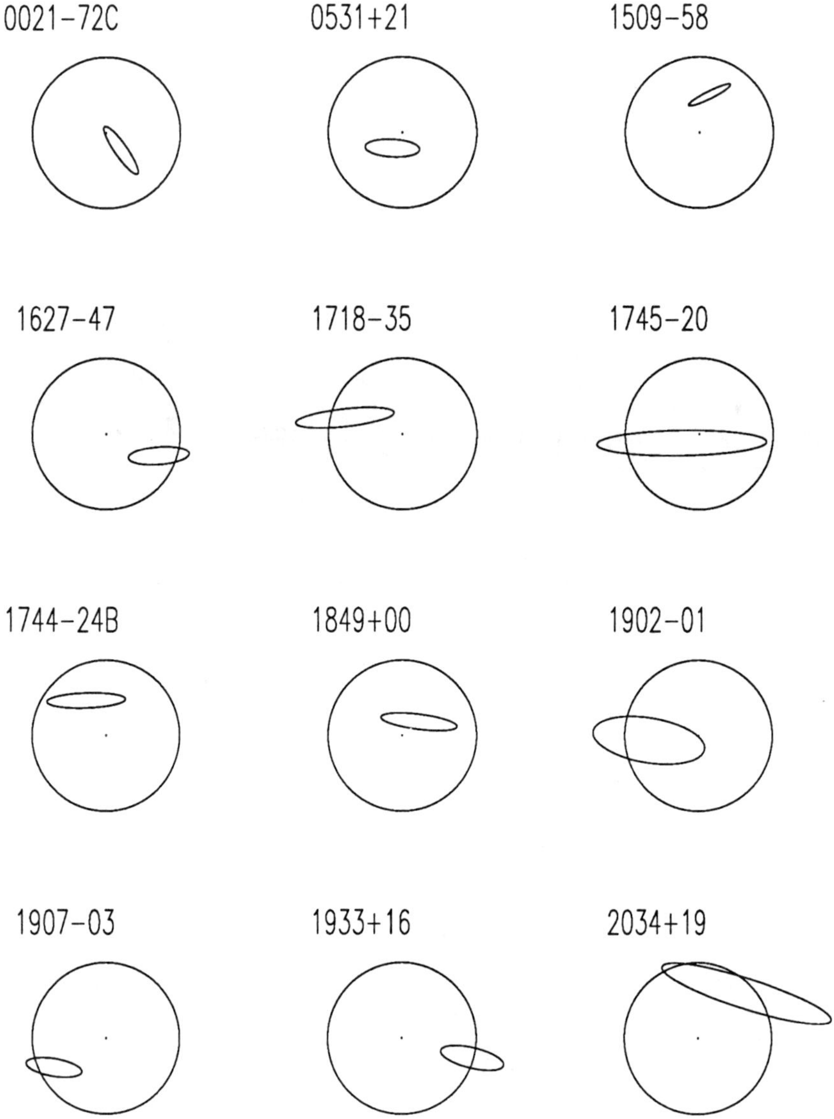

Figure 3. IRAS Point Source Catalog 95% confidence error ellipses superposed on the 1 arcminute radius search regions around the pulsars of figure 2. The sources associated with 0021-72C, 0531+21, 1509-58, 1627-47, 1718-35, 1745-20, 1849+00, and 1902-01 are extended to varying degrees, and the error ellipses underestimate the true positional uncertainties of any point source components.

3.3. 1509-58

This is one of the youngest known pulsars. Its far infrared emission was recently discussed by Arendt (1991) using *IRAS* coadded survey data. Briefly, the far infrared has a color temperature of 50-70 K and a luminosity 2.5×10^{37} erg s^{-1}. This should be compared with the pulsar spin down luminosity of 1.8×10^{37} erg s^{-1} (Weisskopf *et al.* 1983); *i.e.* the two luminosities are equal within the uncertainties. In the simple model for a protoplanetary disk this gives $f \approx 1$ and $r \approx 3 \times 10^3$ AU, rather extreme parameters for a protoplanetary disk. The coincidence between L_{FIR} and $L_{\dot{\omega}}$ poses a strong challenge to models if the association is real.

3.4. 1744-24, the Terzan 5 pulsars

There are two known pulsars in Terzan 5. One is located at the cluster center, near the position of the indicated point source and the other about 30 arcseconds south, coincident with a 12 μm point source visible in the image. Imaging at K is needed to map the locations of AGB stars to eliminate an association of the second source with a known class of far-infrared bright objects. The central source has large excesses in all *IRAS* bands above the fluxes expected from the cluster stars, comparable to or exceeding the spin down luminosities of the of pulsar. The *IRAS* colors resemble most closely the colors of the cirrus, with a color temperature $<$ 20 K, much too low to be a protoplanetary disk.

ACKNOWLEDGEMENTS

DVB thanks the taxpayers of the US for their support through a NASA ADP grant as well as the staff at IPAC for their cheerful assistance.

REFERENCES

Arendt, R. G. 1991, AJ 101, 1260.
Aumann, H. H., Fowler, J. W., and Melnyk, M. 1990, AJ 99, 1674.
Gillett, F. C., de Jong, T., Neugebauer, G., Rice, W. L., and Emerson, J. P. 1988, AJ 96, 116.
IRAS Point Source Catalog, Version 2. 1988, Joint *IRAS* Science Working Group (Washington DC: GPO).
Taylor, J. 1992, anonymous ftp to pulsar@priceton.edu, file pub/psrcat.tar.Z.
Velusamy, T. 1985, MNRAS 212, 359.
Weisskopf, M. C., *et al.* 1983, ApJ 267, 711.
Wolszczan, A. and Frail, D. A. 1992, Nature 355, 145.

Planets Around Pulsars
ASP Conference Series, Vol. 36, 1993
J.A. Phillips, J.E. Thorsett, and S.R. Kulkarni

THE DISTRIBUTION OF COMETS AROUND STARS

SCOTT TREMAINE
Canadian Institute for Theoretical Astrophysics
McLennan Labs, University of Toronto
60 St. George St., Toronto M5S 1A7, Canada

ABSTRACT Stars surrounded by planetary systems that formed from disks are likely to have comets as well. Many of the properties of comet clouds around such stars can be deduced from simple dynamical arguments.

INTRODUCTION

Most plausible formation scenarios for the planets around PSR1257+12 imply that they formed, like the terrestrial planets, by accretion from a disk of planetesimals.

Since planet formation is unlikely to be 100% efficient, there should be some planetesimals left in the disk after formation is complete. Residual planetesimals from the outer regions of our solar system ($r \gtrsim 10$ AU) are identified with comets, so I will call residual planetesimals "comets" even though their mass and composition may well differ from those of solar system comets ("asteroids" would be an equally good name).

Comets remaining in the disk after planet formation have two possible dynamical fates. Comets on near-circular orbits that are well-separated from the planets will survive on these orbits with little or no dynamical evolution. Examples of such objects in the solar system include the asteroids, and comets in a hypothetical Kuiper belt beyond Neptune.

On the other hand, orbits that approach too close to a planet are chaotic. The most important class of chaotic orbits is planet-crossing orbits, which are low-inclination orbits with periastron less than and apastron greater than the planet's semi-major axis. Comets on such orbits eventually collide with the planet or the central star or are ejected from the planetary system. In the solar system, it is believed that this ejection process has led to the formation of the Oort cloud, which contains 10^{11}–10^{12} comets at typical semi-major axes of $(2-3) \times 10^4$ AU.

The aim of this paper is to apply our understanding of the formation of the Oort cloud to deduce the properties of Oort-type comet clouds around other stars. Most of the ideas are well-known in the solar system context (Oort 1950, Hills 1981, Duncan et al. 1987, Fernández and Ip 1991); only the application to other planetary systems is new.

In this paper, an Oort-type cloud is defined to be a roughly spherical distribution of bound comets with typical semi-major axis a_f, formed by ejection of comets from a planetary system with characteristic size $a_p \ll a_f$.

FORMATION OF OORT-TYPE COMET CLOUDS

We consider a system containing a central star of mass $M_\star$, a single planet of mass M_p on a circular orbit of radius a_p, and a number of comets on planet-crossing orbits.

Usually, ejection occurs by a gradual random walk or diffusion of the comet orbit towards escape energy (as is also the case for escape of stars from star clusters). This process leads to highly eccentric orbits, since the comet's semi-major axis a becomes large while its periastron q remains comparable to a_p. The comet receives a kick from the planet's gravity each time it passes through periastron; the orbit is chaotic because the kick depends on the phase of the planet in its orbit.

We describe the comet's energy by the variable $x \equiv 1/a$, and let the diffusion coefficient $D_x = \langle(\Delta x)^2\rangle^{1/2}$ be the rms change in x per periastron passage arising from perturbations by the planet (D_x is independent of a so long as $a \gg a_p$, since all such orbits look like parabolas near periastron). Numerical integrations of highly eccentric, low-inclination, planet-crossing orbits yield (Fernández 1981, Duncan et al. 1987)

$$D_x \simeq \frac{10}{a_p}\frac{M_p}{M_\star}. \tag{1}$$

The characteristic diffusion time, in which the energy x changes by of order itself, is then

$$t_{\text{diff}} \equiv P\frac{x^2}{D_x^2}, \tag{2}$$

where $P = 2\pi a^{3/2}/(GM_\star)^{1/2}$ is the orbital period. Thus

$$t_{\text{diff}}(x) = 1.1 \times 10^9 \text{ y} \left(\frac{M_\star}{M_\odot}\right)^{3/2} \left(\frac{M_p}{M_\oplus}\right)^{-2} \left(\frac{x}{1 \text{ AU}^{-1}}\right)^{1/2} \left(\frac{a_p}{1 \text{ AU}}\right)^2. \tag{3}$$

The diffusion rate speeds up (t_{diff} decreases) as the comet energy increases (x decreases). Initially the comets are on orbits similar to that of the planet, so $x \approx a_p^{-1}$. Thus substantial orbit evolution occurs if and only if

$$t_{\text{ev}} \equiv t_{\text{diff}}(x = a_p^{-1}) \lesssim t_\star, \tag{4}$$

where $t_\star$ is the age of the planetary system. Thus a necessary condition for the formation of an Oort-type comet cloud is

$$\frac{M_p}{M_\oplus} \gtrsim \left(\frac{M_\star}{M_\odot}\right)^{3/4} \left(\frac{t_\star}{10^9 \text{ y}}\right)^{-1/2} \left(\frac{a_p}{1 \text{ AU}}\right)^{3/4}. \tag{5}$$

In an isolated system, most comets on planet-crossing orbits would escape in a few times t_{ev}. However in practice, once the orbit becomes large, the torque from the Galactic tide changes the orbital angular momentum and thus the periastron distance q. Once q is a few times a_p, planetary perturbations become ineffective, and the random walk of the comet's energy ceases: planet-induced diffusion in energy at fixed angular momentum is replaced by Galaxy-induced evolution in angular momentum at fixed energy. For highly eccentric orbits, the specific angular momentum $L = (2GM_\star q)^{1/2} \approx (2GM_\star a_\mathrm{p})^{1/2}$, and the torque per unit mass is $dL/dt = 5\pi kG\rho a^2$ where ρ is the local Galactic mass density and k is a geometrical factor that varies between 0 and 1 depending on the orientation of the orbit (Heisler and Tremaine 1986). For a typical orientation, the time required for the Galactic tide to change the angular momentum by of order itself is then (Duncan et al. 1987)

$$t_\mathrm{tide} = 1 \times 10^{15}\ \mathrm{y} \left(\frac{M_\star}{M_\odot}\right)^{1/2} \left(\frac{\rho}{0.15 M_\odot\ \mathrm{pc}^{-3}}\right)^{-1} \left(\frac{q}{1\ \mathrm{AU}}\right)^{1/2} \left(\frac{x}{1\ \mathrm{AU}^{-1}}\right)^2 . \quad (6)$$

The density $0.15 M_\odot\ \mathrm{pc}^{-3}$ is a compromise between the values obtained for the solar neighborhood by Kuijken (1991) and Bahcall et al. (1992).

Tidal torques freeze the comet out of the random walk in energy at the semi-major axis a_f where $t_\mathrm{tide}(q = a_\mathrm{p}) = t_\mathrm{diff}$, that is,

$$a_\mathrm{f} = 1 \times 10^4\ \mathrm{AU} \left(\frac{M_\star}{M_\odot}\right)^{-2/3} \left(\frac{\rho}{0.15 M_\odot\ \mathrm{pc}^{-3}}\right)^{-2/3} \left(\frac{M_\mathrm{p}}{M_\oplus}\right)^{4/3} \left(\frac{a_\mathrm{p}}{1\ \mathrm{AU}}\right)^{-1} . \quad (7)$$

Many comets escape on their next orbit if $x = 1/a \lesssim D_x$. Thus freezing is an effective barrier to the escape process only if

$$1/a_\mathrm{f} \gtrsim D_x; \quad (8)$$

this condition implies that many of the comets escape unless

$$\frac{M_\mathrm{p}}{M_\oplus} \lesssim 1.7 \left(\frac{M_\star}{M_\odot}\right)^{5/7} \left(\frac{\rho}{0.15 M_\odot\ \mathrm{pc}^{-3}}\right)^{2/7} \left(\frac{a_\mathrm{p}}{1\ \mathrm{AU}}\right)^{6/7} . \quad (9)$$

The frozen comets form an extended disk in the plane of the planets, which is steadily thickened by perturbations from passing stars and the Galactic tide. The timescale to convert the disk into a roughly spherical cloud is simply t_tide, evaluated at $q = a = a_\mathrm{f}$. The condition that this timescale is less than the age $t_\star$ can be shown to be the same as (5). Thus comet clouds formed by this process should be approximately spherical.

The cloud cannot extend beyond the Roche surface of the star, set by the tidal field of the Galaxy. For the Sun the Roche surface has semi-axes 1.47 pc, 0.98 pc, and 0.68 pc, with the shortest axis normal to the Galactic plane (Antonov and Latyshev 1972, Heisler and Tremaine 1986); these values are for $\rho = 0.15 M_\odot\ \mathrm{pc}^{-3}$. The size of the Roche surface can be roughly estimated by comparing the tidal force between two points separated by a vertical distance z, $4\pi G\rho z$, to the star's gravitational force at that distance, $GM_\star/z^2$. If we denote the distance at which these forces are equal as the Roche or tidal radius a_t, we have

$$a_\mathrm{t} = 1.7 \times 10^5\ \mathrm{AU} \left(\frac{M_\star}{M_\odot}\right)^{1/3} \left(\frac{\rho}{0.15 M_\odot\ \mathrm{pc}^{-3}}\right)^{-1/3} . \quad (10)$$

Planets cannot survive outside the tidal radius, so

$$a_\mathrm{p} \lesssim a_\mathrm{t}. \tag{11}$$

A comet cloud can only form if the freezing semi-major axis is less than the tidal radius, $a_\mathrm{f} \lesssim a_\mathrm{t}$, which implies

$$\frac{M_\mathrm{p}}{M_\oplus} \lesssim 8 \left(\frac{M_\star}{M_\odot}\right)^{3/4} \left(\frac{\rho}{0.15 M_\odot\ \mathrm{pc}^{-3}}\right)^{1/4} \left(\frac{a_\mathrm{p}}{1\ \mathrm{AU}}\right)^{3/4}, \tag{12}$$

and the formation process as described here only applies if the freezing semi-major axis exceeds the planet's semi-major axis, $a_\mathrm{f} \gtrsim a_\mathrm{p}$, which implies

$$\frac{M_\mathrm{p}}{M_\oplus} \gtrsim 10^{-3} \left(\frac{M_\star}{M_\odot}\right)^{1/2} \left(\frac{\rho}{0.15 M_\odot\ \mathrm{pc}^{-3}}\right)^{1/2} \left(\frac{a_\mathrm{p}}{1\ \mathrm{AU}}\right)^{3/2}. \tag{13}$$

An additional constraint is that the comets must not impact the planet before they diffuse to the cloud (I am indebted to Al Harris for stressing the importance of this process). Comets initially on orbits with $a \approx a_\mathrm{p}$ make roughly $N = (x_\mathrm{p}/D_x)^2$ orbits before reaching the cloud. If their typical inclination is $\Delta\theta$ and the escape speed from the planet is less than the orbital speed so that gravitational focusing is negligible, the chance of an impact on each periastron passage is roughly $p = (R_\mathrm{p}/a_\mathrm{p})^2/\Delta\theta$, where R_p is the planetary radius. In order that most of the comets do not strike the planet we must have $Np \lesssim 1$, or

$$\frac{M_\mathrm{p}}{M_\oplus} \gtrsim 13 \left(\frac{M_\star}{M_\odot}\right)^{3/2} \left(\frac{a_\mathrm{p}}{1\ \mathrm{AU}}\right)^{-3/2} \left(\frac{\rho_\mathrm{p}}{3\ \mathrm{g\ cm}^{-3}}\right)^{-1/2} \left(\frac{\Delta\theta}{0.1\ \mathrm{rad}}\right)^{-3/4}. \tag{14}$$

where ρ_p is the planet density.

Comets are ejected from the cloud by gravitational perturbations from passing stars and other objects. The half-life of a comet orbiting a star in a region with kinematics similar to the solar neighborhood is (Weinberg et al. 1987, Figure 7a)

$$t_{1/2} = 10^{10}\ \mathrm{y} \left(\frac{M_\star}{M_\odot}\right) \left(\frac{10^4\ \mathrm{AU}}{a}\right) \left(\frac{\rho}{0.15 M_\odot\ \mathrm{pc}^{-3}}\right)^{-1}. \tag{15}$$

This result neglects the uncertain contribution from molecular clouds, and is strictly valid only for $a \ll a_\mathrm{t}$. Replacing a by the cloud radius a_f and requiring that $t_{1/2}$ exceed the age of the system $t_\star$, we obtain the constraint

$$\frac{M_\mathrm{p}}{M_\oplus} \lesssim 6 \left(\frac{M_\star}{M_\odot}\right)^{5/4} \left(\frac{t_\star}{10^9\ \mathrm{y}}\right)^{-3/4} \left(\frac{\rho}{0.15 M_\odot\ \mathrm{pc}^{-3}}\right)^{-1/4} \left(\frac{a_\mathrm{p}}{1\ \mathrm{AU}}\right)^{3/4}. \tag{16}$$

For Population II (halo) stars this constraint must be modified because the relative velocities are higher than in the disk. In this case, the appropriate constraint is strongly dependent on the unknown composition of the dark halo; if the dark matter is in objects of stellar mass or less, the constraint is much weaker than the constraint (16) for disk stars.

APPLICATIONS

The solar system

Figure I shows the constraints (5), (9), (11), (12), (13), (14), and (16) for parameters appropriate for the solar system: $M_\star = 1M_\odot$, $t_\star = 4.5 \times 10^9$ y, and $\rho = 0.15 M_\odot$ pc^{-3}. The positions of the eight major planets are also plotted, and the region of parameter space allowed by the constraints is stippled. The figure exhibits some well-known and some new results:

- Comets crossing the orbits of Jupiter and Saturn cannot create an Oort-type cloud: the planets are so massive that most comets are ejected. Thus Oort cloud comets must come from the Uranus-Neptune region.
- The typical semi-major axis of the comet cloud formed by Uranus or Neptune is $a_\mathrm{f} \approx 10^4$ AU, close to the semi-major axes of new comets (the observed semi-major axes of new comets are somewhat larger, $(2-3) \times 10^4$ AU, because comets with smaller semi-major axes are not visible from Earth until after they have been perturbed by Jupiter).
- Venus or Earth could not efficiently form an Oort-type cloud, even if the massive outer planets were not present: most comets on Venus- or Earth-crossing orbits collide with these planets before reaching the cloud.
- At any planetary semi-major axis a_p there is only a narrow range (less than a factor 5) of planetary masses that are large enough to create an Oort-type cloud within the age of the solar system, yet small enough not to eject most comets. Within this allowed range, the typical size of the comet cloud, a_f, is almost independent of a_p. Outside this range, cloud formation can still occur, but with low efficiency.

PSR1257+12

Figure II shows the same constraints for parameters appropriate to PSR1257+12: $M_\star = 1.4 M_\odot$, $t_\star = 8 \times 10^8$ y, and planetary semi-major axes and masses $a_{p1} = 0.36$ AU, $a_{p2} = 0.47$ AU, $M_{p1} = 3.4 M_\oplus$, $M_{p2} = 2.8 M_\oplus$ (for $\sin i = 1$; if $\sin i < 1$ the masses are larger).

The constraints on the planets that can efficiently form an Oort-type cloud are similar to those in the solar system. However, the allowed range of parameter space is even smaller, mainly because more massive planets are needed to create the cloud within the shorter age of the system.

Comets crossing the orbits of the planets P1 and P2 would mostly be ejected or collide with the planets. Those that did manage to reach the cloud would be found at semi-major axes of a few times 10^4 AU.

HALO STARS

Figure III shows constraints on the formation of a comet cloud for a typical halo star, $M_\star = 0.5 M_\odot$, $t_\star = 10^{10}$ y, $\rho = 0.01 M_\odot$ pc^{-3}. The constraints are qualitatively similar to those for the solar system, but the planets that are most effective in producing a cloud are about a factor of three smaller at a given semi-major axis.

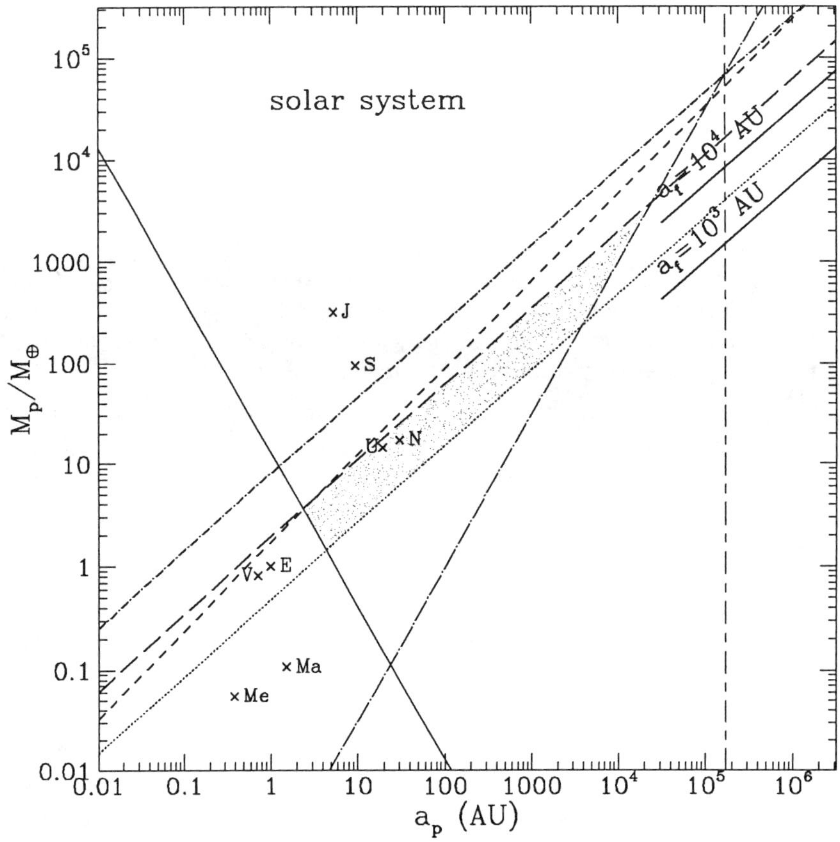

FIGURE I A plot of planet mass against planet semi-major axis showing the regions that are able to create an Oort-type cloud efficiently, for a star of mass $1M_\odot$ and age 4.5×10^9 y that is located in the solar neighborhood. The constraints plotted are: the diffusion time must be less than the age (eq. 5, dotted line); the freezing energy must exceed the rms energy change per orbit (eq. 9, short-dash line); the planet's semi-major axis must be less than the tidal radius (eq. 11, vertical short dash-long dash line); the freezing semi-major axis must be less than the tidal radius (eq. 12, dot-short dash line); the freezing semi-major axis must exceed the planet's semi-major axis (eq. 13, dot-long dash line); the comets must not impact the planet (eq. 14, solid line); the comets must not be ejected by passing stars (eq. 16, long-dash line). Lines have also been marked showing the location of freezing semi-major axes of 10^3 and 10^4 AU; although the lines have been truncated for clarity, they can be extrapolated across the whole figure.

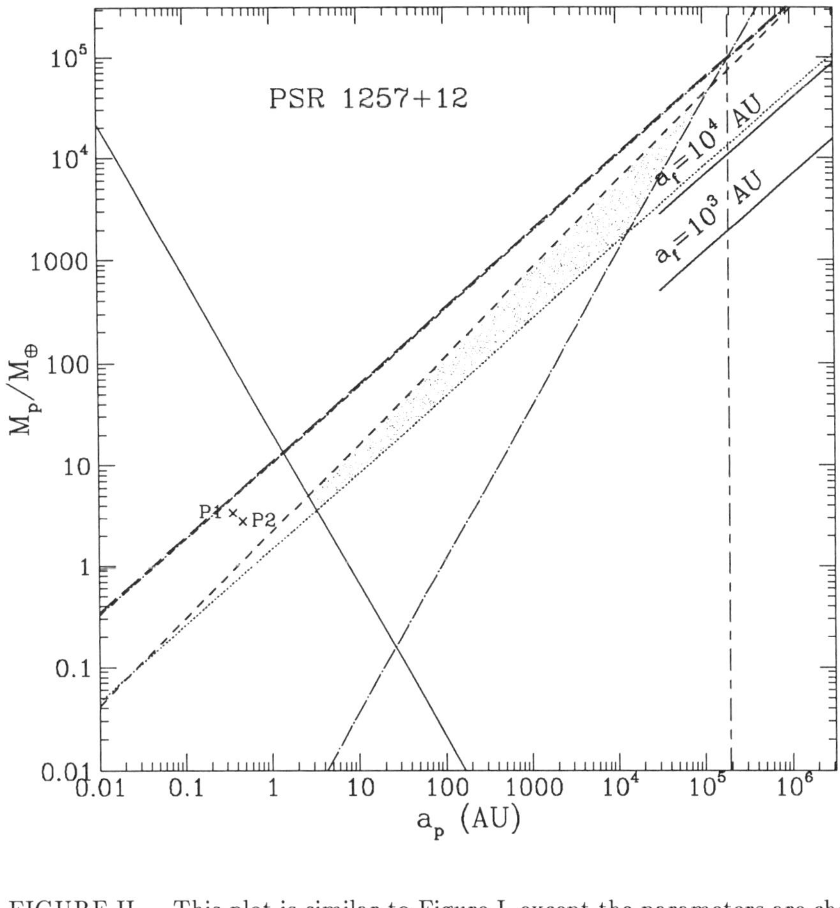

FIGURE II This plot is similar to Figure I, except the parameters are chosen to match those of the pulsar: a $1.4 M_\odot$ star of age 0.8×10^9 y, located in the solar neighborhood.

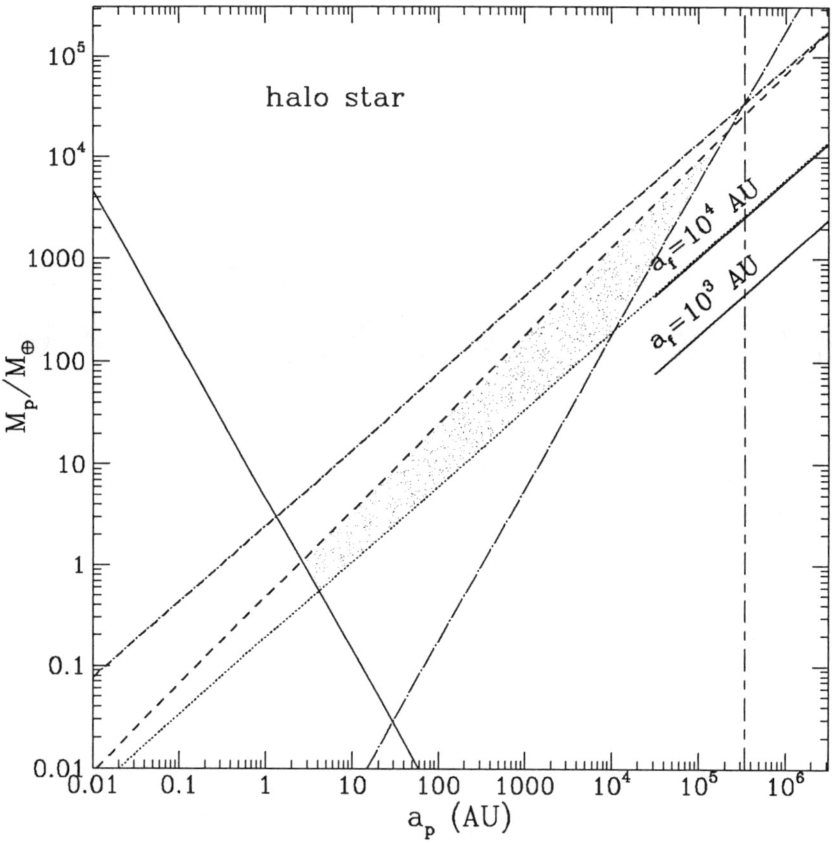

FIGURE III This plot is similar to Figure 1, except the parameters are chosen to represent those of halo stars rather than the solar system: a $0.5M_\odot$ star with an age of 10^{10} y, and a local density of $0.01M_\odot$ pc^{-3}. The constraint that comets are not ejected by passing stars or other objects is not plotted, since the effects of such stars are weak and the number density and mass of the objects composing the halo are not known.

SUMMARY

When planets form from a disk of planetesimals, there is likely to be a residual population of small bodies after planet formation is complete. It is natural to call these residual planetesimals "comets" or "asteroids", although their mass and composition may differ from their namesakes in our own planetary system.

Comets are likely to be found in one of the following configurations:

- A disk or belt of comets on regular (low-inclination, low-eccentricity) orbits that are well-separated from the planets (e.g. the asteroid belt).
- A cloud of comets scattered to large distances by planetary perturbations (e.g. the Oort cloud).

A variety of other configurations is possible. For example, a disk of comets on planet-crossing orbits can be present, if the planetary masses are small enough that the lifetime of such orbits is longer than the stellar lifetime. Another possibility is a spherical distribution of intermingled comets and planets, which can occur if the planets form at large enough radius that their orbits are isotropized by tidal forces after planet formation is complete.

If the configuration of the planets is known, then many of the properties of the comet distribution, in particular of any Oort-type cloud, can be predicted by the same dynamical arguments that successfully explain the structure of the solar system cloud. This paper presents a preliminary outline of the likely properties of Oort-type clouds around stars other than the Sun.

The theory presented here ignores many important issues, such as: the effect of two or more planets (note that comets are injected into the Oort cloud mainly by Uranus and Neptune but removed mainly by Jupiter and Saturn); survival of the comets in the stellar radiation field; the rate of collisions with the star; perturbations by passing molecular clouds; tidal shocks as halo stars pass through the disk; loss of comets when external perturbations bring the periastron back into the planetary system; and so forth.

The stippled areas in the figures show parts of parameter space in which formation of an Oort-type cloud is efficient, that is, in which most comets initially on low-inclination planet-crossing orbits will now be found in the cloud. Clouds may still be present even if the parameters of the star and planets lie outside the stippled region, but in this case cloud formation is inefficient.

The following conclusions indicate the results that can be derived from this type of analysis:

- Stars similar to the Sun can only generate an Oort-type cloud efficiently if they contain planets at semi-major axis $a_{\mathrm{p}} \gtrsim 3$ AU, whose masses lie within a factor of two or so of the relation $M_{\mathrm{p}}/M_{\oplus} = (a_{\mathrm{p}}/1\ \mathrm{AU})^{3/4}$. Thus Oort-type clouds are not expected to be found around every star, and may in fact be rather rare.
- The typical semi-major axis in an Oort-type cloud around a solar-type star will be $a_{\mathrm{f}} \approx 10^4$ AU.
- Halo stars require planets that are smaller by about a factor of 3 at a given semi-major axis, and produce a somewhat larger cloud, $a_{\mathrm{f}} \approx 2 \times 10^4$ AU.

- From equations (5), (9), and (11) we may deduce that no Oort-type cloud can be formed if $t_\star \lesssim 3 \times 10^7$ y($\rho/0.15 M_\odot$ pc^{-3}). Hence young, massive stars ($M \gtrsim 7 M_\odot$) do not have Oort-type comet clouds.

REFERENCES

Antonov, V. A., and Latyshev, I. N. 1972, in *The Motion, Evolution of Orbits, and Origin of Comets*, G. A. Chebotarev, E. I. Kazimirchak-Polonskaya, and B. G. Marsden, eds., Dordrecht: Reidel, 341

Bahcall, J. N., Flynn, C., and Gould, A. 1992, *ApJ*, **389**, 234

Duncan, M., Quinn, T., and Tremaine, S. 1987, *AJ*, **94**, 1330

Fernández, J. 1981, *A&A*, **96**, 26

Fernández, J. A., and Ip, W. H. 1991, in *Comets in the Post-Halley Era*, R. L. Newburn, M. Neugebauer, and J. Rahe,eds., Dordrecht: Kluwer, 487

Hills, J. G. 1981, *AJ*, **86**, 1730.

Heisler, J., and Tremaine, S. 1986, *Icarus*, **65**, 13.

Kuijken, K. 1991, *ApJ*, **372**, 125

Oort, J. H. 1950, *BAN*, **11**, 91

Weinberg, M. D., Shapiro, S. L., and Wasserman, I. 1987, *ApJ*, **312**, 367

Planets Around Pulsars
ASP Conference Series, Vol. 36, 1993
J.A. Phillips, J.E. Thorsett, and S.R. Kulkarni

THE ROTATION CURVE CONSPIRACY AND NEUTRON STAR/ASTEROID MODELS FOR GAMMA RAY BURSTS

EDWIN E. SALPETER AND IRA WASSERMAN
Center for Radiophysics and Space Research, Cornell University, Ithaca, NY 14853

INTRODUCTION

Many Gamma Ray Bursts (GRB hereafter) have a complicated time development with structure from less than a msec to more than 10 sec. The short end of this range is about the dynamic time scale of a neutron star. There are various possible models for the energy source and for the relatively long total duration, but one old model is particularly relevant to this conference: Harwit and Salpeter (1973) postulated comets in bound orbits around neutron stars which occasionally get disrupted and accreted, releasing their gravitational infall energy. In 1973, GRB's were generally thought to occur nearby; distances $d \sim$ 20pc to 100pc, say, seemed plausible and therefore cometary masses m of order 10^{16} to 10^{18} gm were required to give the observed GRB fluences. The energy spectra of burst photons usually range from X-rays to γ-rays above 1 MeV, sometimes with line features at $\sim 10 - 50$ keV and $\lesssim 511$ keV, and even evolve in the course of a burst (cf Higdon and Lingenfelter 1990). As discussed in papers during the intervening 19 years [see especially Colgate and Petschek (1981), Katz (1986), Tremaine and Zytkow (1986)], the physics of the infall process is also complex, so one can expect a complex and evolving spectrum of photons, but it is difficult to explain why more of the emission is observed above 100 keV than below.

In Section 2 we discuss how the new GRO/BATSE results show, in conjunction with older PVO and KONUS data, that the distribution in space of the GRB sources must have an outer bounding surface which is approximately a sphere centered on our location. The Galactic disk has a bounding surface, but it is a pair of planes and not a sphere. Cosmological redshifts at $z \sim 1$ naturally provide a bounding sphere. If gravitational infall energy powers GRB's the required mass $m \sim M_\odot$ cosmological distances; models for the merger of a neutron star binary have been published since 1986 [see, e.g. Paczynski 1986, Goodman 1986, Narayan, Piran and Shemi 1991]. We do not discuss this class of models at all, but only neutron stars in some kind of extended halo around our own Galaxy. The required mass of an infalling object is then of order $m \sim 10^{21}$ to 10^{23} gm, and we shall use the term "asteroid" for any solid object in this mass range, irrespective of its formation history or orbit.

THE BATSE DIPOLE MOMENT AND HALO MODELS

Several hundred GRB's have been recorded ($\gtrsim 300$ by GRO) and BATSE observations have enabled the detection of fainter bursts. We now have detailed data on the distributions of observed gamma ray burst count rates and positions. To interpret the results, we shall oversimplify by neglecting variations in burst duration, spectrum and detection threshold; with these simplifications we can discuss observed flux $S = L/4\pi d^2$ (where L and d are the luminosity and distance of the source) instead of peak count rate. Consider first some small area on the sky and let $f(S)$ be the observed distribution function on the logarithmic scale, i.e. $d(\text{number}) = f(S)d\ln S$. If the distribution of sources in the cone towards that area of the sky were uniform all the way to infinity, then $f(S) \propto S^{-1.5}$ would hold, for any luminosity function $\Psi(L)$. Now consider an idealized case where $\Psi(L) \to 0$ for $L > L_{max}$ and the source density is zero at $d > d_{max}$ but uniform for $d > d_{max}$. With $S_{kink} \equiv L_{max}/4\pi d_{max}^2$ this gives $f(S) \propto S^{-1.5}$ for $S > S_{kink}$ and $f(S) \propto \Psi\left(4\pi S d_{max}^2\right)$ for $S > S_{kink}$ if $d(\text{number}) = \Psi(L)\, d\ln L$, and $\Psi(L)$ decreases with L. (If $\Psi(L)$ increases with L, then $f(S) \sim \Psi(L_{max})$ at $S < S_{kink}$.) For this uniform, bounded model, the variation of S_{kink} with direction in the sky gives the variation of d_{max} with direction (assuming L_{max} to be independent of source direction), except for an unkown scale-factor (since L_{max} is unkown).

Kinks will also appear in $f(S)$ if the density of sources is nonuniform. If the number of sources within distance d is $N(< d)$ in some direction — with $N(< d) \propto d^3$ if $d \ll d_c$ and $N(< d) \propto d$ if $d \gg d_c$ for a Galactic halo — then if GRB's were standard candles ($\Psi(L) \propto L_0\delta(L - L_0)$) we would find $f(S) \propto N(< \sqrt{L_0/4\pi S})$ (cf Brainerd 1992, Paczynski 1991). This would imply $f(S) \propto S^{-0.5}$ for $S < S_{kink}$. A non-trivial luminosity function $\Psi(L)$ alters $f(S)$ at $S < S_{kink} \equiv L_{max}/4\pi d_c^2$, but if $\Psi(L)\sqrt{L}$ decreases with L at large L, then roughly speaking $f(S) \propto \Psi\left(4\pi S d_c^2\right)$. Thus d_c plays the same role as d_{max} in realistic halo models (at least as long as $d_c \gtrsim R_0 \approx 8.5$ kpc, the galactocentric radius of the Sun); the "effective core radius" is a function of direction on the sky for observers not exactly at the halo center (see Mao and Paczynski 1992 and Loredo and Wasserman 1992, hereafter LW92).

Earlier data from PVO and KONUS (for which $f(S) \propto S^{-1.5}$ or so; Epstein and Hurley 1988, Chuang et al. 1992) plus the more sensitive BATSE data (for which $f(S) \propto S^{-0.8}$; Meegan et al. 1991) imply a kink in the distribution function and moreover that S_{kink} is roughly independent of direction. We thus have the important conclusion that the (effective) outer bounding surface of the source distribution is, at least approximately, a sphere centered on our location.

For models involving some kind of Galactic halo, one might therefore hope that the *form* of $f(S)$ at $S < S_{kink}$ uniquely determines $\Psi(L)$. For a homogeneous halo (or at least one in which $\rho(d)$ truncates rapidly outside $d_{max} = d_c$), $f(S) \propto S^{-0.8}$ implies $\Psi(L) \propto L^{-0.8}$ (Wasserman 1992). For a more realistic halo $\rho(d) = \rho_0\left(d^2 + d_c^2\right)^{-1}$ the situation is murkier. Although $f(S) \to \Psi\left(4\pi S d_c^2\right)$ for $S \ll L_{max}/4\pi d_c^2$, the distribution flattens at $S \lesssim S_{kink}$ for any $\Psi(L)$ and various luminosity functions near $\Psi(L) \propto L^{-0.8}$ may fit the source distribution (which is not precisely a powerlaw) equally well (LW92). However, the fact that $f(S) \propto S^{-0.8}$ falls off more steeply than $f(S) \propto S^{-0.5}$, the relation for standard candles, suggests that most observed bursts come from $d \sim d_{max}$ rather than $d > d_{max}$, and have $L < L_{max}$, so that decreasing $\sqrt{L}\Psi(L)$ (which includes $\Psi(L) \propto L^{-0.8}$) is favored. Note that the contribution to integrated luminosity $L\Psi(L) \propto L^{0.2}$ increases slowly to its peak at $L \approx L_{max}$. While $\Psi(L)$ may not be exactly zero for $L > L_{max}$, the fact that $f(S) \propto S^{-1.5}$ at large S shows that contamination from $L \gg L_{max}$ is unimportant.

The isotropy of the source distribution detected by BATSE strongly constrains d_c or d_{max}. If Θ is the angle between the source position and the Galactic Center, then according to Meegan et al. 1991, $D \equiv \langle \cos\Theta \rangle \lesssim 0.1$ (see also Brainerd 1992). Since $D \propto R_0/d_c$ for small R_0/d_c, an upper bound on D implies a lower bound on d_c. For a uniform halo, $D \lesssim 0.1$ implies $d_c \gtrsim 40$ kpc for $\Psi(L) \propto L^{-0.8}$ (Wasserman 1992) and for a nonuniform halo, $d_c \gtrsim 25$ kpc (LW92; see also Paczynski 1991, Brainerd 1992, Mao and Paczynski 1992). By contrast, most models for the dark matter in the Milky Way based on velocity data and the distribution of stars give $d_{core} \lesssim R_0$ (e.g. Bahcall and Soneira 1980). There are two ways out of this dilemma, if one wants to associate GRB's with halo stars: (i) if neutron stars only constitute a small fraction of the total mass of the dark matter, they could have a distinct spatial distribution (see Woosley's contribution); (ii) alternatively, GRB sources are distributed with the dark matter except for "a hole in the middle". The rationale for such models is given in Section 3, but the dipole moment D is clearly lower than for conventional halos. A partially hollow distribution can be thought of as a sum of the "normal" density profile and a compensating $\rho(d) < 0$ at $d < d_{hole}$, where d_{hole} may be larger than R_0 (but not by very much). The second term tends to cancel the contribution of the first to D. These "hollow halo" models must satisfy a number of observational constraints simultaneously in order to be viable: (1) $f(S) \propto S^{-1.5}$ at $S > S_{kink}$; (2) $f(S) \propto S^{-0.8}$ or so at low S; (3) dipole moment within acceptable bounds ($D \lesssim 0.1$). (See LW92.)

THE ROTATION CURVE CONSPIRACY AND HOLLOW HALOS

The gravitational potential produced by a "standard" spherical halo has two parameters, the amplitude ρ_0 and the core radius d_{core}. The potential produced by the mass of the visible stars in the Galactic disk also has an amplitude parameter and a length scale for the exponential decrease of density with d. For general values of the two pairs of parameters one would expect a very complicated rotation curve $V_{rot}(d)$, varying from one galaxy to another. This is not the case and $V_{rot}(d)$ is in fact a simple, smoothly varying function for most spiral galaxies, which seems to require a "conspiracy" among the various parameters (e.g. Sancisi and van Albada 1987). Bahcall and Casertano (1985) have suggested that a natural explanation may be that *all* the dark matter associated with individual spiral galaxies is baryonic. Assume first that the proto-galaxy had a single distribution of gas $\sim (d^2 + d_{core}^2)^{-1}$. Assume second that the early star formation in the *outer* regions of the galaxy led preferentially to dark objects (or objects whose remnants are dark after $\sim 10^{10}$ years), but preferentially to ordinary stars with the usual IMF in the *inner* regions. If the distribution of total density, visible plus dark, remained unchanged during the evolution of the proto-galaxy, the rotation curve would have a simple shape, since the distinction between the two components is irrelevant for V_{rot}.

In Section 4 we shall postulate that GRB's are associated with present-day dark matter in the Galactic halo, so we need to speculate on the redistribution of matter. We shall consider original orbits for "clumps" of matter and postulate that dispersed gas after clump-clump collisions will form the Galactic disk (plus spheroid), whereas gas in an intact clump eventually makes neutron stars plus small solid objects. Even with this extreme assumption, the final distribution of matter depends not only on the original density distribution, but also on the original distribution of angular momentum and of orbit eccentricities. Since the rotating disk must be formed, there must have been some net preference for the angular momentum vector $\mathbf{\Omega}$ originally, and possibly some preference for nearly circular orbits (even for other $\mathbf{\Omega}$ directions) over nearly radial ones. As one illustration, consider two clumps originally on circular orbits of radius 12 kpc with $\mathbf{\Omega}$ vectors in the same hemisphere but deviating by an angle α in two opposite directions from the disk's spin axis $\mathbf{\Omega_0}$. If the clumps collide, the resulting angular momentum of the merged gas is $\mathbf{\Omega_0} \cos\alpha$ and the decrease in rotational kinetic energy appears as heat. If one waits for the heat to be radiated away, with orbits spiralling in, the end result is a slightly smaller disk orbit, e.g. $\sim$ 8 kpc. The original 12 kpc orbits are now missing and the new disk orbits will make visible stars, not dark objects. Clumps on highly eccentric (almost radial) orbits which are lost at small distances from the center are partially replaced by clumps coming in on similar orbits from further out (but there is no overcompensation). The missing circular orbits are not replaced at all, leaving a hole. This hole in the middle will only be a partial one, with the typical hole radius and density contrast depending on (a) the eccentricity distribution of the initial orbits and (b) the collision history (see Section 4).

Besides the division into a visible and a dark component, the postulates we outlined also result in some redistribution of the total density as well. The gas which has suffered collisions only forms stars slowly (as observed in present-day galactic disks) and dissipates much of its random velocity (down to present-day disk velocity dispersions of about $20\,\mathrm{km\,s^{-1}}$). As a result the final galactic disk is not only highly flattened, but typical radii are also slightly smaller than the projected radii of the orbits which are missing from the dark matter distribution. This model will imply a "hump" in $V_{rot}(d)$ but it is not yet clear whether the predicted "hump" will be larger than found observationally [Kent (1987); Casertano and Van Gorkom (1991); Begeman, Broeils and Sanders (1991)].

THE METAL EXCESS EMBARRASSMENT AND NEUTRON STAR CLUSTERS

We are exploring models where $\gtrsim 10\%$ of the total dark matter mass is now in the form of neutron stars. Presumably they were formed early in the development of the Galaxy (stellar population III), so a large amount of heavy elements was produced in supernova explosions before much of stellar populations I and II formed. If the progenitors of the neutron stars were distributed individually throughout the halo, the supernova debris would have been spread out rather uniformily and may have excessively enriched the interstellar medium with heavy elements. We therefore postulate that most of the interstellar gas in the proto-

Galaxy was not distributed uniformly, but occurred in clusters of rather large mass M_{cl} orbiting in the halo.

The allowed values of M_{cl} and $v_{cl} \sim \sqrt{GM_{cl}/R_{cl}}$, the cluster velocity dispersion, are constrained by a host of requirements. Here we shall only give a partial listing, and selected numerical examples; for more detail see Salpeter and Wasserman (1992).

(1) Heating of Disk Stars: Neutron star clusters, on passing through the Galactic disk, will heat the stellar populations, producing velocity dispersions $\sigma_\star \approx (Dt)^{1/2}$ at sufficiently late times; the diffusion coefficient $D \propto \rho_h M_{cl}/v_h$, where ρ_h and v_h are the mass density and velocity dispersion of halo clusters. Based on observations of stars in the solar neighborhood, Ostriker and Lacey (1985) have estimated that $\rho_h M_{cl} \approx 2\times10^4 M_\odot^2\,\mathrm{pc}^{-3}$ suffices to produce the *entire* velocity dispersion of the oldest nearby stars. Taking $\rho_h \approx f_{cl} \times 0.01 M_\odot\,\mathrm{pc}^{-3}$ (with the diminution factor f_{cl} introduced to take into account the hollowing of the halo, which already may be non-negligible at R_0), we find $M_{cl} \lesssim 2 \times 10^6 f_{cl}^{-1} M_\odot$. For the numerical examples given below, we adopt $M_{cl} \approx 10^7 M_\odot$ as an estimate, so that $M_{cl} = 10^7 M_{cl,7} M_\odot$. For $M_{cl} \sim 10^7 M_\odot$, the number of clusters out to $d_{max} \sim 10-20$ kpc is of order 10^4, which is much larger than the number of bursts observed so far by BATSE. Thus, it is unlikely that BATSE has observed more than a single burst from any given cluster, which is consistent with the absence of repeaters in the burst sample.

(2) Tidal Radius: A cluster at radius R has a tidal radius $R_T \approx R[M_{cl}/M(R)]^{1/3}$. Using $M(R) \approx V_c^2 R/G$, where $V_c \approx 220\,\mathrm{km\,s^{-1}}$ is the (flat) Galactic rotation speed, we get $R_T \approx 400\,\mathrm{pc}\, M_{cl,7}^{1/3}(R/R_0)^{2/3}$ or $v_{cl} \approx 10\,\mathrm{km\,s^{-1}} M_{cl}^{1/3}(R/R_0)^{-1/3}$.

(3) Protocluster collisions: A family of protoclusters in the halo must not collide with one another while still predominantly gaseous, actively forming stars and undergoing supernovae, since collisions will disperse the (possibly metal-rich) gas, thereby destroying the cluster. Indeed, we shall want clusters to be destroyed at $R \lesssim R_0$ to produce the "hole in the middle" of halo. Clusters of radius R_{cl} can avoid collisions during their entire "formation timescale" t_F if $4\pi R_{cl}^2 f_{cl}\rho_h(R) v_h t_F/M_{cl} \lesssim 1$, or, equivalently, if the column density of clusters, $\pi R_{cl}^2 f_{cl}\rho_h(R) R/M_{cl} \lesssim R/4v_h t_F$. We adopt a plausible formation timescale, $t_F \sim 10^9$ yrs, implying a nominal column depth of order $R_0/4v_h t_F \approx 0.01 t_{F,9}$. Avoiding collisions during t_F therefore implies $R_{cl} \lesssim 20\,\mathrm{pc}\,(M_{cl,7}/f_{cl}t_{F,9})^{1/2}(R/R_0)$ and $v_{cl} \gtrsim 50\,\mathrm{km\,s^{-1}}(M_{cl,7} f_{cl} t_{F,9})^{1/4}(R/R_0)^{-1/2}$. Below, we adopt $50\,\mathrm{km\,s^{-1}}$ as an estimate for cluster velocity dispersion. Collisions between clusters after all of their gas has been consumed by stars, asteroids, etc. are not destructive if v_{cl}/v_h is small, and in any case result in little or no ejection of metal-rich gas.

(4) Retention of Neutron Stars: Various studies of pulsar proper motions (e.g. Lyne, Anderson and Salter 1982, Bailes et al. 1990) and interstellar scintillations (e.g. Cordes 1986) indicate a fairly broad distribution of pulsar velocities in space with mean speeds $\sim 100\,\mathrm{km\,s^{-1}}$ but with a considerable fraction somewhat below and a non-negligible tail well above $100\,\mathrm{km\,s^{-1}}$. From these studies we surmise that if $v_{cl} \ll 100\,\mathrm{km\,s^{-1}}$ a substantial fraction of all neutron stars formed in the cluster would end up escaping, but if $v_{cl} \lesssim 100\,\mathrm{km\,s^{-1}}$ (for example, if $v_{cl} \sim 50\,\mathrm{km\,s^{-1}}$) then a substantial fraction could be retained. In any case, it is likely that a distributed population of neutron stars would be spawned by a primordial halo filled with massive clusters of the type we envision. This population could in itself be sufficient to produce the properties of the observed gamma ray burst distribution unless cluster membership is somehow central to triggering burst events (see Section 5, and also Woosley's contribution).

(5) Containing Supernova Ejecta: A basic requirement for our hypothetical clusters is that they be capable of confining their supernova blast waves. For the rather dense clusters we envision (i.e. $\rho_{cl} \approx 3M_{cl}/4\pi R_{cl}^3 \approx 3v_{cl}^6/4\pi G^3 M_{cl}^2 \approx 2 \times 10^4 m_p \, \mathrm{cm}^{-3} v_{cl,50}^6/M_{cl,7}^2$) we must scale up from numerical results for supernova explosions (cf. Mansfield and Salpeter 1974, Chevalier 1974, Cioffi, McKee and Bertschinger 1988). For $M_{cl,7} \approx v_{cl,50} \approx 1$ individual blast waves with initial energy $E_0 \approx 10^{51}$ ergs may be confined by a swept up mass $M_s \approx 2000 M_\odot \zeta^{-0.15} f_g^{-0.10} \ll M_{cl}$, where ζ is the metallicity in solar units and f_g is the fraction of the cluster mass in gaseous form.

(6) Avoiding Superbubbles: If $\sim 10^7$ supernovae occur in $\sim 10^9$ yrs within an extremely compact cluster ($R_{cl} \lesssim 20\,\mathrm{pc}$) then it is conceivable that the supernova blast waves from individual explosions percolate before merging with the interstellar gas in the cluster, thereby forming a "superbubble" (e.g. Mac Low and McCray 1988) which could burst into the surrounding Galaxy, polluting it with metal-rich gas. For $v_{cl,50} \approx M_{cl,7} \approx 1$ we find that a supernova blast wave merges with the general interstellar medium of the cluster after expanding to a radius $r_s \approx 1\,\mathrm{pc}\,\zeta^{-0.05} f_g^{-0.36}$, which happens at a time $t_s \approx 6000\,\mathrm{yrs}\,\zeta^{-0.05} f_g^{-0.36}$ after the explosion. During a time t_s, the nearest supernova to a randomly selected point in the cluster should take place at a distance $r(t_s) \approx R_{cl}(Mt_F/M_{cl}t_s)^{1/3}$ where $M \approx M_\odot$ is the cluster mass per supernova. For our standard parameter values, we get $r(t_s) \approx 5\,\mathrm{pc}\, t_{F,9}^{1/3} \zeta^{0.02} f_g^{0.1}$, which is considerably larger than r_s. These values depend sensitively on v_{cl} ($r(t_s)/r_s \sim v_{cl}^{1.8}$) so that avoiding superbubbles places a fairly strong lower bound on v_{cl} (i.e. clusters with $v_{cl} \lesssim 20\,\mathrm{km\,s^{-1}}$, or $R_{cl} \gtrsim 100\,\mathrm{pc}$ are disfavored).

(7) Internal Dynamics: The dynamical timescale for the clusters we envision is $t_d \approx 2\pi GM/v_{cl}^3 \approx 2 \times 10^6\,\mathrm{yrs}\, M_{cl,7}/v_{cl,50}^3$. If, once star formation and its aftermath cease, the cluster is dominated by neutron stars, then its two-body collisional relaxation timescale is $t_R \approx 2 \times 10^{10}\,\mathrm{yrs}\, M_{cl,7}^2/v_{cl,50}^3(\Lambda/15)$, where $\Lambda \equiv \ln(0.4N)$ (see Spitzer 1987 and references therein). If the cluster eventually becomes dominated by low mass stars, planetary debris or asteroids, the relaxation timescale is correspondingly larger, but the dynamical friction timescale for neutron stars to be dragged in to the cluster center is roughly the same (cf. Binney and Tremaine 1987). Thus, collisional effects are marginally significant, at best, in the entire lifetime of the cluster. In particular, we do not expect significant mass segregation to occur in the cluster.

The hope for baryonic dark matter in galaxy halos stems in part from the idea of "bimodal star formation" (Shu, Adams and Lizano 1987). At present there is a tendency for quiet conditions to produce mainly low mass stars, violent conditions to produce massive stars which themselves induce violent conditions via supernovae, etc. Conditions must certainly have been different for early star formation and it is possible that it favored either only low masses (e.g. Salpeter 1992), or only high mass stars leading to neutron stars and black holes. We mainly want the latter, but we also need solid objects of quite small mass ("asteroids" for short). These may be formed during the outflow from individual supernova (see contribution by Woosley) or independently, e.g. where two supernova blast waves in their snowplow phases compress each other. There is yet another possibility, if it is the metal abundance in the ISM which mainly controls the stellar IMF: at an early stage, when only $\lesssim 10\%$ of the cluster gas has been used up, the star formation may have favored massive stars, making supernovae and metals, but then the IMF shifted fairly rapidly and drastically towards very low mass brown dwarfs and asteroids, etc., once the metal abundance in the gas has built up to some critical value. In such a model less of the total mass of a cluster would now be in neutron stars than in very low mass objects, but even less would be in visible main sequence stars ($0.1M_\odot$ to $2M_\odot$). Nevertheless there would have to be *some* visible stars present, born in the transition period between the two different extremes of the IMF. We have argued for a projected filling factor $\gtrsim 0.01$ for clusters on the sky and a search in the near infrared for patches of low surface brightness emission in the sky would be useful. (See also Silk 1992.)

ASTEROID IMPACTS WITH CLUSTER NEUTRON STARS

If clustered neutron stars in the Galactic halo are the sites of GRB's then the burst rate per neutron star must be $\sim 1\,\mathrm{day}^{-1}/10^{12} f_{NS}$ or $\sim 10^{-9} f_{NS}^{-1}\,\mathrm{yr}^{-1}$, where f_{NS} is the fraction of the halo in (clustered) neutron stars. We imagine that GRB's result when the distance of closest approach of an asteroid to a neutron star is $\leq q_B$. If asteroids (or fragments from a tidally disrupted asteroid) are dragged in by emitting Alfven waves in the neutron star's magnetosphere (Drell, Foley and Ruderman 1965, Neubauer 1980, van Buren 1981, Tremaine and Zytkow 1986), then we estimate $q_B \lesssim 100$ km, which is much smaller than the tidal disruption distance, $q_L \approx 10^{10}$ cm for asteroidal material of tensile strength $\approx 10^{10}$ dyne cm^{-2} and density $\rho \sim 1$ gm cm^{-3}. Note that $\sim 10\%$ of the total infall energy may be injected directly into the magnetosphere, and subsequently turned into nonthermal (gamma-ray?) emission, by this process.

How often do asteroids try to come within $q_B \lesssim 100$ km of a neutron star? One possibility is that the general interstellar medium of a cluster contains considerable low mass debris — including asteroids or planetesimals — which occasionally accretes onto a neutron star. The flux of interstellar asteroids within $q_B = 10^7 q_{B,7}$ cm of a given neutron star (mass $M = 1.4 M_{1.4} M_\odot$) is $\Phi \approx 2\pi G M q_B \rho_{cl}/M_A v_{cl}$, where M_A is the mass per ISM asteroid in the cluster ($\geq m$, the typical asteroid mass). Thus we require
$M_A \approx 2 \times 10^{24}\,\mathrm{gm}\, q_{B,7} v_{cl,50}^5 M_{1.4}/\Phi_9 M_{cl}^2$, where $\Phi = 10^{-9}\Phi_9$ yr^{-1}, which is comparable to the mass of Ceres, the heaviest asteroid in the solar system. A substantial fraction of the mass of a cluster must be in the form of ISM asteroids for a significant flux.

If, on the other hand, most cluster asteroids are bound to individual neutron stars, then passing neutron stars could induce asteroid showers analogous to comet showers from the Oort cloud (Hills 1981, Tremaine and Zytkow 1986). Orbiting asteroids are only susceptible to perturbations by passing stars at $a \geq a_I \approx 20\,\mathrm{AU} M_{1.4}^{3/2}/q_{L,10}^{1/2} v_{cl,50}^3$, where the impulse approximation holds for encounters capable of filling the loss cone, $q_L \sim 10^{10}$ cm. Orbits widen substantially over a timescale $T = 10^{10}T_{10}$ years for $a \geq a_{diss} \approx 40\,\mathrm{AU}M_{cl,7}^2/v_{cl,50}^5 T_{10}$, so showers can only be triggered in a rather narrow range of a for nominal parameter values. (Note, however, that $a_I/a_{diss} \sim v_{cl}^2$.) It turns out that the loss cone empties much faster than it fills for this range of a, so that the average flux of asteroids to $\lesssim q_B$ is $\Phi \approx \pi \rho_{cl} a q_B N(a)(2G/Mq_L)^{1/2}$, where $N(a)$ is the number of asteroids at $a_I \lesssim a \lesssim a_{diss}$. For a given flux we must therefore have $aN(a) \approx 3 \times 10^7\,\mathrm{AU}\Phi_9 M_{cl,7}^2 (M_{1.4} q_{L,10})^{1/2}/q_{B,7} v_{cl,50}^6$, or $N(a) \approx 10^6 \Phi_9$ at $a \approx 30\,\mathrm{AU}$ if $M_{cl,7} \approx v_{cl,50} \approx q_{B,7} \approx q_{L,10} \approx 1$. Since $q_B N(a)/a \approx 0.02$, few showers into the loss cone produce GRB's, and repeated bursts during a single shower are extremely rare. The flux of asteroids into the loss cone is $q_L \Phi/q_B \approx 10^{-6}\Phi_9\ \mathrm{yr}^{-1}$, which means that only $\sim 1\%$ of the asteroids at ≈ 30 AU are disrupted in 10^{10} years. If asteroids are distributed uniformly in binding energy to ~ 1 AU, then there must be $\sim 3 \times 10^7 \Phi_9$ asteroids around a typical cluster neutron star. If the loss cone is much larger (e.g. $q_L \sim$ AU if planets scatter asteroids from $a \lesssim a_{diss}$ into either short period or unbound orbits), then $aN(a) \sim q_L^{1/2}$. The fraction captured in 10^{10} yrs is also $\sim q_L^{1/2}$, and may become appreciable (i.e. $\sim 30\%$ for $q_L \sim$ AU and $a \sim a_{diss}$).

Asteroids in a planetary system orbiting a cluster neutron star may also be scattered inward by a combination of gravitational interactions (particularly resonances with giant planets like Jupiter) and physical collisions with other asteroids (e.g. Wisdom 1985, 1987, Wetherill 1985, and Greenberg and Nolan 1989). It is not yet clear to us that these processes are capable of producing asteroid orbits with periastrons as small as q_B at the rates required by GRB observations. However, it is interesting to note that the cumulative distribution of asteroid volumes $\propto V^{-a}$ with $a \approx 0.8$ for relatively large asteroids (sizes $\sim 1-100$ km; see Dermott et al. 1984). This exponent is consistent with steady state collisional fragmentation theory (e.g. Dohanyi 1969), and is also just what is neeeded to get $\Psi(L) \propto L^{-0.8}$ if gravitational energy release by asteroid infall powers GRB's.

ACKNOWLEDGMENTS

This research was supported in part by NSF grant AST89-13112 and NASA grant NAGW-666.

REFERENCES

Bahcall, J.N. and Casertano, S. 1985, *ApJ*, **293**, L7

Bahcall, J.N. and Soneira, R.M. 1980, *ApJS*, **44**, 73

Bailes, M., Manchester, R.N., Kesteven, M.J., Norris, R.P. and Reynolds, J.E. 1990, *MNRAS*, **247**, 332

Begeman, K., Broeils, A., and Sanders, R. 1991, *MNRAS*, **249**, 523

Binney, J. and Tremaine, S. 1987, *Galactic Dynamics* (Princeton: Princeton University Press)

Brainerd, J. 1992, *Nature*, **355**, 522

Casertano, S. and Van Gorkom, J. 1991, *ApJ*, **101** 1231

Chevalier, R. 1974, *ApJ*, **188**, 501

Chuang, K.W., White, R.S., Klebesadel, R.W. and Laros, J.G. 1992, *ApJ*, **391**, 242

Cioffi, D.F., McKee, C.F. and Bertschinger, E. 1988, *ApJ*, **334**, 252

Colgate, S.A. and Petschek, A.G. 1981, *ApJ*, **248**, 711

Cordes, J.M. 1986, *ApJ*, **311**, 183

Dermott, S.F., Nicholson, P.D., Burns, J.A. and Houck, J.R., *Nature*, **312**, 305

Drell, S.D., Foley, H.M. and Ruderman, M.A. 1965, *JGR*, **70**, 3131

Dohanyi, J.S. 1969, *JGR*, **74**, 2531

Epstein, R.I. and Hurley, K. 1988, *Ast. Lett. and Comm.*, **27**, 229

Goodman, J. 1986, *ApJ*, **308**, L47

Greenberg, R. and Nolan, M.C. 1989, in *Asteroids II*, eds. R.P. Binzel, T. Gehrels, and M.S. Matthews (Tucson: University of Arizona Press)

Harwit, M. and Salpeter, E.E. 1973, *ApJ*, **186**, L37

Higdon, J.C. and Lingenfelter, R.E. 1990, *ARA&A*, **28**, 401

Hills, J.G. 1981, *AJ*, **86**, 1730

Katz, J.I. 1986, *ApJ*, **309**, 253

Kent, S.M. 1987, *ApJ*, **93**, 826

Loredo, T.J. and Wasserman, I. 1992, in preparation

Lyne, A.G., Anderson, B. and Salter, M.J. 1982, *MNRAS*, **201**, 503

Mac Low, M. and McCray, R. 1988, *ApJ*, **324**, 776

Mansfield, V.N. and Salpeter, E.E. 1974, *ApJ*, **109**, 305

Mao, S. and Paczynski, B 1992, *ApJ*, **389**, L13

Meegan, C.A., Fishman, G.J., Wilson, R.B., Paciesas, W.S., Pendleton, G.N., Horack, J.M., Brock, M.N. and Kouveliotou, C. 1992, *Nature*, **355**, 143

Narayan, R., Piran, T., and Shemi, A. 1991, *ApJ*, **379**, L17

Neubauer, F. 1980, *JGR*, **85 A3**, 1171

Ostriker, J.P. and Lacey, C.G. 1985, *ApJ*, **299**, 633

Paczynski, B. 1986, *ApJ*, **380**, L43

Paczynski, B. 1991, *Acta Ast.*, **41**, 157

Salpeter, E.E., and Wasserman, I. 1992, in preparation

Sancisi, R. and van Albada, T. 1987, in *Dark Matter* eds. G. Knapp and J. Kormendy (Dordrecht: Reidel)

Shu, F.H., Adams, F.C. and Lizano, S. 1987, Ann. Rev., **25**, 23

Silk, J. 1992, *Phys. Rep.*, in press.

Spitzer, L. 1987, *Dynamical Evolution of Globular Clusters* (Princeton: Princeton University Press)

Tremaine, S. and Zytkow, A.N. 1986, *ApJ*, **301**, 155

van Buren, D. 1981, *ApJ*, **249**, 297

Wasserman, I. 1992, *ApJ*, in press.

Wetherill, G.W. 1985, *Meteoritics*, **20**, 1

Wisdom, J. 1985, *Nature*, **315**, 731

Wisdom, J. 1987, *Icarus*, **72**, 241

Planets Around Pulsars
ASP Conference Series, Vol. 36, 1993
J.A. Phillips, J.E. Thorsett, and S.R. Kulkarni

GAMMA-RAY BURSTS FROM PLANETESIMAL ACCRETION

S. E. WOOSLEY[1]
Board of Studies in Astronomy and Astrophysics, University of California, Santa Cruz, CA 95064

ABSTRACT Planetesimals that form following the explosion of supernovae provide a reservoir of material that can be accreted and generate high energy transient emission long after the neutron star has ceased being a pulsar. Blue supergiants, which would have been the predominant progenitor of supernovae during galaxy formation, may be particularly efficient in producing such systems. Owing to their birth at an early epoch and the kick velocity given the neutron star during the explosion, these sources would lie in a very extended halo (R $\sim$ 50 kpc). In this picture, some or most gamma-ray bursters would be 10^{10} year old relics of galaxy formation. At a minimum a very thick disk population would be expected.

INTRODUCTION

The study of cosmic gamma-ray bursts is currently experiencing a creative chaos that, if historical precedents such as the debates over the cosmological nature of quasars and "spiral nebulae" apply, may lead to a major advance in our understanding of the universe. Unlike these earlier precedents, however, it may turn out that gamma-ray bursts are both - local (i.e., "Galactic") and cosmological (more distant than about 1 Gpc). In favor of the local hypothesis one can cite compelling evidence for nearby magnetic neutron stars such as the 8 s pulsing tail (22 cycles) on the light curve of the March 5, 1979 event (Mazets et al., 1979; Barat et al., 1979; Cline et al., 1980; Fenimore et al., 1992); the location of at least two additional so called "soft repeaters" in the direction of the Galactic disk (B1900+14, Mazets et al., 1982; Golenetskii et al. 1987; SGR 1806-20, Atteia et al., 1987; Laros et al., 1987); the observation of cyclotron features in a fraction of bursts (Murakami et al., 1988; Yoshida et al., 1991); the soft approximately thermal tail of x-ray emission seen following several gamma-ray bursts (Yoshida et al., 1989); the stringent beaming requirements if the burst is situated at great distances (Harding 1991 and references therein); and the existence of long complex gamma-ray bursts lasting over one minute that may prove difficult to explain in cosmological models. On the other hand, recent results from the Compton Observatory (Meegan et al., 1992) have extended the restric-

[1]General Studies Group, Physics Department, Lawrence Livermore National Laboratory, Livermore, CA 94550

tions provided by earlier missions on both the isotropy and the near complete sampling of sources. Taken at face value these observations imply a gamma-ray burst source distribution either in a very extended halo or at cosmic distances (Paczynski 1991ab; Mao and Paczynski 1991). While compelling models have yet to be offered in the literature, it is clear that the proponents of cosmological models are far from exhausted and complete models have never been a *forte* of the local hypothesis either. Recent suggestions by Meszaros and Rees (1992) offer encouragement to the cosmological school.

Here a model for producing gamma-ray bursts in an extended halo, or at a minimum, a very thick disk, is briefly described. It is based upon the premise that macroscopic solid condensations (henceforth "planetesimals") should routinely form around all neutron stars, especially those coming from Type II supernovae in a low metallicity population. These provide a reservoir of material for accretion which then gives rise to bursts either from direct accretion or thermonuclear runaway. The exact mechanism for making gamma-rays, as opposed say to x-rays, remains elusive though non-linear Alfven wave dissipation can be invoked.

THE SUPERNOVA EXPLOSION AND FALL BACK

Considerable progress has been made in our understanding of the Type II supernova mechanism and of important multidimensional effects - "mixing" - that occur during and immediately after the explosion. Naively one might think that if matter is going to fall back to the neutron star, that material would just be the small mass immediately above the "mass cut" that failed to achieve escape velocity. Such a view would overlook some very important physics. First, the matter does not just fall back, it is pushed. The hydrodynamic interaction of the dense helium core with the low density hydrogen envelope, or more specifically of a region of low ρR^3 with a region of higher ρR^3, gives rise to a "reverse shock" that propagates inwards in mass (Chevalier 1990; Bethe 1990). This shock decelerates the outgoing material, the extreme inner portion of which is slowed to less than the escape speed. This deceleration happens from two hours after the explosion of a blue supergiant, such as the progenitor of SN 1987A, to two days after the explosion of the red supergiants believed to make modern Type II plateau supernovae.

The distinction is important because the amount of material that falls back is sensitive to the density near of the neutron star when the shock arrives there (Chevalier 1990). During homologous expansion the density declines as t^{-3}. Thus a shock that arrives back at the neutron star ten times earlier will encounter a density 1000 times greater leading to an enormously enhanced fall back mass. The material that falls back in a blue supergiant like SN 1987A may be orders of magnitude greater than in a red supergiant. The exact value is difficult to estimate in either case, but may be $\sim 0.01\ M_\odot$ in the blue case (Chevalier 1990). Fall back has also been seen in multidimensional calculations (e.g., Müller, Fryxell, and Arnett 1991), though again quantitative estimates of the mass are lacking. An important point is that almost all Type II supernovae with metallicity less than about 0.1 solar will occur in blue supergiant progenitors (Woosley, Pinto, and Weaver 1989). Thus when the Galaxy was forming, the typical supernova would have been a blue supergiant; nowadays it would not. It is important to know at just what age, and more importantly, just what stage of galaxy formation, this transition occurred. Unfortunately current calculations are not clear about this.

Another important area of uncertainty is the "kick" imparted to the neutron star during its formation. The origin of this kick, observationally $\sim$200 km s^{-1} (Lyne 1987) or more, is probably the minute asymmetry in neutrino emission from one hemisphere of the neutron star to the other (Woosley 1987; Herant, Benz, and Colgate 1991), rotationally averaged. Colgate and Epstein (1992) have also noted that high velocity neutron stars kicked by this mechanism might make an interesting gamma-ray burst population. This paper differs by invoking kicks at an earlier Galactic epoch which may be necessary to obtain good agreement with observations (D. Hartmann, private communication 1992). Otherwise, much of the following is common to both models.

Yet another area of ignorance is the angular momentum of the material that falls back. This is complicated by an uncertain distribution of angular momentum in the progenitor star, the mixing that goes on during and after the explosion, and the fact that the neutron star moves an appreciable distance during the two hours (or two days in a red supergiant) that it takes for the material to fall back. Some of these issues have been discussed by Lin, Woosley, and Bodenheimer (1991). Within the considerable uncertainties involved, especially the redistribution of angular momentum that occurs during convection in the late stages of stellar evolution and the mixing that occurs during and after the explosion (Herant and Benz 1992; Herant, Benz, and Colgate, 1992; Fryxell, Müller, and Arnett 1991; Müller, Fryxell, and Arnett, 1991), it is not unreasonable to invoke a specific angular momentum of $\sim 2 \times 10^{17}$ erg s^{-1} g^{-1}. This is the angular momentum of a Keplerian orbit at 10^8 cm, so the material clearly cannot accrete directly onto the neutron star. The initial formation of a disk appears likely.

Beyond this point, the evolution is governed by the viscous spreading of the disk and its interaction with the pulsar. Early on the pulsar may be quite luminous, but then the surface density of the disk is large, $\sim 0.01\ M_{\odot}/\pi r^2$ $= 10^{15}$ g cm^{-2}. It is quite optically thick. Material will be ablated from the surface, but the initial spreading time of a disk of this density is very short. As the disk spreads under the action of viscosity, mass moves in, probably to be lost from the system as it is spun off by the rotating neutron star's magnetic field, and angular momentum moves out. Conserving angular momentum, the surface density declines as $r^{-5/2}$.

FORMATION AND ACCRETION OF PLANETESIMALS

As the disk spreads out to about 1 AU, the condensation of solids becomes feasible (Lin et al., 1991). This material lacks hydrogen, though it may contain appreciable helium (both from mixing with the helium shell and from explosive silicon burning), carbon, silicon, iron, and, especially, oxygen. Its opacity should be high; even at 1 AU the disk remains optically thick. At this radius the internal heating from viscous dissipation becomes low enough that solids can form. Unless the disk is entirely ablated from its surface, this happens in a region that has never experienced the pulsar radiation.

Grains form first and then grow by cohesive collisions to planetesimals of uncertain size. A typical value may be 10^{-12} $M_\odot$ (Lin et al., 1991). The original 0.01 $M_\odot$ that fell back has by this point eroded to $\sim 10^{-5}$ - 10^{-4} $M_\odot$, or approximately one million planetesimals. What happens after this is important, not so much because a certain mass is necessary to make a gamma-ray burst, but because it determines how the planetesimals interact with one another. If large bodies form, they may act as scatterers to eject the remaining planetesimals into an extended Oort cloud (Tremaine et al., this volume). The evolution of a large number of planetesimals starting at 1 AU and interacting only with one another and without the influence of massive planets has not been studied. It is practical, and would be interesting to do.

Because these bodies are composed of heavy elements, they would have a tensile strength larger than the fluffy comets considered by Tremaine and Zytkow (1986). We assume $T \approx 10^7$ dyne cm^{-2}. Tidal disruption then occurs at a radius, $r_b \approx 5 \times 10^{10}$ cm for a planetesimal of 2×10^{21} g (10^{-12} $M_\odot$) while an object of 10^{-6} $M_\odot$ would disrupt at about 10^{12} cm. In between there will be a cascade to smaller scales. The key quantity is the rate at which mass is fed inside the disruption radius for the characteristic planetesimal size. We assume that each time a planetesimal is disrupted, a substantial fraction of its mass moves in to a smaller perigee. A single large object might power bursts for centuries or more, 10^{26} g breaking down to 10,000 objects of 10^{22} g and so on. Bursts themselves would enhance accretion by vaporizing the surfaces of planetestimals out to several AU.

In order to explain a substantial fraction of the $\sim$1000 gamma-ray bursts known to occur each year (Meegan et al., 1992), a certain mass must accrete in each system. We assume that 10^8 to 10^9 neutron stars have formed in the halo with planetesimal systems as described here. We take a mean distance of 50 kpc as required to explain the observed isotropy and confinement (Paczynski 1990; 1991ab). The typical burst, 10^{-6} erg cm^{-2} at 50 kpc, then requires an energy of 10^{41} erg. If the event is accretion powered, this translates into 10^{21} g. If the events are nuclear powered, as we shall discuss, 100 times more mass is needed. There have been 10^{13} bursts in the life of the Galaxy which for 10^8 neutron stars implies one event per system, on the average, every 100,000 years. These numbers may be changed by beaming, which reduces the mass requirement, but increases the number of events required. It may also be that the events come in "swarms" with recurrences temporarily as frequent as every century. The observations do not say; so far we have sampled only $\sim$700 of the 10^8 - 10^9 proposed sources.

This translates into an average accretion rate of 5×10^{-18} $M_\odot$ yr^{-1}, if the bursts are accretion powered, and 5×10^{-16} $M_\odot$ yr^{-1} if they are nuclear powered. That makes a total of 5×10^{-8} (or 5×10^{-6} $M_\odot$) which must be accreted in a Hubble time, a small fraction of the estimated 10^{-5} - 10^{-4} $M_\odot$ total mass of the planetesimals, though in the nuclear case one may need to invoke a larger number of neutron stars, beaming, or more angular momentum in the fall back. The fundamental issue is whether the evolution of the planetesimal cloud can provide such accretion rates, say of planetesimals passing inside 10^{12} cm, 10^{10} years after the neutron star is born.

Alternatively, one might build a similar model in which the accretion lasts for a shorter period of time, say 10^8 - 10^9 yr, but uses younger neutron stars blasted out of the disk (Colgate and Epstein 1992). Then with a smaller number of neutron stars, say $\sim 10^6$ produced in the last 10^8 yr, one would need accretion rates 100 times larger, but the same total mass accreted per neutron star. One would also need larger kick velocities to eject the neutron star from the Galaxy.

In either case accretion will be resisted while the neutron star remains an active pulsar, Though the formulae usually employed for gaseous accretion at a constant rate are not directly applicable, the accreted material will probably be evaporated or magnetically disrupted outside the Alfven radius, $r_a \approx 3 \times 10^{10} \dot{M}_{10}^{-2/7}$ cm where $\dot{M}_{10}$ is the accretion rate in units of 10^{10} g s^{-1}. This in turn is outside the light cylinder for periods less than 6 s. The centrifugal radius, $(GM/\Omega^2)^{1/3}$, will also be inside the Alfven radius until the neutron star slows down. Since the accretion is transitory, none of these formulae can be used with confidence, but the implication is that a pulsar is unlikely to accrete for the kinds of mass fluxes we envision. As time passes, the luminosity and rotation rate decline appreciably. 10^8 to 10^{10} years after the supernova, the rotation of a strongly magnetic neutron star will have slowed to at least several seconds and the luminosity will have declined to a small fraction of the sun's. Solid objects can then come quite close to the neutron star before attaching to the magnetic field. The instantaneous accretion rate, which depends on how much tidal disruption and orbital spreading has occurred by this point, could be very high. Obtaining 10^{21} g of accretion in 10 s, a typical gamma ray burst time scale, may prove difficult, but that is a problem generic to all gamma-ray burst models that invoke solid body accretion (Harwit and Salpeter 1973; Colgate and Petschek 1981; Tremaine and Zytkow 1986).

PRODUCTION OF THE GAMMA-RAY BURST

Providing the requisite amount of energy for a gamma-ray burst is just (the easy) part of the story. Why does it come out in gamma-radiation? This too is a historic problem and the solutions have always been speculative, typically relying on magnetic interaction to produce non-thermal radiation. The situation following the sudden accretion of 10^{21} g of material onto the surface of a neutron star with magnetic field 10^{12} gauss is similar to that in the "quake" model of Blaes et al., (1989; 1990). The initially solid nature of the accreted material may allow it to accrete at a vastly super-Eddington rate (though see below). This material will be heated by compression to billions of K, and is unlikely to cool simply by thermal diffusion. The temperature gradient is probably superadiabatic and overstable oscillations will be an efficient source of Alfven waves that can propagate out into the magnetosphere and damp by accelerating electrons (Bonazzola et al. 1984; Hameury and Lasota 1986). The flux carried by these waves is $\sim (\delta B^2/8\pi)v_a$, with v_a the Alfven speed, $\sim 10^9$ cm s^{-1}. For $\delta B \sim 10^{10}$ gauss, i.e., about 1% of B, $j \sim 10^{38}$ erg s^{-1} km^{-2}. The waves would dissipate where $\delta B \sim B$, several neutron star radii.

Given the potential difficulties in getting 10^{21} g to accrete all at once, one might want to consider an alternative in which a much greater mass is accreted over a long period of time. Suppose the cascade of planetesimal masses to smaller scales under the action of tidal disruption proceeds to its logical limit, complete disintegration to dust and gas. Then there will be periods when the accretion rate is high and others when it is essentially zero, but we not longer attribute the burst to accretion itself. Rather accretion provides the fuel for a thermonuclear flash.

Assume that 10^{23} g accretes in 10^5 years. The *minimum* steady accretion rate at which this can occur is 5×10^{-16} $M_\odot$ yr^{-1}. Bilsden, Salpeter, and Wasserman (1992) have pointed out that at such low accretion rates, spallation reactions in the accreting material will strip down heavy elements such as CNO (and presumably iron and silicon) to helium and even to hydrogen. Many studies of thermonuclear instability in accreting neutron stars then become applicable (Lamb and Lamb 1978; Ergma and Tutukov 1980; Hameury et al., 1982, 1983; Woosley and Wallace 1982; Woosley 1982). The work by Hameury et al. in particular sets a limit on the accretion rate below which hydrogen will burn stably, $\sim 10^{-15}$ $M_\odot$ yr^{-1} km^{-2} (*nb.*, per km^2). The critical mass is about 2×10^{20} g km^{-2}. Given the very long time scale and thick accreted layers involved, it might be reasonable to assume that the accreted material spreads over a significant fraction of the entire surface, $\sim$1000 km^2, so the total mass would be 10^{23} g. At this mass and low accretion rate, hydrogen would begin to burn by electron capture (at $\rho = 1.45 \times 10^7$ g cm^{-3}). Thus for this and lower accretion rates, a thick but potentially explosive layer of helium would accumulate.

It is difficult to estimate just how far this accumulation might go if the accretion rate is highly variable. During a high state, a hydrogen shell flash could ignite that not only consumed all the hydrogen, but touched off the helium layer as well. If the density of the helium layer has reached more than about 10^8 g cm^{-3}, a detonation might occur with a very fast rise time (Fryxell and Woosley 1982). For each gram of hydrogen-rich material that burns, $\sim 5 \times 10^{18}$ erg is produced; for helium, or a mixture of helium, carbon, and oxygen, the value is close to 10^{18} erg g^{-1}. Obviously a large diversity of results - total energy, time scale, and duration - is possible. The mechanism for producing gamma-radiation would probably be similar to that described above for accretion events - non-linear Alfven wave damping.

It is also possible that the slow accumulation of material produces starquakes (Blaes et al., 1989, 1990). Basically the planetesimals provide a reservoir of matter for old, isolated neutron stars that is needed in a wide variety of models.

CONCLUSIONS

Planetesimal formation may occur around all neutron stars (Lin et al., 1992), the condensible material falling back from the partially mixed heavy elements and helium made in the explosion and presupernova star. While occurring at some level for all supernovae, these planetesimals form much more copiously around the remnants of supernovae that occur in blue supergiants and thus may be more plentiful around very old neutron stars in the halo. A limit on planetary masses in systems that are presently active pulsars would therefore not rule out the gamma-ray burst model proposed here. On the other hand, a positive detection would confirm at least part of the story.

Since it is unlikely that either supernovae in blue supergiants or planetesimal formation around neutron stars ceased entirely after the Galaxy was born, there may be a thick disk population (d $\sim$ few kpc) of such sources, as well as the extended halo population emphasized here. Lingenfelter and Higdon (1992) have shown that such a dual distribution would be consistent with the Compton Observatory observations (though see Paczynski 1992). The thick disk population would need to be characterized by a lower characteristic energy which might be reasonable if the younger neutron stars have a higher accretion rate.

Gamma-ray bursts are produced as these planetesimals accrete onto the neutron star. The energy release may come directly from accretion, or in a delayed form from a thermonuclear runaway or starquake. If the planetesimals accrete either too slowly or too quickly, the model is ruled out. In the halo, one needs to have time averaged accretion rates that after 10^{10} years remain at about 10^{16} - 10^{18} g yr^{-1}. Accretion is inhibited at early times when the neutron star is rapidly rotating and luminous.

The manner in which the planetesimal is captured, and in particular the extent to which it is tidally disrupted, are important, highly uncertain aspects of the model that might be addressed by numerical calculation. In both limits, complete disruption to gas or accretion of a nearly intact planetesimal, a burst is possible, the former by thermonuclear explosion, the latter by direct conversion of accretion energy. The latter requires 100 times less mass. In between these two limits, a burst sufficiently brilliant to be visible at 50 kpc is unlikely. The individual fragments would not give adequate accretion energy and the temporary high accretion rate would pre-ignite nuclear burning before a large critical mass was reached.

It remains to be seen if the many speculations presented here will bear up under detailed study. It must be acknowledged that the need to move bursts out to 50 kpc poses severe problems for all models based on comet accretion or thermonuclear instability. Perhaps, in the end, these ideas will prove applicable to only a small fraction of all bursts (which may be significantly closer than 50 kpc), but even that would be very exciting.

ACKNOWLEDGMENTS

This work has been supported by the National Science Foundation (AST -91 15367) and NASA (NAGW 2525).

REFERENCES

Atteia, J. L., and 15 others, 1987, *ApJ*, **320**, L105.

Barat, C. et al. 1979, *A&A*, **79**, L24.

Bethe, H. A. 1990, *Rev. Mod. Phys.*, **62**, No. 4, 801.

Bilsden, L., Salpeter, E. E., and Wasserman, I. 1992, *ApJ*, in press.

Blaes, O., Blandford, R., Goldreich, P., and Madau, P. 1989, *ApJ*, **343**, 839.

Blaes, O., Blandford, R., Madau, P., and Koonin, S. 1990, *ApJ*, **363**, 612.

Bonazzola, S., Hameury, J. M., Heyvaerts, J., and LaSota, J. P. 1984, *A&A*, **136**, 89.

Chevalier, R. A. 1990, *ApJ*, **346**, 847.

Cline, T. et al. 1980, *ApJ*, **237**, L1.

Colgate, S. A., and Petschek, A. G. 1981, *ApJ*, **248**, 771.

Colgate, S. A., and Epstein, R. 1992, *BAAS*, **23**, No. 4, p. 1356.

Ergma, E. V., and Tutukov, A. V. 1980, *A&A*, **84**, 123.

Fenimore, E., Epstein, R. I., Ho, C., Klebesadel, R. W., Laros, J. G., & Ulmer, A. 1992, *BAAS*, **23**, No. 4, p. 1322.

Fryxell, B. A., Müller, E., & Arnett, W. D. 1991, *ApJ*, **367**,619.

Fryxell, B., and Woosley, S. E. 1982, *ApJ*, **261**, 332.

Golenetskii, S., Aptekar, R., Guryan, Yu, Ilinskii, V., and Mazets, E. 1987, *Sov. Astron. Lett.*, **13**(3), 166.

Hameury, J. M., Bonazzola, S., Heyvaerts, J., and Ventura, J. 1982, *A&A*, **111**, 242.

Hameury, J. M., Heyvaerts, J. and Bonazzola, S. 1983, *A&A*, **121**, 259.

Hameury, J. M., and Lasota, J. P. 1986, in *Gamma-Ray Bursts*, AIP Conf. Proc. 141, eds. E. Liang and V. Petrosian, (AIP, New York), p. 164.

Harding. A. 1991, *Phys. Rept.*, **206**, 327.

Harwit, M., and Salpeter, E. E. 1973, *ApJ*, **186**, L37.

Herant, M., & Benz, W. 1992, *ApJ*, **387**, in press.

Herant, M., Benz, W., and Colgate, S. A. 1992, *ApJ*, submitted. Also Chap 3 of Herant, M. 1992, PhD thesis, Harvard Univ.

Lamb, D. Q., and Lamb, F. K. 1978, *ApJ*, **220**, 291.

Laros, J., and 16 others 1987, *ApJ*, **320**, L111.

Lin, D., Woosley, S. E., and Bodenheimer, P. 1991, *Nature*, **353**, 827.

Lingenfelter, R. E., and Higdon, J. C. 1992, *Nature*, **356**, 132.

Lyne, A. 1987, in *The Origin and Evolution of Neutron Stars*, eds. D. Helfand and J. Huang, (D. Reidel: Dordrecht), p. 23.

Mao, S., and Paczynski, B. 1991, *ApJ*, **374**, L37.

Mazets, E. P. et al. 1979, *Nature*, **282**, 587.

Mazets, E. S., Golenetskii, S., Guryan, Yu, and Ilinskii, V. 1982, *Astrophys. Spac. Sci.*, **84**, 173.

Meegan, C. A., Fishman, G. J., Wilson, R. B., Paciesas, W. S., Pendleton, G. N., Horack, J. M., Brock, M. N., and Kouveliotou, C. 1992, *Nature*, **355**, 143.

Meszaros, P., and Rees, M. J. 1992, *ApJ*, in press for Oct. 1 issue.

Murakami, T. et al. 1988, *Nature*, **335**, 234.

Müller, E., Fryxell, B., & Arnett, W. D. 1991, *Max Planck Institut für Physik und Astrophysik, Preprint MPA 586*, submitted to *A&A*.

Paczynski, B. 1990, *ApJ*, **348**, 485.

Paczynski, B. 1991a, *Acta Astron.*, **41**, 157.

Paczynski, B. 1991b, *Acta Astron.*, **41**, 257.

Paczynski, B. 1992, *Acta Astron.*, **42**, 1.

Tremaine, S., & Zytkow, A. 1986, *ApJ*, **301**, 155.

Woosley, S. E. 1982, in *Accreting Neutron Stars*, eds. W. Brinkman and J. Trümper, (Max Planck Institute für Astrophysik: Garching), p. 189.

Woosley, S. E. 1987, in *The Origin and Evolution of Neutron Stars*, eds. D. Helfand and J. Huang, (D. Reidel: Dordrecht), p. 255.

Woosley, S. E., Pinto, P. A., and Weaver, T. A. 1989, *Proc. Astron. Soc. Australia* **7**, No. 4, 355.

Woosley, S. E., and Wallace, R. K. 1982, *ApJ*, **258**, 716.

Yoshida, A. et al. 1989, *Publ. Astron. Soc. Japan*, **41**, 509.

Yoshida, A., Murakami, T., Nishimura, J., Kondo, I., and Fenimore, E. E. 1991, *Publ. Astron. Soc. Japan*, in press.

Section 7:

Conferencc Summary

Planets Around Pulsars
ASP Conference Series, Vol. 36, 1993
J.A. Phillips, J.E. Thorsett, and S.R. Kulkarni

CONCLUDING COMMENTS

MARTIN J. REES
Institute of Astronomy, Madingley Road, Cambridge, CB3 0HA, UK

This has undoubtedly been a most stimulating and informative meeting. The chosen topic – plainly a novel and exciting one – is of interest to specialists in pulsars, protostars, binary stars, planet formation and planetary dynamics. The programme included excellent presentations on all these subjects, and I am sure that all of us, whatever our prime specialities, have broadened our perspective and will go away much better informed. I shall not attempt a comprehensive summary of contributions which were often, in themselves, broad reviews. Instead, I will offer just a few impressionistic remarks, trying to be very brief in view of the short time that remains.

Theorists often think, even if they do not openly say, "with hindsight we could have predicted that". So many theories have already been offered that it now seems in no way implausible – on the contrary, it almost seems natural – that planets should exist around some pulsars. The difficulty is, rather, to decide among the many options that have been proposed. But the imaginations of theorists are so impoverished that they need the stimulus of observations. And that is why, even though pulsars have been known for 25 years, discussions of planets around them date almost exclusively from within the last year.

The prime question in my mind when I came to this meeting was, are the periodic variations in PSR 1257+12 real? Alex Wolszczan's opening talk, and the reaction to it by the other pulsar experts present, left little doubt that the answer was yes, and that we had a real phenomenon to discuss during the meeting.

The next question posed by the 1257+12 data is "Do the periods imply orbiting bodies?" Could the data alternatively be attributed to internal oscillations, precession, etc.? There are three requirements of any such interpretation, which I believe are very hard to meet.

(1) Because the mean slowing down rate $\dot{p}$ for this particular pulsar is small, the periodic variations involve a very large *fractional* change in $\dot{p}$. Indeed, $\dot{p}$ changes sign, so that 1257+12 is actually *speeding up* for part of the time. This is hard to reconcile with any model in which precession causes the slowing down torque to vary.

(2) There is no readily discernible change in the pulse profile, as would be expected if the magnetic axis were wobbling through a large angle owing to free precession.

(3) The phase shift, measured by the ratio of the amplitude of the residuals to the period, is a large angle, of order π. This is a constraint on ideas involving, for instance, Tkachenko modes – variations in the surface rotation rate caused by

periodic redistribution of interior angular momentum due to mutual repulsion of neutron superfluid vortex lines in the core. (This third option otherwise has the advantage that it can surmount the problems of (1) and (2) above.)

While I personally remain slightly openminded, to the extent of assigning perhaps ten per cent probability to one of the above possibilities, I believe that at the moment it is wise to bet firmly on the hypothesis that 1257+12 has planetary-mass bodies orbiting around it. It is, moreover, very gratifying that, as we heard from Renu Malhotra, it will only be a few years before the tell-tale effects of orbital perturbations should be detectable in the timing data. If these are found, that would clinch the case for orbiting bodies completely.

The many scenarios for planetary system formation have been discussed by other speakers. In particular, Philipp Podsiadlowski offered an elaborate classification scheme, which I will not reproduce here. However, a key demarcation between the different theories depends on the answer to the question: did the formation involve a binary system?

It is widely believed that binary companions were implicated in spinning up millisecond pulsars. Moreover, in some cases a low mass companion still exists around such systems. Given that 1257+12 is itself a millisecond pulsar, it therefore seems natural to attribute the planets to some consequences of binary evolution. My personal preference is for the idea that ablation of a low mass companion eventually leads to its sudden disruption on a dynamical timescale, and thus to formation of a disc containing $> 0.1 M_\odot$ of material. (An alternative to this is the idea that a circum-binary disc is produced before the companion disappears.) Other scenarios, which would again produce planets around specifically millisecond pulsars, involve mergers between binary white dwarfs, or a white dwarf and neutron star binary.

Let me just mention two other very different models, neither of which involve binaries, and which could both equally well apply to ordinary slow pulsars. The first, discussed by Stan Woosley, involves disc formation from the infall in the first few years after a supernova explosion. The problem here is that the angular momentum is only marginally sufficient; the mass also is likely to be modest (though to some extent this is compensated by the fact that it would be almost 100% heavy elements.) A quite different idea is that a pre-existing neutron star collided with an ordinary solar system. The original "sun" would form a Thorne-Zytkow object around the neutron star, the material being eventually either accreted or expelled. This scenario, discussed further by Philipp Podsiadlowski, involves a rather improbable type of collision, and would therefore only be tenable if solar systems were very common. If it were correct, this idea would therefore have the consequence, welcome no doubt to our NASA sponsors, that our Galaxy must be teeming with solar systems like our own.

Ed van den Heuvel emphasised that millisecond pulsars are themselves rare and exotic systems, so that one need not feel too inhibited about invoking *a priori* unlikely models to explain them. This therefore adds interest to the question of whether there are planets around ordinary pulsars as well. Jim Cordes and Matthew Bailes gave very interesting assessments of what we can conclude from the present data. At the moment, one cannot rule out low mass planets around many other pulsars. On the other hand, the tentative data on 1828-10 seemed, at least to my naive eyes, quite convincing evidence that there may be things in orbit around that pulsar too.

If planets are discovered around other pulsars, or if continuing observations reveal further longer period planets in 1257+12, we will be in a better position to discriminate among the models. In particular, it is important to know what the total angular momentum of the planetary system is, and whether outer planets are also in circular orbits. (Some of the scenarios – for instance, the white dwarf coalescence model – predict that planets could in principle exist around some high-mass white dwarfs. While one cannot in general get timing data of comparable precision for white dwarfs, it was interesting to learn that in one or two cases it might not be completely hopeless to set interesting limits.)

The proto-planetary discs that can result from, for instance, disruption of a low mass companion, have parameters not grossly different from those postulated for our own solar nebula. But two distinctions are perhaps important. First, in most of the pulsar scenarios, the material is injected at a radius of 10^{10} cm. Viscosity then causes the disc to *expand*: the process is *excretion*, not accretion. The angular momentum budget permits ten per cent of the matter to move out to a hundred times the injection radius, and one per cent to attain out to 10^4 times that radius.

The disc would be irradiated by the central pulsar, or by an accretion-powered X-ray source. However, as Steve Ruden and Sterl Phinney emphasised, plausible estimates of viscosity suggest that planets form about 10^6 years after creation of the disc; by that time, the accretion-driven luminosity would have fallen to a level not substantially above the output from our proto-sun.

As we have heard from several speakers, the modelling of proto-stellar discs, planetesimal evolution, and planet formation is a lively area of research. Progress in this relatively "conventional" subject will, as a byproduct, help us to sift out which of the various ideas for circum-pulsar planet formation are most promising. Let me just highlight, however, one interesting theoretical issue which arises specifically in the pulsar context. This is the structure and evolution of *Thorne-Zytkow objects* – neutron stars surrounded by an extended giantlike atmosphere containing a mass of stellar order. They were first studied here at Caltech in the 1970s. Ed van den Heuvel explained that such systems arise when neutron stars collide with ordinary stars, or during the common envelope phase of close binary evolution. When these objects are puffed up to a radius of several a.u., they may, to a first approximation, be spherical. They would, however, contain enough angular momentum to form a planetary system. The mass loss rate, which determines the lifetime of the Thorne-Zytkow phase, is at the moment only estimated by analogy with winds from supergiants, and this is a current uncertainty in all modelling. Another uncertainty concerns what happens in the later phases, when the mass in the extended envelope is substantially depleted. Does it remain quasispherical? Or does it deflate into an extended rotationally supported disc? We already a have surfeit of models, so perhaps an extra one is not very welcome. However, if (as Krolik, for instance, has speculated) the Thorne-Zytkow envelope eventually produced an extended disc, a collision between a neutron star and a main sequence star might produce a secondary planetary system, even if the ordinary star did not previously have a solar system around it.

If the planetary systems around neutron stars are at all analogous to our own, we would expect there to be a lot of low-mass debris particles, in addition to the planets themselves. This raises a number of questions. Are asteroidal bodies or molecules detectable around pulsars? If planetary systems form around ordinary pulsars, and take more than 10^6 years to do so, is there evidence for proto-planetary systems around the youngest ordinary pulsars? Are there huge numbers of comets around some pulsars, as in our own Kuiper belt or Oort cloud?

Planetary systems around pulsars may offer clues to another phenomenon where theorists are completely flummoxed, namely the cosmic gamma-ray bursts. Some speakers at the meeting did not resist the temptation to link these two mysteries, by devising (or reviving) the idea that the bursts are due to impacts of comets or asteroids onto old neutron stars. The data from GRO imply an isotropic distribution, plus a number-versus-intensity relationship that is too flat for a homogenous "Euclidean" distribution of sources. These two facts, in combination, would require the sources to be far out in the Galactic Halo: the "core radius" of the parent population must be > 30 kpc – larger than for any other galactic population, even the dark matter. The bursts must either involve an old halo population of neutron stars formed during the collapse of our Galaxy (and given a transverse "kick" on formation), or there must be some reason why passage through or near the Galactic disc inhibits the gamma-ray burst mechanism (e.g. by altering the properties of an Oort cloud). But unfortunately (or perhaps, fortunately!) the clock prevents me from speculating further along these lines today.

In conclusion, I am sure all the other participants would join with me in wishing to thank the organisers for arranging this splendid meeting. It was advertised as the *first* meeting on planets around pulsars. It will certainly not be the last, and when the next one takes place I am confident that there will be great progress to report. This progress will be manifested by more observations, and perhaps the discovery of new systems. In parallel, we should expect (and, indeed, hope for) a *reduction* in the number of tenable theories, coupled with an improvement in the sophistication and convincingness of those that survive.

Planets Around Pulsars
ASP Conference Series, Vol. 36, 1993
J.A. Phillips, J.E. Thorsett, and S.R. Kulkarni

THE PULSAR PLANET PRODUCTION PROCESS

E. S. PHINNEY AND B. M. S. HANSEN
Theoretical Astrophysics, 130-33 Caltech, Pasadena, CA 91125, USA

ABSTRACT Most plausible scenarios for the formation of planets around pulsars end with a disk of gas around the pulsar. The supplicant author then points to the solar system to bolster faith in the miraculous transfiguration of gas into planets. We here investigate this process of transfiguration. We derive analytic sequences of quasi-static disks which give good approximations to exact solutions of the disk diffusion equation with realistic opacity tables. These allow quick and efficient surveys of parameter space. We discuss the outward transfer of mass in accretion disks and the resulting timescale constraints, the effects of illumination by the central source on the disk and dust within it, and the effects of the widely different elemental compositions of the disks in the various scenarios, and their extensions to globular clusters. We point out where significant uncertainties exist in the appropriate grain opacities, and in the effect of illumination and winds from the neutron star.

1. INTRODUCTION

Like most things in astronomy, scenarios for the formation of planets and planetary systems around pulsars can be divided into two classes:

i) *Salamander scenarios* (recall that the mythical salamander could survive, even thrive in fire). In these, the planets were formed as part of the formation of an ordinary main sequence star, and survived its violent transformation to a rapidly spinning neutron star. In this category we may place planets which: survive pre-supernova evolution and the supernova explosion and recoil; spiral into an extended envelope; are the remains of the ablation of a much more massive companion; or are captured from a passing star. These scenarios (reviewed in these proceedings by Podsiadlowski) require unlikely events or physical circumstances of controversial verisimilitude. More importantly, they have nothing new to teach us about planet formation, since such planets would have formed in circumstances similar to the solar system, and the challenge is to get them to survive, not to make them. We shall thus consider these scenarios no further, as they fall outside the statutory limits of our title.

ii) *Memnonides scenarios* (recall that according to Ovid (8), the Memnonides were birds which annually constructed themselves from the ashes of the warrior Memnon, and circled his funeral pyre). In these the planets are formed from the scattered remains of a disrupted, evaporated, ablated, or otherwise dismembered star (for a comprehensive description of the proposed scenarios, see the review by Podsiadlowski in these proceedings).

Neutron stars surrounded by gaseous stellar remains are ubiquitous in standard models of binary evolution, supernovae, and globular cluster pulsar formation. White dwarf analogues are also ubiquitous. Understanding the evolution of such gaseous remnants around degenerate stars is consequently of general importance, whether or not they ultimately transfigure themselves into planets.

2. SYNOPSIS OF MEMNONIDES SCENARIOS

In table 1,we summarize the Memnonides scenarios which have been proposed, listing for each the physical properties of the initial gaseous disk (our personal favorite, number 5 in the table, a suggestion of C. Thompson, is illustrated in figure 1). There are many significant differences between the conditions in the gaseous disks produced by these scenarios and the conditions in the Solar nebula.

i) *The large luminosity of the central source.* In all the scenarios, $\gtrsim 10^{-2} M_\odot$ is accreted toward the neutron star. Averaged over a typical planet formation timescale of 10^6 y (see below), this corresponds to an accretion rate of $\gtrsim 10^{-8} M_\odot\,\mathrm{y}^{-1}$, giving a luminosity $\sim 10^{38}\,\mathrm{erg\,s}^{-1}$, about the Eddington luminosity for the neutron star. At 0.4 AU from such a luminosity, the black body temperature is $\sim 6,000\,\mathrm{K}$, well above the sublimation temperature for even the most refractory solids. To make planets it is therefore crucial either to shield any coagulating dust and planetesimals from this radiation, or more likely, to wait until the rate of accretion onto the central source has dropped below $10^{-10} M_\odot\,\mathrm{y}^{-1}$. In scenarios 2–5, naïve application of standard disk models (see below) would predict that most of the mass would try to accrete in $\lesssim 10^4$ y. Since this would lead to an enormously super-Eddington luminosity, the bulk of the mass evidently cannot be accepted by the neutron star, and must be ejected from the system. It is energetically possible for the resulting wind to carry off the incipient outer disk, and thus prevent planet formation altogether.

ii) *The planets must be formed from the expanding, not the contracting parts of the nebula* in all the scenarios of table 1, except for no. 6. The planets inferred around PSR 1257+12 (Wolszczan & Frail 1992; Wolszczan, these proceedings) lie at radii much larger than the initial radii of the gaseous disk. In standard viscous disk models, this does not present an insuperable difficulty, since the total angular momentum J of the matter in the disk is nearly conserved (negligible angular momentum is accreted onto the point-like neutron star), so $J_d = M_d(t)\sqrt{GMa(t)} = \mathrm{const}$. Since for most reasonable viscosities, the bulk of the mass of the disk is near its outer edge, of radius $r_e(t)$, it follows that

$$r_e(t) \simeq r_e(0) \left[\frac{M_d(0)}{M_d(t)}\right]^2 . \tag{2.1}$$

As the mass of the disk drops through accretion, it must move outwards to conserve angular momentum. The fact that at most (see worries about super-Eddington winds in (i) above) $\sim [r_e(0)/0.4\,\mathrm{AU}]^{1/2}$ of the total initial

Table 1. Memnonides scenarios for pulsar planet formation

	J_{disk} ($g\ cm^2\ s^{-1}$)	R_{disk} AU	$Z_{CO}/Z_{\odot}$	$Z_{Si,Fe}/Z_{\odot}$	V_{sys} ($km\ s^{-1}$)	Age of NS at Planet form. (y)	Occur in glob clust?	WD analogue?
Solar nebula	$> 10^{52}$	> 10	1	1	20			
Memnonides Scenario		initial						
1. Supernova fallback	10^{49}	10^{-5}	$\sim 100?$	$\sim 100?$	any	10^6	No	No
2a. NS disrupt WD	10^{50}	10^{-4}	~ 100	1	any	$> 10^8$	Yes	—
2b. WD disrupt WD	10^{50}	10^{-4}	~ 100	1	~ 20(Pop I)	AIC	Yes	Yes
3. Unstable xfer disrupt compan. $M_c \sim 0.1 M_{\odot}$	10^{50}	10^{-3}	1	1	any	$> 10^9$	Yes	Yes
4. Nonconservative xfer/evap forms circumbinary disk	10^{50}	10^{-2}	1	1	any	any	Yes	No
5. Supernova recoil into companion	10^{51}	$10^{-2.5}$	1	1	> 100	10^6	(tidal cap)	(GC only)
6. Disk captured from Be* wind	10^{52}	1	1	1	~ 20	$< 10^7$	No	Yes
PSR 1257+12	$> 10^{50} \frac{Z_{\odot}}{Z}$	$a \sim 0.4$	—	—	$\sim 200?$	lo B $\rightarrow > 10^7?$		

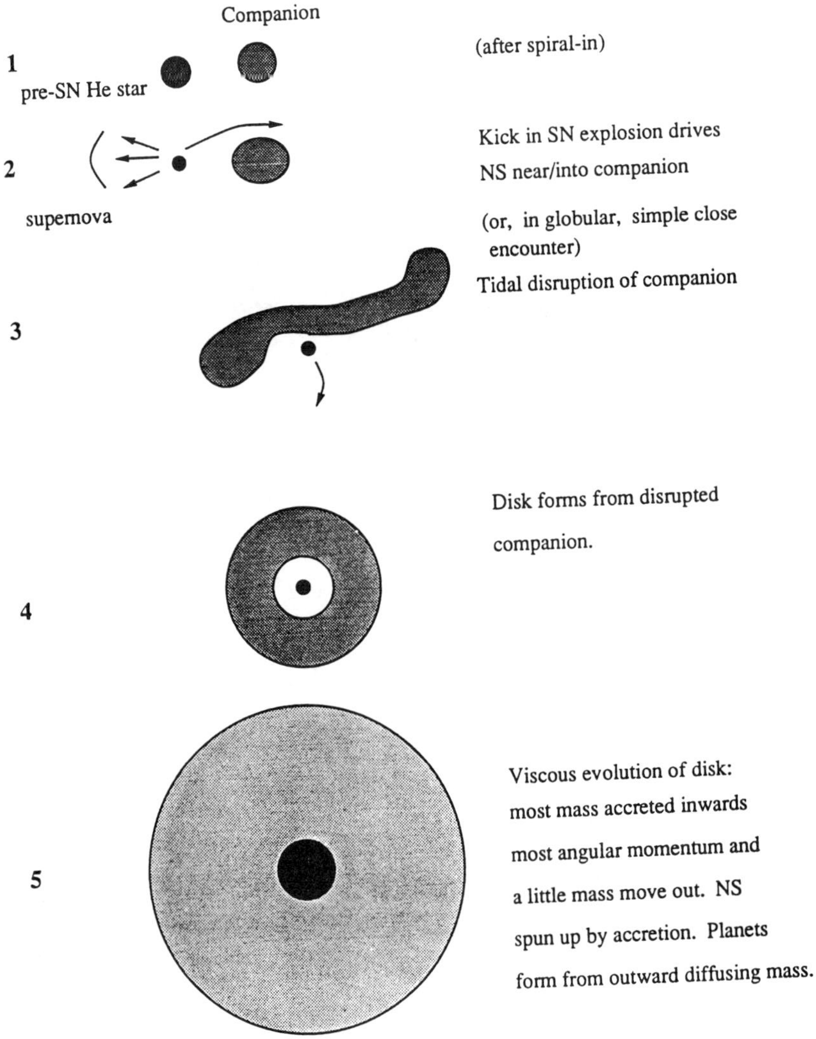

FIGURE 1 Cartoon of Memnonides scenario no. 5 of Table 1. Steps 3–5 could also occur in a globular cluster, during the close encounters which are believed to produce many of the cluster pulsars and X-ray binaries. Steps 3–5 are also representative (with modest changes of scale) of all the other scenarios in Table 1.

disk mass reaches the radii of PSR 1257+12's planets means that in scenarios 1–4 (and to a lesser extent in scenario 5), mass at the desired planetary radii is a dangerously scarce commodity, requiring the planet formation to be quite efficient (see also figure 5). In some of the models which have been proposed for the ionized parts of accretion disks in active galactic nuclei (Blandford & Payne 1982, Blandford 1976) and adapted to protostellar disks (Königl 1989), all of the disk's angular momentum is removed in a magnetised wind, and the disk mass simply accretes *without* any outward expansion of the disk. If these models applied to the early evolution of the circumpulsar disks, they would lead to the failure of all of scenarios 1–5 [no. 6 would survive, but in this scenario it is difficult to explain the apparently high proper motion of PSR 1257+12 (Wolszczan, these proceedings)]. Those (including the authors) with an attachment to these scenarios might then take the existence of planets around PSR 1257+12 as an indication that disks solve their angular momentum problem by viscous transport (e.g. via small-scale magnetic cells), not by angular momentum ejection along a global magnetic field.

iii) *The grain opacities could be quite different from those in the ISM and the protosolar nebula.* In Memnonides scenarios there is no dust initially present in the disk. In the solar nebula, aboriginal dust nuclei from the interstellar medium (from the winds of carbon and other extreme giant stars, and perhaps from supernovae) can simply grow by sticking atoms they encounter to their surfaces. In a pulsar nebula, by contrast, the gas must first supercool and nucleate grains. The high densities and long evolution times (respectively thousands of times higher and longer than in red giant photospheric winds —cf. Frenklach *et al.*, 1989*a, b* for discussion of formation of PAHs and silicon carbide grains in red giants) suggest that this should present no difficulty. However, the different environment is likely to yield a different density of nucleation sites and rate of grain growth. Grain radius a grows as $\dot{a} \sim 30\,\mu\mathrm{m}\ \mathrm{y}^{-1}$ in solar metallicity disks with the properties shown in figure 4. Consequently the grain size distribution, and therefore the opacity κ could be very different from those assumed in the opacity tables used in solar nebula models (and in the rest of this paper). Simple classical nucleation theory (Salpeter 1974; note however that this is rarely applicable! —cf. Frenklach *et al.* 1989b) would predict that the density of nucleated grains scales roughly as the inverse of the cooling time, so that the grains in the circumpulsar disk would be $\lesssim 10^{-3}$ times as numerous as in the protosolar disk, but $\gtrsim 10^{3}$ times more massive -i.e. $\gtrsim 10$ times larger. To see the consequences for the circumpulsar disk, define the viscous stress as αp (hence viscosity $\nu = (2/3)\alpha c_s^2 \Omega^{-1}$ for a Keplerian disk of angular velocity $\Omega = (GM/r^3)^{1/2}$ and isothermal sound speed c_s), and equate the dissipation per unit area of the disk (see §4) to the radiation flux diffusing out. This gives an implicit equation for the disk surface density Σ as a function of temperature and radius:

$$\Sigma(T) \simeq \left(\frac{32\sigma_b \mu}{9\alpha k_b}\right)^{1/2} \left[\frac{T^3}{\kappa(\rho, T)}\right]^{1/2} \Omega(r)^{-1/2}, \qquad (2.2)$$

where σ_b is the Stefan-Boltzmann constant, T the temperature at the midplane of the disk, and ρ the density there, $\rho(r, \Sigma, T) = \Sigma\Omega/(2c_s)$. Since grain opacity is only weakly dependent on ρ, it follows that $\Sigma(T) \propto a^{1/2}$ (a^1) for $ak_bT/(hc) \gg 1$ $(\ll 1)$, where a is the typical grain size, k_b the Boltzman constant, and h Planck's constant. The grain size distribution thus determines the structure and evolution of the disk once it cools below $\sim 3000\,\mathrm{K}$.

iv) *The heavy element abundances may be quite different from solar.* In scenarios 1 and 2, the disk may be formed almost entirely of heavy elements, with a mixture quite different from cosmic (e.g., nearly pure C, O or Si). This dramatically increases the mean mass per particle μ in the neutral gas phase (recall that the thickness of a thermally supported disk $h \propto (T/\mu)^{1/2}$), and the opacity in almost all phases, bottling up the radiatin and increasing the midplane temperature of the disk. Scenarios 3–5 can operate in globular clusters which have heavy element abundances as low as 10^{-2} of solar. Over most of their evolution such disks are much colder at given surface density and radius than the solar nebula. But for the high central source luminosity, these could form planets at smaller radii than Galactic systems.

In the particular application to PSR 1257+12, scenarios 1, 2b, 5 and 6 require the pulsar to have evolved in a way which would before 1988 have been considered ridiculous, but is now more fashionable (on phenomenological, not theoretical grounds). PSR 1257+12 belongs to the distinct group of pulsars with low magnetic dipole moments and high spin frequencies which are found (with the sole exception of PSR 1937+21) only in binary systems and in globular clusters (see review by van den Heuvel in these proceedings). These differ from isolated Galactic pulsars both in being much older ($\gtrsim 10^8\,\mathrm{y}$ vs. $\lesssim 10^7\,\mathrm{y}$) and in having accreted from companions. Before 1988, it was generally accepted that the significant difference was the age difference, and that neutron star magnetic fields decayed spontaneously (Narayan & Ostriker 1990). This belief was not hindered by an absence of theoretical justification (cf. review by Goldreich & Reisenegger 1992). More recent data and analyses of pulsar kinematics (Bhattacharya *et al.* 1992) and X-ray binary evolution (Verbunt *et al.* 1990) suggest that the fields may not spontaneously decay. It has thus become fashionable to argue that the significant difference between the high and low field pulsars is not their age, but the fact that the low field pulsars have accreted (cf. Romani 1990). Though the observational and theoretical justification for this view are far from compelling, they would permit a young, high-field neutron star to be turned into a low field pulsar like 1257+12, thus saving scenarios 1, 2b, 5 and 6.

3. VICIOUS ASSUMPTIONS

In this section, we examine the transfiguration of the hot gas of a dismembered star into a disk of gas large and cool enough to begin forming planetesimals. In doing so, we make three major assumptions whose validity can be questionned.

i) *Neglect self-gravity in the early stages of the disk evolution.* In scenarios 1,2,3 and 5, the disrupted companion initially forms (in a few dynamical times) a thick radiation-pressure supported torus with an Eddington-limited luminosity (Evans & Kochanek 1989, Frank 1978, Abramowicz *et al.* 1978), which must last for at least

$$\tau_{KH} \sim \frac{GM_c^2}{R(\text{init})L_{\text{Edd}}}, \tag{3.1}$$

which is of order 10 y for a disrupted main sequence star, and of order 10^3 y for a disrupted white dwarf. In scenarios 2 and 5, the resulting torus is massive enough that it is susceptible to the $m = 1$ and 2 fission instability (the I-mode of Christodoulou & Narayan 1992), and may gather itself in a few orbits into a single orbiting lump, rather than an axisymmetric disk. In scenarios 1 and 3, the torus is less massive, but as it loses internal energy and ceases to be supported by internal radiation pressure, it becomes Jeans unstable (the J-mode of Christodoulou & Narayan 1992) and breaks up into many orbiting lumps. Gravitational interactions between these lumps will spread the lumps over a wider range of radii, where on a longer timescale viscosity might be able to smear them back into an axisymmetric disk, now of much wider radial extent, which would then be Jeans stable (see below). We assume that this occurs, and begin our calculations with an axisymmetric disk which has lost its initial internal energy (i.e. we start at $t \gg \tau_{KH}$).

ii) *Neglect the effect of radiation from the neutron star and inner disk on the outer disk.* In scenarios 1, 2, 3, 5 and 6, the neutron star initially accretes at a highly super-Eddington rate. As discussed above, the wind from the inner disk could then have a dynamical effect on the outer disk, perhaps blow it away. Even as the luminosity drops below Eddington, the neutron star and inner disk can still affect the evolution of the entire disk. The surface density distribution and rate of radial spreading of the disk are controlled in part by the thermal structure of the disk. If the temperature drops more slowly than $T \propto r^{-1}$, then the disk half-thickness $h \propto T^{1/2}r^{3/2}$ increases faster than r^1, i.e., h/r increases with r. This occurs, for example, in the inner regions of the disk where bound-free ($h/r \propto r^{0.08}$) and electron scattering opacity ($h/r \propto r^{0.05}$) dominate, and in the optically thin outer ice grain regions. When

$$L_c \frac{d\ln(h/r)}{d\ln r} > \frac{3}{2}\frac{GM\dot{M}}{r}\,;\text{i.e.}\quad \frac{d\ln(h/r)}{d\ln r} \gtrsim 15\frac{GM}{rc^2}\,, \tag{3.2}$$

which is satisfied for all the regions mentionned above for $r \gtrsim 10^8$ cm, the flux absorbed from the vicinity of the neutron star exceeds the flux generated locally by viscous dissipation in the disk. The temperature and vertical temperature structure of the disk are then substantially different from those in standard disk models. Even if h/r decreases with r, radiation scattered down from a wind above the disk can provide significant surface heating. The far infrared spectra of protostellar nebulae appear to require such scattering (or some other means of heating the outer disk far more than standard disk models predict —Kenyon & Hartmann 1987). The most serious effects occur in the outer (ice-coated grain) regions. The surface heating can stabilize the disk against convection, which is sometimes argued to provide the turbulent eddy viscosity justifying an α-viscosity prescription (e.g. Ruden *et al.* 1988; see discussion under (iii) below, however). Furthermore, even a small fraction f ($f \sim \max[d\ln(h/r)/d\ln r, \tau_w]$, where τ_w is the scattering optical depth between r and $2r$ in the wind above the disk) of the central source luminosity L can sublimate the solid grains crucial for planetesimal formation, if incident on the disk. Consider grains with a sublimation temperature $T_{\rm sub} = 10^3 T_{s3}$ K. Graphite grains sublimate at $T_{s3} \sim 2$ in the solar nebula, and at $T_{s3} \sim 4$ in the denser, C-rich environment of a disrupted white dwarf disk. Ice grains sublimate at $T_{s3} \sim 0.2$ in the solar nebula, and at $T_{s3} \sim 0.25$ in the disk of a disrupted main sequence star. Grains of the given sublimation temperature will not form inside

$$r_{\rm sub} = 2(fL/10^{36}\,{\rm erg\ s^{-1}})^{1/2} T_{s3}^{-2}\,{\rm AU}\,. \tag{3.3}$$

Note that grains smaller than $\sim 10\mu$m are transparent to the keV X-rays from the neutron star, but also radiate inefficiently at frequencies $\nu \sim kT/h$, so both heating and cooling rates scale with volume for small grains; for larger grains, they scale with area. Equation (3.3) applies roughly to both. We see that at the ~ 0.4 AU radius of the planets around PSR 1257+12, refractory grains cannot form until $fL \lesssim 10^{36}\,{\rm erg\ s^{-1}}$ (i.e., $f\dot{M} \lesssim 2 \times 10^{-10} M_\odot\,{\rm y^{-1}}$), and ice grains until $fL \lesssim 2 \times 10^{33}\,{\rm erg\ s^{-1}}$ ($f\dot{M} \lesssim 5 \times 10^{-13}\,M_\odot\,{\rm y^{-1}}$), less than the current pulsar spin-down luminosity $I\Omega\dot{\Omega} \simeq 2 \times 10^{34}\,{\rm erg\ s^{-1}}$ (as we discuss in §7, the pulsar wind luminosity is likely to be in such a pentrating form that equation (3.3) does not apply to it). In gas of solar metallicity, the X-rays penetrate only $\sim 0.1\,{\rm g\ cm^{-2}}$, less than 10^{-5} of the total disk surface density $\gtrsim 10^4\,{\rm g\ cm^{-2}}$ at 0.4 AU required to make the planets of PSR 1257+12. The shielded inner layers of the disk will then be heated by the inward-diffusing infrared reradiation from the directly heated gas. When the infrared optical depth is high, the temperature of the shielded layers can be somewhat cooler than (3.3) would indicate. Any illuminated but not sublimated grains with radii $a < 0.4(L/10^{38}\,{\rm erg\ s^{-1}})$ cm will feel a radiation pressure force exceeding the force of gravity, and can drive a wind from illuminated dusty surfaces of the disk.

iii) *We adopt an α viscosity prescription.* We assume the viscous stress in the disk to be give by αp, with α a constant of order 10^{-2}–10^{-1}, and p the gas pressure. This gives a viscosity

$$\nu = \frac{2}{3}\alpha c_s^2 \Omega^{-1}\,. \tag{3.4}$$

In the ionized portions of the disk this popular prescription has been lent new respectability by the demonstration by Hawley & Balbus (1991) and Balbus & Hawley (1992) that any weak magnetic field in the disk will grow exponentially on the orbital timescale until it becomes dynamically important (forcing reconnection, or rising buoyantly out of the disk).* A dynamically important field gives $\alpha \sim B^2/(8\pi p)$, and models of flux loops (Coroniti 1981, Sakimoto & Coroniti 1989) suggest $\alpha \sim 10^{-2}$. In the outer neutral regions of the disk, turbulent convection has been suggested to provide an $\alpha \sim 10^{-2}$ in the optically thick regions of the disk (Ruden *et al.* 1988, though in the linear regime convective modes carry angular momentum *inwards*, not outwards, and thus do not have the characteristics of viscosity —Ryu & Goodman 1992). Ambipolar diffusion through a global magnetic field (Königl 1989) can also remove angular momentum, but as discussed in §2, this does not act like a true viscosity, and the disk does not expand outwards as required in scenarios 1–5.

4. VISCOUS DEVELOPMENT

A thin viscous accretion disk operates as follows. Relative to disk material at some distance r from the neutron star, matter closer to the neutron star is moving faster, and matter on the outside moving slower. Across the mean free path of the viscosity (e.g., magnetic cell or convection cell size, both of order the disk thickness h if $\alpha \sim 1$), matter on the inner fast track and the outer slow track share momentum. Thus the inner matter on average loses momentum (hence angular momentum) and the outer matter gains it. The inner mass, losing angular momentum, spirals inwards, accreting, while the outer matter will move outward, unless there is matter beyond it to which it can in turn transfer its angular momentum.

In the standard model of a neutron star accreting from a companion star orbiting it, mass is lost from the companion at the L1 point, and forms an accretion disk around the neutron star. Initially the disk is quite small ($\sim$ 0.1 the Roche lobe size), but as discussed above, the outward transport of the angular momentum lost by matter accreting toward the neutron star makes the disk expand outward until its edge is close enough to the companion that gravitational torques can transfer the angular momentum back to the companion star. The disk thus has a sharp outer edge, close to a resonance (Goldreich & Tremaine 1982, Priedhorsky & Verbunt 1988, Whitehurst & King 1991). In the Memnonides scenarios (save no. 6), there is no outer companion, and the disk is free to expand forever.

The time evolution of a disk of surface density Σ and viscosity ν (given, for our purposes, by equation (3.4)) is thus described by a diffusion equation

$$\frac{\partial \Sigma}{\partial t} = \frac{3}{r}\frac{\partial}{\partial r}\left(r^{1/2}\frac{\partial}{\partial r}\left[\nu \Sigma r^{1/2}\right]\right) , \tag{4.1}$$

* The objection of Knobloch (1992) that the linear instability is strictly an overstability for non-vertical fields is of purely semantic interest. With *any* realistic seed field expected in a disrupted star, the exponential growth of the "overstability" for any field orientation will have reached the nonlinear phase simulated by Hawley *long* before the linear overstability would have begun oscillating.

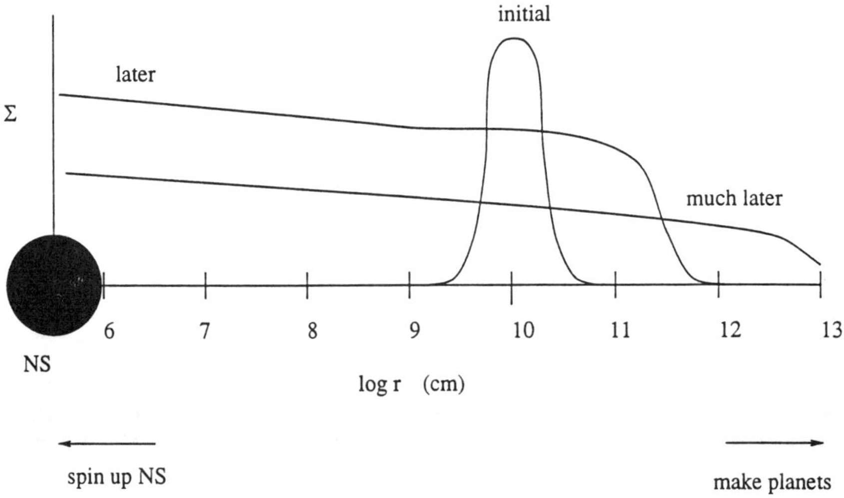

FIGURE 2 General features of the evolution of a viscous disk.

(cf. Frank, King & Raine 1992) provided that we ignore winds from the disk, and interpret ν as the vertically-averaged viscosity. The energy per unit area produced by viscous dissipation on each side of the disk midplane is

$$F_H = \frac{9}{8}\Sigma\nu\Omega^2 \,, \tag{4.2}$$

which must be balanced by radiative losses if the disk is to remain thin. Substituting (3.4) and equating to the radiation flux in the diffusion approximation (large optical depth) then gives us equation (2.2), with κ a Rosseland mean opacity. Equations (4.1), (3.4), and (2.2), together with an opacity table $\kappa(\rho, T)$, and $\rho = \Sigma\Omega/(2c_s)$ from vertical hydrostatic equilibrium then give a complete set of equations for the evolution of a disk. Given Σ at some r, (2.2) determines T and hence all the other disk properties, including $\nu(\Sigma, r)$ which can then be substituted into (4.1) and evolved.

The general character of the evolution from a torus of radius R(init) (see table 1) is as follows (see figure 2). Most of the mass rapidly accretes inwards, but a small fraction of the mass (usually given by (2.1)) moves outwards to conserve angular momentum. At some particular radius $r > R$(init), the surface density of the disk will initially rise (as the outer edge of the disk approaches r), and then fall (as the outer edge moves beyond r, and the mass of the disk continues to drop by accretion). The surface density at $r > R$(init) thus reaches a maximum at a time $t_m(r)$. At this time, the maximum mass is available to form planets near r, provided that the disk is there cool enough to form grains. If it is not, one must wait until it has cooled further, but the surface density and available mass will then be much less.

5. MATCHED QUASI-STATIC SIMILARITY SOLUTIONS

We have integrated the differential equation (4.1) directly (see §6), but we have found that its solutions, even for disks with realistic opacities, can be represented well by analytic (algebraic) solutions which we now describe. These allow us to quickly survey parameter space, and the effects of changes in opacity, composition, etc.

As described following (4.2), using the α prescription, the thermal and vertical hydrostatic equilibrium conditions for a given opacity, determine the midplane temperature T, viscosity ν and disk thickness h all as functions of the surface density and radius. For an opacity law

$$\kappa = \kappa_0 \rho^m T^n , \tag{5.1}$$

we have that

$$T = \left(\frac{9\alpha\kappa_0}{32\sigma_b}\right)^{\frac{2}{6-2n+m}} \left(\frac{k_b}{\mu m_p}\right)^{\frac{2-m}{6-2n+m}} \Sigma^{\frac{2m+4}{6-2n+m}} \Omega^{\frac{2m+2}{6-2n+m}} \tag{5.2}$$

$$\nu = \frac{2}{3}\alpha^{\frac{8-2n+m}{6-2n+m}} \left(\frac{9\kappa_0}{32\sigma_b}\right)^{\frac{2}{6-2n+m}} \left(\frac{k_b}{\mu m_p}\right)^{\frac{8-2n}{6-2n+m}} \Sigma^{\frac{2m+4}{6-2n+m}} \Omega^{\frac{m+2n-4}{6-2n+m}} \tag{5.3}$$

$$h = \left(\frac{k_b}{\mu m_p}\right)^{1/2} \frac{T^{1/2}}{\Omega} \tag{5.4}$$

where $\Omega = (GM/r^3)^{1/2}$ is the Keplerian angular velocity, μ is the mean particle mass in units of m_p and σ_b is the Stefan-Boltzmann constant.

Thus, we have the viscosity in the functional form

$$\nu = \nu_0 \Sigma^q r^p . \tag{5.5}$$

Since for most opacities the viscous timescale (inflow time) decreases inwards, the inner parts of the disk must have reached a quasi-steady state, with $\dot{M} = 3\pi\nu\Sigma$ nearly independent of r. Since even at a maximal spin period $\sim 0.7\,\mathrm{ms}$, a neutron star cannot have angular momentum $> 10^{49}$erg s, loss (or gain) of angular momentum to it will not have a significant effect on the total disk angular momentum in any scenario (except possibly no. 1). Thus the total disk angular momentum $J \sim \Sigma(r_e) r_e^4 \Omega(r_e)$ where r_e is ~ 0.7 times the radius of the outer edge of the accretion disk. The viscous timescale there is $t_\nu \sim r_e^2/\nu$, which is also of order the time t for the outer edge of the disk to reach r_e. Combining these equations, we have

$$r_e \propto t^{2/(4-2p+5q)} J^{2q/(4-2p+5q)} , \tag{5.6}$$

$$\dot{M} \propto t^{-1-1/(4-2p+5q)} J^{1-q/(4-2p+5q)} . \tag{5.7}$$

These suggest the existence of more rigorous similarity solutions, which indeed there are. Pringle (1991; see also Cannizzo *et al.* 1990) has shown that similarity solutions of the Zel'dovich-Raizer type can be found for (4.1) when ν is of the form (5.5), i.e. (5.3). Two different types of similarity solution can be found depending on whether the mass or the angular momentum of the accretion disk is conserved. The solution for the angular momentum conserving case is given below.

$$\Sigma = \Sigma_0 x^{-3} \tau^{-2\lambda} \xi^c (1 - k\xi^b)^a \,, \tag{5.8}$$

where $x = (r/r_0)^{1/2}$, $\tau = \frac{3}{4}\nu_0 r_0^{p-2} \Sigma_0^q t$, $\xi = x\tau^{-\lambda}$ and

$$a = 1/q \tag{5.9}$$

$$b = 1 + c \tag{5.10}$$

$$c = \frac{3q + 3 - 2p}{1 + q} \tag{5.11}$$

$$\lambda = 1/(5q + 4 - 2p) \tag{5.12}$$

$$k = q\lambda/(4q + 4 - 2p)\,. \tag{5.13}$$

The opacity table used here (all quantities in cgs units) is based on that of Lin & Papaloizou (1985), with an electron scattering opacity added for high densities.

1. Electron scattering: $\kappa = \kappa_1$, where $\kappa_1 = 0.2(1 + X)$, and X is the mass fraction of hydrogen. This opacity is valid for $T \geq 4.55 \times 10^8 \rho^{2/5}$.
2. Free-Free/Bound-Free: $\kappa = \kappa_2 \rho T^{-2.5}$, where $\kappa_2 = 1.5 \times 10^{20}$. This opacity is bounded from below by $T = 1.76 \times 10^4 \rho^{1/21}$ Instead of the partially ionized opacity favoured by Lin and Papaloizou, we approximate the ionization front as a discontinuous jump into the molecular dissociation and grain evaporation regime, which is characterised by the following two opacities.
3. $\kappa = \kappa_3 \rho^{2/3} T^3$, $\kappa_3 = 2 \times 10^{-8}$. In the temperature and density regime used here, this opacity rarely comes into play, the transition usually taking place from opacity 2 to:
4. $\kappa = \kappa_4 \rho^{2/3} T^{-9}$, $\kappa_4 = 2 \times 10^{34}$. This opacity is bound from above by $T = 3000$ K, and from below by $T = 4.6 \times 10^3 \rho^{1/15}$
5. The next three opacities are due to various types of grains: $\kappa = \kappa_5 T$, $\kappa_5 = 5 \times 10^{-3}$, and bounded from below by $T = 210$ K. This is representative of the high-Z refractory grains thought to be important in terrestrial planet formation.
6. $\kappa = \kappa_6 T^{-7}$, $\kappa = 2 \times 10^{16}$. This is bounded below by $T = 170K$.
7. Below 170 K, the opacity is represented by the ice grain opacity $\kappa = \kappa_7 T^2$, $\kappa_7 = 2 \times 10^{-4}$.

These opacities map out a "phase diagram" in the T-ρ plane, as shown in Figure 3.

By associating a single opacity with a given range of density and temperature in the disk, we can build an analytic model of the disk by linking the similarity solutions for different opacities. Each similarity solution has two free parameters, Σ_0 and τ. The τ value of the outermost similarity solution is fixed by the disk outer radius, where $\Sigma = 0$. The τ's of the inner solutions are fixed such that each similarity solution would have its edge at the same place if it were

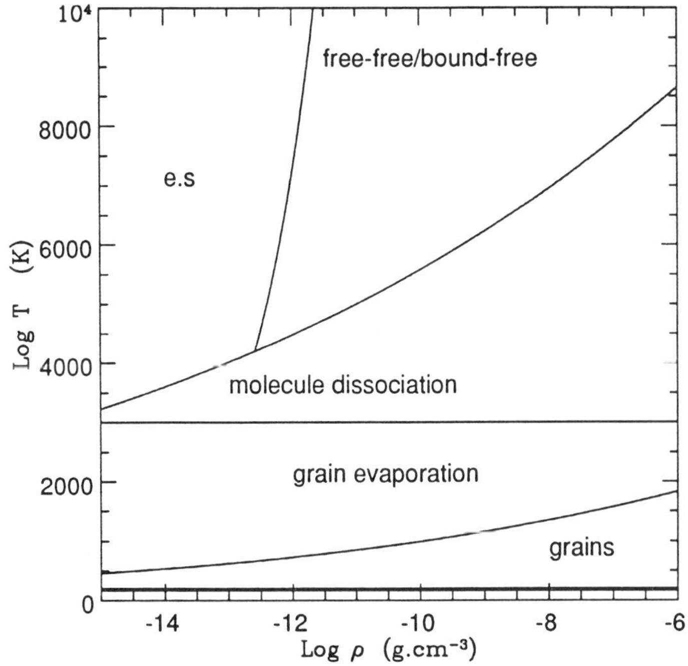

FIGURE 3 The phase diagram in T-ρ space, showing the dominant opacity source in each region. 'e.s.' is electron-scattering. The opacity below the double lines at $T = 170\,\mathrm{K}$ is due to ice-coated grains.

extended to $T = 0$. The other boundary condition, i.e. that which determines Σ_0, is that the accretion rate $\dot{M}$ remains constant across the boundary between two similarity solutions. The accretion rate at radius r can be calculated from $\dot{M} = \partial g/\partial h$ (Lynden-Bell & Pringle 1974), where g is the torque exerted by the outer annulus on the inner annulus and h is the specific angular momentum at r. This leads to an expression

$$\dot{M} = 6\pi\nu_0 r^{1/2}\frac{\partial}{\partial r}\left[r^{p+1/2}\Sigma^{1+q}\right] . \tag{5.14}$$

Except at the extreme edge of the accretion disk, each similarity solution may be characterised by a power law of the form $\Sigma \propto r^{-p/(1+q)}$, and thus the accretion rate is constant throughout most of the disk. Thus the outer radius and the accretion rate at the inner edge define the patchwork accretion disk completely.

Although a number of opacities are required for a full description, one can obtain an idea of the behaviour of the disk at early and late times by considering just a single opacity. The first practical early time approximation is to assume that the disk is dominated by the free-free/bound-free opacities. The timescales over which this remains valid are of the order of a year. While this is still not applicable to planet formation, one might expect this solution to bear the main characteristics of the disk structure at early times. This solution may also be used to calculate a quick lower bound on the mass in a given disk, since its surface density drops faster at large radii than the properly self-consistent solutions. Using opacity 2 from the above table, we obtain $m = 1$, $n = -2.5$, hence $q = 0.5$, $p = 1$, and so from (5.6) and (5.7),

$$r_e \propto t^{4/9} J^{2/9} \; ; \quad \dot{M} \propto t^{-11/9} J^{7/9} \quad (\mathrm{bf/ff}) . \tag{5.15}$$

For disks with $r_e \gtrsim 5 \times 10^{13}$ cm, other opacities are required for a self-consistent description. Furthermore, to obtain significant disk mass for r_e less than this, one requires super-Eddington accretion rates. This means that matter arrives at the neutron star Alfvén radius more quickly than it can be accepted. Hence there is likely to be a build-up of material at the inner edge of the disk. How much of this is "blown-off" by the pulsar radiation and how much builds up in a thick boundary layer outside the magnetosphere is a matter that can only be decided by detailed modelling. With the magnetic field of PSR 1257+12, the Alfvén radius $r_A \sim 4R_{NS}$. Hence the mass expelled is likely to be only ~ 4 times the mass accreted. Thus, a significant fraction of the mass could pile up in a boundary "shield" outside the magnetosphere.

The analytic solution from which the most concrete conclusions can be drawn is that corresponding to the ice grain opacity, the low temperature regime that dominates the outermost reaches of the disk. The approximation that the disk is dominated by this opacity improves at later times and it is this approximation that is valid on timescales required for planet formation. For this disk, $m = 0$, $n = 2$, hence $q = 2$, $p = 0$, and

$$r_e \propto t^{1/7} J^{2/7} \; ; \quad \dot{M} \propto t^{-15/14} J^{6/7} \quad (\text{ice grain}) . \tag{5.16}$$

$$r_e = 3.0 \times 10^{14}\,\mathrm{cm} \left(\frac{\alpha}{0.1}\right)^{1/3} \left(\frac{\mu}{0.6}\right)^{-1/3} \left(\frac{\mathrm{M_{disk}}}{10^{-3}\mathrm{M_\odot}}\right)^{1/2} \left(\frac{\dot{\mathrm{M}}}{\dot{\mathrm{M}}_{\mathrm{Edd}}}\right)^{-1/6} \tag{5.17}$$

$$J = 3.7 \times 10^{50}\,\mathrm{erg\ s} \left(\frac{\alpha}{0.1}\right)^{1/6} \left(\frac{\mu}{0.6}\right)^{-1/6} \left(\frac{M_{disk}}{10^{-3}M_\odot}\right)^{5/4} \left(\frac{\dot{M}}{\dot{M}_{\mathrm{Edd}}}\right)^{-1/12} \tag{5.18}$$

$$t_\nu = 3.6 \times 10^{3}\,\mathrm{y} \left(\frac{M_{disk}}{10^{-3}M_\odot}\right) \left(\frac{\dot{M}}{\dot{M}_{\mathrm{Edd}}}\right)^{-1} . \tag{5.19}$$

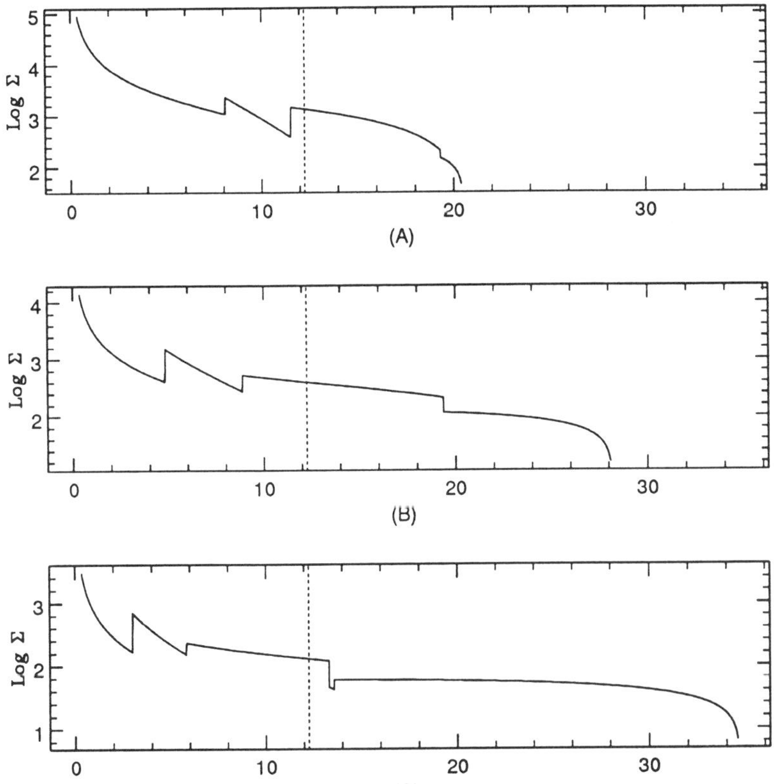

FIGURE 4 The sequence of quasi-static disks representing the evolution of a disk with total angular momentum J$\sim$ 2×10^{50} gm cm^2 s^{-1}. The dashed line lies at 1 AU. The abscissa is $x \equiv (r/10^{11}\,\mathrm{cm})^{1/2}$. A: $t = 10^3$ y, B: $t = 10^4$ y, C: $t = 10^5$ y (see figure 5). The temperature and physical state at each radius can be estimated by noting that each discontinuity in Σ corresponds to crossing one of the opacity boundaries in figure 3, and the temperature increases from right to left (to smaller x). The outermost part of each disk shown here is dominated by the opacity of ice-coated grains.

The inner regions of this disk will, of course, be hotter and described by different opacities. Performing matchings at constant $\dot{M}$ as described above equation (5.14), we can derive an evolutionary sequence for the complete disk, shown in figure 4. We note that except at their very edges, these disks are stable against self-gravitational instabilities: $Q = \Omega c_s/(\pi G \Sigma) \gg 1$ (cf. Goldreich & Tremaine 1982).

To apply these models to the PSR1257+12 system, we note that the total mass of the two observed planets, $M = 2 \times 10^{-5} M_\odot / \sin i$, requires a disk mass $M_{\rm disk} > 2 \times 10^{-5} M_\odot$ and $J \gtrsim 2 \times 10^{48}$ erg s. Since the planets are most likely to be formed from refractory elements (see discussion following (3.3)), more realistic minimum disk mass and angular momentum are larger than these by the reciprocal of the abundance of refractory elements (C, O, Si, Fe).

For a given disk profile, the independent variables are the outer radius, the inner radius and the accretion rate. There are two further parameters, the

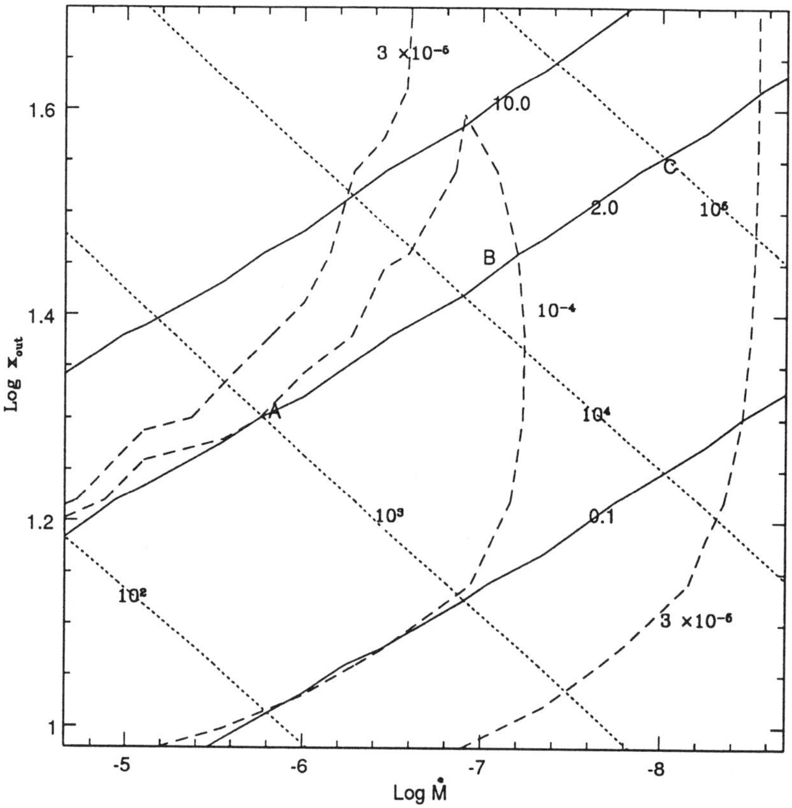

FIGURE 5 The parameter space spanned by our quasi-static disk models. The ordinate $x_{\rm out} = (r_e/10^{11}\,{\rm cm})^{1/2}$, where r_e is the radius of the outer edge of the disk. The abscissa is the accretion rate through the inner parts of the disk, in $M_\odot\,{\rm y}^{-1}$. To good approximation, a real disk evolves upwards to the right along the solid line of constant total angular momentum $J = 10^{50} J_{50}\,{\rm g\,cm^2\,s^{-1}}$. The solid lines of constant J are labelled in units of J_{50}. The dotted lines are the viscous time at the edge of the disk, roughly the disk age, and are labelled in years. The dashed lines represent the mass in the cold, refractory grain disk that lies inside 1 AU, and are labelled in $M_\odot$. The letters A,B and C are the positions of the three profiles in figure4.

value of α and the value of μ. The latter parameter will be determined by the composition of the original material. For the following discussion, $\alpha = 10^{-2}$ and the composition of the material is solar (μ will vary with the degree of ionization).

Figure 5 shows the distribution of disk properties in the $\dot{M}$–r_e plane, the natural co-ordinates of the disk model. The solid lines indicate curves of constant total disk angular momentum. These are the evolutionary tracks for disks with a given total angular momentum. The dotted lines are lines of constant inspiral time from the edge of the disk, a rough indication of the disk age. The important curves are the dashed lines. These lines represent contours of mass in the region of the disk where grains have formed, and which lies within 1 AU. This mass was calculated by summing all the mass in that part of the disk dominated by opacity number 5 above and which lay at radii less than 1 AU. This is (a generous estimate of) the mass which would be available for the formation of terrestrial-type planets, the most likely mode of formation for the planets of PSR 1257+12 (see above).

The most favourable region of parameter space is thus that where there is the most mass in grains inside 1 AU. For example, in the evolutionary sequence shown in figure 4, the point at which the viscous evolution ends will depend on the grain growth in the disk. However, contours of constant mass in the grain region in figure 5 show that planetesimal formation would probably have to be completed before 2×10^4 y, although one could stretch this timescale to 2×10^5 y if one assumed 100% efficiency in converting disk mass into planets.

Notice that the mass constraint outlined above restricts the amount of disk angular momentum to $J \sim 10^{50}$ erg s. If it is much higher than this, then there is never enough mass in heavy grains to make planets. If it is less than this, then the ice grain disk forms too early, leaving little time for planet formation.

6. VALIDITY OF THE MATCHING SOLUTIONS

As a check on the accuracy of the matched quasi-static solutions described above, we have also calculated the disk evolution using a finite difference scheme. We solve equation (4.1) as transformed defining $x = (r/R(\mathrm{init}))^{1/2}$, and $S = x^3\Sigma$, $\tilde{\nu} = \nu/x^2$: $\partial S/\partial\tau = \partial^2/\partial x^2(S\tilde{\nu})$, where $\tau = 3t/(4R^2(\mathrm{init}))$. To provide convenient stepsizes, we transform to a new radial variable $y = x^{1/a}$. The following difference scheme conserves both angular momentum and mass explicitly:

$$S_j^{n+1} = S_j^n + \frac{\Delta t}{a\Delta y} y_j^{1-a}\left[\frac{(S\tilde{\nu})_{j+1}^n}{(y_{j+1}^a - y_j^a)} - \frac{(y_{j+1}^a - y_{j-1}^a)(S\tilde{\nu})_j^n}{(y_j^a - y_{j-1}^a)(y_{j+1}^a - y_j^a)} + \frac{(S\tilde{\nu})_{j-1}^n}{(y_j^a - y_{j-1}^a)}\right]. \tag{6.1}$$

This scheme has the stability criterion $\Delta t < (a\Delta y)^2/(2\tilde{\nu}_j)y^{2a-2}$, and thus is second order in space, despite appearances. With a single opacity law, the solutions of our scheme rapidly approach the similarity solution (5.8)to very high precision. With multiple opacity laws, comparison with the results of §5 is not quite straightforward, since the definitions for opacities used in §5completely cover the ρ–T plane, but not the Σ-x plane. Thus for the numerical integration, we used a simplified version of the opacity table, with 5 regions which cover the Σ-x plane. These regions are shown in figure 6, along with the corresponding analytical patchwork solution. Such comparisons lead us to believe that the results of §5 are reliable to within a factor of two —much better than the uncertainties in the physics!

7. CONCLUSIONS

In the particular case of PSR 1257+12, scenarios 1, 2, 5, and 6 are constrained by the requirement that the pulsar be spun up to its present 6 ms period. To spin it up requires accreting

$$\Delta M \gtrsim 0.02 P_{6\mathrm{ms}}^{-4/3} I_{45}\, M_\odot\,, \tag{7.1}$$

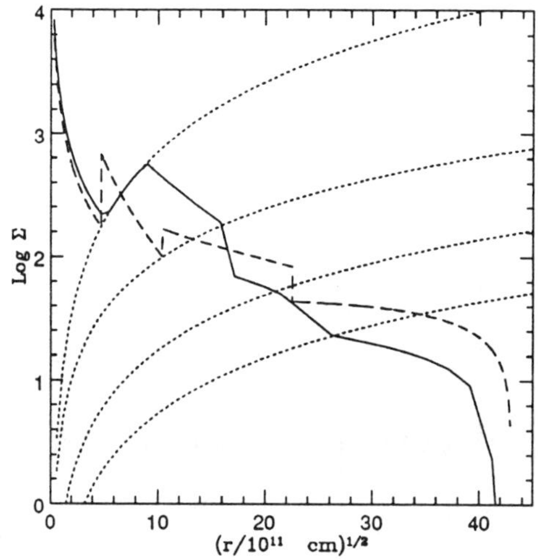

FIGURE 6 The state at $t = 10^6$, y of a $J = 2.3 \times 10^{50}$ erg s disk, started as an annulus at $R(\text{init}) = 10^{11}$ cm. The solid curve is the numerical solution of (4.1). The dashed curve is the quasi-static analytic solution with the same accretion rate. The dotted curves represent the transition boundaries between opacities for the numerical solution. The definitions for opacities in §5 completely cover the ρ–T plane, but not the Σ-x plane. Thus for the numerical integration, we used a simplified version of the opacity table, with 5 regions which cover the Σ-x plane. The modest differences in the opacities used in the two solutions exaggerates their differences. Nevertheless, the track each other well.

where the moment of inertia of the neutron star is $10^{45} I_{45}$ g cm^2. This result is insensitive to whether or not the field was initially low, or decayed as a result of accretion, since most of the mass is required to make the last factor of two change in P: $\Delta M \gtrsim I\Omega/j$, where $j = (GMr_A)^{1/2}$ is the specific angular momentum at the Alfvén radius, and $r_A < (GM/\Omega^2)^{1/3}$ as long as spin up continues. The time to spin up to this period is bounded below by the fact that the neutron star cannot accept matter at a super-Eddington rate, so

$$t_{\text{spin-up}} \sim 2 \times 10^6 I_{45} P_{6\,\text{ms}}^{-4/3} (\dot{M}_{\text{Edd}}/\dot{M})\,\text{y}\,. \tag{7.2}$$

With standard disk physics, as shown in figure 5, only disks with $J > 10^{51}$ erg s maintain the accretion this long. This would favor scenario 5 (or 3 and 4, in which the pulsar could have been spun up at leisure before the disruptive event).

At the other extreme of the evolution, one may enquire what happens as the accretion rate tails off at late times. For a pulsar with magnetic dipole moment $\mu_B = 10^{27}\mu_{27}$ (PSR 1257+12 has $\mu_{27} \sim 1$), the Alfvén radius will be pushed beyond the pulsar's light cylinder when

$$\dot{M} < \dot{M}_* \simeq 10^{-12}\mu_{27}^2(P/6\,\mathrm{ms})^{-7/2}M_\odot\,\mathrm{y}^{-1}\,. \tag{7.3}$$

Since beyond the light cylinder, the ram pressure of the pulsar wind falls as r^{-2}, more slowly than the ram pressure of accretion, roughly $\propto r^{-5/2}$ (Illarionov & Sunyaev 1975, Shaham & Tavani 1991), the pulsar wind will then be able to break free, and prevent any further accretion. The pulsar will turn off as an X-ray source, and turn on as a radio pulsar.

In the late stages of evolution of the solar nebula, Poynting-Robertson drag is often invoked to clear dust from the young planetary system (Lin & Papaloizou 1985). In the context of the pulsar nebula, this is much less effective once $\dot{M} < \dot{M}_*$ (7.3). The drag timescale is

$$t_{\mathrm{PR}} \simeq \frac{4\pi c^2 r^2 a_{gr}\rho_{gr}}{f_{abs}L_{psr}}\,. \tag{7.4}$$

The difference here is that the penetrating part of the pulsar wind, mainly ultrarelativistic ($\gamma \gtrsim 10^5$, Kulkarni *et al.* 1992) electrons and ions, has a stopping column $\gtrsim 10^2\,\mathrm{g\ cm^{-2}}$. Thus grains smaller than baseball size are transparent, and the fraction of the incident pulsar flux absorbed, $f_{abs} \propto (a_{gr}\rho_{gr})$, so all such grains sink at roughly the same rate,

$$t_{PR} \sim 10^9 r_{AU}^2\,y\,, \tag{7.5}$$

so slowly as to be probably irrelevant.

In conclusion, we find that the scenarios listed in table 1 are just barely capable of producing planets of the masses and semi-major axes appropriate to PSR 1257+12, if the disk approximations employed in solar nebula modelling are valid in these circumstances. Such model disks are not capable of producing a massive planet at large radii, as is inferred around the globular cluster PSR 1620–26 in M 4 from its apparently large $\ddot{P}$ (Backer, Sigurdsson, these proceedings). However, the effects of illumination by the central source, its super-Eddington winds, self-gravity in early stages, and the problem of grain nucleation in the nebulae bear investigation beyond our cursory discussion.

ACKNOWLEDGEMENTS

Supported in part by NASA grant NAGW-2394 and the Sloan Foundation.

REFERENCES

Abramowicz, M., Jaroszyński, M. & Sikora, M. 1978, A&A, 63, 221.

Balbus, S.A. & Hawley, J.F. 1992, ApJ, in press.

Blandford, R.D. 1976, MNRAS, 176, 465.

Blandford, R.D. & Payne, D.G. 1982, MNRAS, 199, 883.

Bhattacharya, D., Wijers, R., Hartman, J. & Verbunt, F. A&A 254, 198.

Cannizzo, J.K., Lee, H.M. & Goodman, J. 1990, ApJ, 351, 38.

Christodoulou, D.M. & Narayan, R. 1992, ApJ, 388, 451.

Coroniti, F.V. 1981, ApJ, 244, 587.

Evans, C.R. & Kochanek, C.S. 1989, ApJL, 346, L13.

Frank, J. 1978, MNRAS, 184, 87.

Frank, J., King, A.R. & Raine, D.J. *Accretion Power in Astrophysics*, 2nd Ed. Cambridge: Cambridge Univ. Pr., 69.

Frenklach, M. & Feigelson, E.D. 1989a, ApJ, 341, 372.

Frenklach, M., Carmer, C.S. & Feigelson, E.D. 1989b, Nature, 339, 196.

Goldreich, P. & Reisenegger, A. 1992, ApJ, 395, 250.

Goldreich, P. & Tremaine, S. 1982, ARAA, 20, 249.

Hawley, J.F. & Balbus, S.A. 1991, ApJ, 376, 223.

Illarionov, A.F. & Sunyaev, R.A. 1975, A&A, 39, 185.

Kenyon, S.J. & Hartmann, L. 1987, ApJ, 323, 714.

Knobloch, E. 1992, MNRAS, 255, 25P.

Königl, A. 1989, ApJ, 342, 208.

Kulkarni, S.R., Phinney, E.S., Evans, C.R. & Hasinger, G. 1992, Nature 359, 300.

Lin, D.N.C. & Papaloizou, J. 1985, in *Protostars & Planets II*, D.C. Black & M.S. Matthews, eds, Tucson: Univ. Arizona Pr., 981.

Lynden-Bell, D. & Pringle, J.E. 1974, MNRAS, 168, 603.

Narayan, R. & Ostriker, J.P. 1990, ApJ, 352, 222.

Ovidius Naso, P. 8, *Metamorphoses*, book XIII.

Priedhorsky, W.C. & Verbunt, F. 1988, ApJ, 333, 895.

Pringle, J.E. 1991, MNRAS, 248, 754.

Romani, R.W. 1990, Nature, 347, 741.

Ruden, S.P., Papaloizou, J.C.B. & Lin, D.N.C. 1988, ApJ, 329, 739.

Ryu, D. & Goodman, J. 1992, ApJ, 388, 438.

Sakimoto, P. & Coroniti, F.V. 1989, ApJ, 342, 49.

Salpeter, E.E. 1974, ApJ, 193, 579.

Shaham, J. & Tavani, M. 1991, ApJ, 377, 588.

Verbunt, F., Wijers, R.A.M.J. & Burm, H.M.G. 1990, A&A 234, 195.

Whitehurst, R. & King, A. 1991, MNRAS, 249, 25.

Wolszczan, A. & Frail, D.A. 1992, Nature, 355, 145.

Author Index